Key Challenges in Geography

EUROGEO Book Series

This book series addresses relevant topics in the wide field of geography, which connects the physical, human and technological sciences to enhance teaching, research, and decision making. Geography provides answers to how aspects of these sciences are interconnected and are forming spatial patterns and processes that have impact on global, regional and local issues and thus affect present and future generations. Moreover, by dealing with places, people and cultures, Geography explores international issues ranging from physical, urban and rural environments and their evolution, to climate, pollution, development and political economy.

Key Challenges in Geography is an initiative of the European Association of Geographers (EUROGEO), an organization dealing with examining geographical issues from a European perspective, representing European Geographers working in different professional activities and at all levels of education. EUROGEO's goal and the core part of its statutory activities is to make European Geography a worldwide reference and standard. The book series serves as a platform for members of EUROGEO as well as affiliated National Geographical Associations in Europe, but is equally open to contributions from non-members.

The book series addresses topics of contemporary relevance in the wide field of geography. It has a global scope and includes contributions from a wide range of theoretical and applied geographical disciplines.

Key Challenges in Geography aims to:

- present collections of chapters on topics that reflect the significance of Geography as a discipline;
- provide disciplinary and interdisciplinary titles related to geographical, environmental, cultural, economic, political, urban and technological research with a European dimension, but not exclusive;
- deliver thought-provoking contributions related to cross-disciplinary approaches and interconnected works that explore the complex interactions among geography, technology, politics, environment and human conditions;
- publish volumes tackling urgent topics to geographers and policy makers alike;
- publish comprehensive monographs, edited volumes and textbooks refereed by European and worldwide experts specialized in the subjects and themes of the books;
- provide a forum for geographers worldwide to communicate on all aspects of research and applications of geography, with a European dimension, but not exclusive.

All books/chapters will undergo a blind review process with a minimum of two reviewers.

An author/editor questionnaire, instructions for authors and a book proposal form can be obtained by contacting the Publisher.

Aikaterini Klonari ·
Maria Luisa De Lázaro y Torres · Athanasios Kizos
Editors

Re-visioning Geography

Supporting the SDGs in the post-COVID era

euro
geo

Springer

Editors
Aikaterini Klonari
Department of Geography
University of the Aegean
Mytilene, Greece

Athanasios Kizos
Department of Geography
University of the Aegean
Mytilene, Greece

Maria Luisa De Lázaro y Torres
Department of Geography
Universidad Nacional de Educación
a Distancia and EUROGEO
Madrid, Spain

ISSN 2522-8420 ISSN 2522-8439 (electronic)
Key Challenges in Geography
ISBN 978-3-031-40746-8 ISBN 978-3-031-40747-5 (eBook)
https://doi.org/10.1007/978-3-031-40747-5

This Springer imprint is published by the registered company Springer Nature Switzerland AG
The registered company address is: Gewerbestrasse 11, 6330 Cham, Switzerland

Contents

Geospatial Technologies

Promotion of Geography Student Teachers' Abilities to Diagnose Pupils' Written Argumentation Skills with the Help of a Diagnostic Tool

Alexandra Budke and Kimberley Hindmarsh

Abstract Argumentation competencies are essential in order to participate in social discourse. Teaching these skills is one of the aims of geography lessons. However, since pupils often find it difficult to produce correct, complete and convincing arguments, there is a need for support from the teacher in developing these skills. In order to be able to support pupils individually, their existing argumentation competencies must first be determined by an educational diagnosis. In the study presented here, a diagnostic tool was developed in a digital learning unit and used and evaluated in a seminar with geography student teachers (This study was conducted with student teachers of the subject geography. We will further refer them as "students") at the University of Cologne in Germany. The study showed that the students were mostly able to use the diagnostic tool correctly and to apply it competently. The results of the evaluation also showed a high level of acceptance of the OER and a good rating of the diagnostic tool by the students, especially due to its practical relevance.

Keywords Open educational resource (OER) · Diagnosis · Argumentation competency · Teacher education · Geography

Introduction

In geography lessons, central social issues are dealt with, such as climate change, migration, poverty, urban and regional development, sustainable resource use, etc. In order to understand these topics, pupils necessarily need argumentation skills to be able to comprehend the respective social discourse. In addition, the aim of geography lessons is to contribute to the education of mature citizens who not only

A. Budke · K. Hindmarsh (✉)
Institute of Didactics of Geography, University of Cologne, Gronewaldstraße 2, 50931 Cologne, Germany
e-mail: kimberley.hindmarsh@uni-koeln.de

A. Budke
e-mail: alexandra.budke@uni-koeln.de

A. Klonari et al. (eds.), *Re-visioning Geography*, Key Challenges in Geography,
https://doi.org/10.1007/978-3-031-40747-5_1

understand the arguments of the social groups involved in the discourse, but can also individually assess and evaluate them and justify their own position on this basis (Budke 2016). Therefore, the teaching of argumentation skills is an objective of geography education defined in the national educational standards and curricula specific to the federal states (DGfG 2014).

However, there are various studies that show that pupils and also students often find it difficult to formulate complete, technically correct and convincing arguments (e.g. Budke and Kuckuck 2020; Chase 2011; Lytzerinou and Iordanou 2020; Uhlenwinkel 2015). Pupils should be supported by their teachers in geography lessons in developing the competence to formulate technically good arguments. In order to have the necessary didactic competencies, it is, therefore, necessary in teacher training that students learn and practise the diagnostics of the argumentation competencies of the pupils.

In order to promote these competencies, a tool for the diagnosis of students' written argumentation competence was developed and integrated into a digital learning unit. This is available as an open educational resource (OER).

Since the Corona pandemic, the need to include digital teaching and learning materials in higher education has become obvious and i.a. there has been an unexpected surge in new didactic concepts and materials developed by lecturers and published as OERs. This surge in practical development in higher education should now be followed by an examination of the exact learning effects and a discussion about the transferability of the particularly positive approaches to other higher education institutions and subjects.

In this context, the focus of the chapter is the presentation of an OER that was developed within the framework of the BMBF project "DiGeo"[1]—Generalisability and transferability of digital subject concepts using the example of the responsible use of digital geomedia in teacher training. The aim of DiGeo is to develop a subject concept for building up competence in the responsible use of digital geomedia in geography teacher training. For this purpose, exemplary digital learning formats are developed and tested as open educational resources (OER). In the sub-area of argumentation/communication, ten OERs have been created so far, which are intended on the one hand, to expand the argumentative competencies of student teachers of geography and on the other hand, to develop their skills for the didactic advancement of the argumentation competencies of their future students.[2] One of these OERs deals with the topic of the didactic diagnosis of argumentation competencies and will be presented in more detail below.[3] First, the theoretical foundations on the basis of which the OER was created will be explained (section "Theoretical Background" of this chapter), then the didactic tool will be introduced (section "Diagnostic Tool for Argumentative Text Products"), the didactic conception of the learning unit will

[1] DiGeo website: https://digeo-oer.net/doku.php?id=start.

[2] Website of the DiGeo sub-project Argumentation/Communication: https://www.ilias.uni-koeln.de/ilias/goto_uk_cat_3758292.html.

[3] Website for the OER "Diagnosis of argumentation skills": https://www.ilias.uni-koeln.de/ilias/goto_uk_lm_4679282.html.

be presented (section "The Didactical Concept of the OER"), the methodological procedure for the evaluation in the context of a seminar will be explained (section "Methodological Approach to the Evaluation") and then the results on the following research questions will be presented (section "Results" and section "Summary and Conclusion"):

- How interesting, meaningful, appropriate, understandable and relevant did the students find the work with the digital learning unit to diagnose written argumentation skills?
- How competently did the students use the diagnostic tool from the learning unit to diagnose a pupil's argumentation competence based on his or her text?

The chapter concludes with a summary of the results and an outlook (section "Summary and Conclusion").

Theoretical Background

According to Witt (2015, p. 2), pedagogical diagnostics is understood as follows: "The term 'diagnostics' includes all activities that determine the prerequisites and conditions for successful teaching and learning processes of a learner. Furthermore, learning processes are analysed and their effectiveness, which is reflected in the learning outcome, is determined. The aim of diagnostics is to optimise the individual learning process".

The diagnosis of pupils' argumentation competencies is very demanding, as a look at the broad scientific discussion on the topic proves (e.g. von Aufschnaiter et al. 2008; Benetos and Bétrancourt 2020; Gronostay 2017; Maier and Budke 2018). As part of the SpiGu-project,[4] an analysis grid for evaluating argumentative pupil texts was developed in cooperation between geography and German didactics and on the basis of already developed approaches (Budke et al. 2020).[5] Overall, the analytical grid contains the following four main categories and is thus based on an interdisciplinary combination of theories from German and geography didactics: (1): Argumentative organisation of the text through (subject-)linguistic action schemata; (2): Linguistic and structural organisation of the text; (3): Material reference; (4): Content quality of the argumentation. The focus in the first two parts of the analysis grid is on the linguistic quality and structural organisation of the argumentation, for which approaches to specific "text procedures" in written argumentation and material-based writing were the basis (e.g. Feilke 2014; Steinseifer 2018). For this purpose, the (technical) linguistic action schemata used, such as "positioning" or "justifying", are examined, among other things. Since geographical argumentation

[4] SpiGu: Language-sensitive teaching and learning in inclusive geography teaching: formats of support in material-based argumentative writing.

[5] Link to the analysis grid: https://geodidaktik.uni-koeln.de/multimedia/analysebogen-zur-beurteilung-von-schriftlichen-argumentationen.

is usually written with the support of material (Budke 2021), which means that pupils write their argumentation on the basis of maps, diagrams, texts and pictures, the third part of the grid examines whether the pupils used the material provided by the teacher or researched by themselves and the content quality of the material reference in the argumentative text. In the last part of the grid, the structural quality of the argumentation is examined, referring to Toulmin's (1996) approach, which has been used successfully in many studies on argumentation quality measurement (e.g. Lam et al. 2018; Stapleton and Wu 2015). In addition, the content quality of the argumentation is determined, whereby the theoretical approach of Kopperschmidt (2016, pp. 62–64) was used as well as geography-specific approaches (Budke et al. 2015, p. 276).

Based on these theoretical considerations, a simplified diagnostic instrument was developed that teachers can use to diagnose their students' written argumentation skills (Fig. 1). This consists of five parts. Relevant criteria are used to assess the number of arguments in the text (1), the quality of the structure of the text (2), the quality of the presentation of the conflict in the text (3), the inclusion of relevant actors involved in the argumentation (4) and the quality of the integration of materials (maps, diagrams, texts, etc.) into the text to derive evidence for the arguments (5). In contrast to the scientific questionnaire, the assessment of the fulfilment of the quality criteria described above is not carried out precisely in the diagnostic tool for teachers, but is assessed with the categories "completely true", "partly true" and "not true".

Diagnostic Tool for Argumentative Text Products[6]

The following diagnostic tool (Fig. 1) can be used to diagnose the students' competencies for the written formulation of material-based argumentations. It can be used to give feedback to the students about their level of proficiency, but also to adapt to their own lesson planning to this level of proficiency. The tool is divided into five sections (number of arguments, general structure, argumentation on the conflict/problem, actors, material and material reference).

The Didactical Concept of the OER

The OER for the diagnosis of written argumentation skills of pupils in geography lessons was developed and implemented together with Sarah Schwerdtfeger, Laura Nieß and Laura Maßmann. The target group is student teachers of geography. It can

[6] The diagnostic tool is based on the questionnaire from the SpiGU project at the University of Cologne under the direction of Prof. Dr. Alexandra Budke, Prof. Dr. Alexandra L. Zepter and Dr. Diana Gebele. This is available under the following link: https://geodidaktik.uni-koeln.de/multimedia/raumnutzungskonflikt-innerer-gruenguertel-koeln.

		Examples/Explanation				Comments:
1. Number of arguments						
	Number of complete arguments formulated by the student.	Complete arguments contain an opinion, a proof and a validity relation that logically connects opinion and proof. These are marked in the text and then counted. The smileys on the right can be used to evaluate the number of complete arguments against the background of the topic, the materials and the class level.	Number of complete arguments: _____ ☺ appropriate number 😐 partly appropriate number ☹ inappropriate number			
2. General structure			☺ completely true	😐 partly true	☹ not true	
2.1	Student organises and structures his/her own text. Recognisable and text-progressively compre-hensible sections of meaning that appropriately support the intention of the presentation.	Introduction Main part Conclusion e.g. by paragraphs				
2.2	Student uses structuring terms in his/her text.	e.g. first; second, next; last; first				
2.3	Student concludes the text with a conclusion.	e.g. My conclusion is; for these reasons; as a result; I conclude that...).				
3. Argumentation on the conflict/problem			☺ completely true	😐 partly true	☹ not true	
3.1	Student correctly describes the conflict/problem being argued.	Describes the problem/issue being argued about, on which social groups have different views/interests correctly				
3.2	Student correctly reflects the spatial conditions of the conflict.	Student correctly describes the spatial relevance of the problem/issue being argued about. e.g. localisation of the place of conflict, effects on different scales, spatial action and reach of the actors.				
3.3	Student correctly reflects the temporal conditions of the conflict.	When did the conflict arise and how did it develop in time? e.g. Since MM/YYYY it has been discussed; by the year YYYY it should,...; By YYYY, it will be decided whether....				
3.4	Student positions him/herself on the conflict and gives his/her opinion on the conflict.	e.g. in my opinion; I am for/ for/ against/ against....				

Fig. 1 Diagnostic tool for argumentative text products, authors: Alexandra Budke, Sarah Schwerdtfeger, Laura Nieß and Rica Maßman

3.5	The student justifies his/her position in a meaningful way.	e.g. because; since; therefore; due to;				
3.6	Student formulates counter-arguments to his/her thesis.	e.g. against...; although... ;however....				
3.7	Student weighs up or invalidates counter-arguments.	e.g. although...but; both..., and; nonetheless; nevertheless				
4. Actors			☺ complete ly true	😐 partly true	☹ not true	
4.1	Student names the relevant actors for the issue/conflict.	e.g. local residents, business representatives, environmentalists				
4.2	Student correctly reflects the positions of the actors.	The opinions/views/arguments of the actors in the conflict are correctly reproduced and used in one's own argumentation. e.g. The business representatives are of the opinion that...,				
5. Material and material reference			☺ complete ly true	😐 partly true	☹ not true	
5.1	Student uses information from provided or researched material in own text and presents the information correctly.	Information contained in the materials is used to form the arguments and is used correctly. Correct means that the information used for the arguments corresponds to that of the material or that the information beyond the material is verifiably correct.				
5.2	The student makes a linguistic reference to the source in his or her text.	*-As described in material X, ...* *-Material X shows/evidences/illustrates/...* *-On the basis of the map/diagram from material X it can be read that ... - M1; x says/writes etc. that.*				
5.3	Student compares materials and links information in a meaningful way.	*e.g. Both material X and material Y show that ...* *e.g. Compared to; comparable with; ...*				
Total:						
Comment:						

Fig. 1 (continued)

be accessed through the following link: https://www.ilias.uni-koeln.de/ilias/goto_uk_lm_4679282.html.

The OER was technically implemented on the learning platform "ILIAS" used by the University of Cologne, whereby many interactive learning elements (H5P elements) were integrated.

For the orientation of the users, information on the target group, the time required, the learning objectives, the structure and the authors is provided on the starting page (Chapter 1[7]). Further information such as literature, didactic commentary for lecturers on the use of OER in events and documents to download can be found at the end of the learning unit (Chapters 7–9).

The core content of the OER are Chapters 2–4. In Chapters 2 and 3, student teachers of geography learn about the theoretical basis of diagnosis in general and the diagnosis of pupils' argumentation skills using a diagnostic tool. The tool can be used to determine the quality of a pupil's argumentative,[8] material-based text and to give individual feedback to the learners on this basis. These two chapters are designed as self-learning units in which new knowledge content is not only presented in a multimedia format, but also practised and repeated. In contrast, Chapter 4 consists of a complex application task for which no standardised solution is offered. The students' results should be discussed during university courses. Rounding off the content part, Chapter 5 contains questions for reflecting on learning gains. The exact didactic implementation of the self-learning unit is explained below.

Self-Learning Unit

The self-learning unit (Chapters 2–3) consists of materials for knowledge acquisition on the one hand and interactive practice and repetition elements on the other. The knowledge acquisition begins with the fictitious case study of a teacher who does not know how to give competent feedback to his student on her argumentative text (see Fig. 2).

This introduction should sensitise the students to the importance of pedagógical diagnosis in general and the diagnosis of argumentation competencies through the evaluation of argumentative texts in geography lessons and motivate them to work on the self-study unit. The situation in the comic is taken from everyday classroom life and invites the prospective teachers to identify with the comic figure of the teacher. The subject content is then offered as an interactive presentation. In addition, the diagnostic tool with all its criteria is presented and explained in detail. One criterion that can be used to diagnose the argumentation quality of the student text is, for example, its general structure (see Fig. 3).

The application of each criterion is practised using examples from one pupil texts. The students have to use the respective criterion to examine and evaluate the texts. For example, they have to mark in the text where the respective criterion is to be applied (see Fig. 4). The examination of the learning gain is done, i.e. by multiple choice answers.

[7] By the term "chapter", we refer to the chapters in the OER.

[8] A text written by a pupil of an 8th grade of a grammar school (ger.: *Gymnasium*) in Germany was used. His text was not written within a lesson at school, but as a test run at home.

Fig. 2 Comic as an introduction and to raise awareness for the diagnosis of argumentation skills (screenshot)

General structure

Now turn to the second part of the diagnostic sheet.

2. General structure			completely true	partly true	not true	
2.1	Student organises and structures his/her own text. Recognisable and text-progressively comprehensible sections of meaning that appropriately support the intention of the presentation.	Introduction Main part Conclusion e.g. by paragraphs				
2.2	Student uses structuring terms in his/her text.	e.g. first; second, next; last; first				
2.3	Student concludes the text with a conclusion.	e.g. My conclusion is; for these reasons; as a result; I conclude that...).				

Fig. 3 The individual parts of the diagnostic questionnaire are presented and explained in the learning unit (screenshot)

Application Task

The learned knowledge is to be applied in Chapter 4 in a challenging, authentic and creative task. The students are to assess the quality of an authentic pupil's argumentation using the entire diagnostic tool and give feedback to the pupil. Thus, this task ties in with the classroom situation illustrated by the comic strip in the introduction, in which teachers want to give their pupils qualified feedback on their texts. The

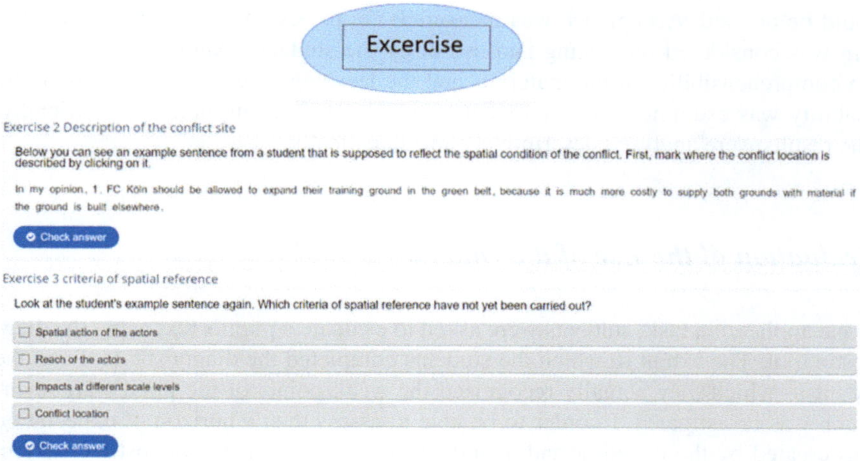

Fig. 4 Exercises in the self-study unit

students now have to apply the individually learned criteria in context and weight them in their feedback. The diagnostic tool is provided as a Word document so that students can fill it out digitally and also use it in their future teaching practice.

Methodological Approach to the Evaluation

The evaluation and investigation of the students' use of the diagnostic questionnaire to determine their diagnostic competence was carried out with undergraduate student teachers for Hauptschule, Realschule, Sekundarschule and Gesamtschule at the Institute for Didactics of Geography at the University of Cologne. The 18 students had attended the bachelor's seminar "Teaching and learning argumentation with digital geomedia" of the subject geography, in which the learning unit was carried out and the application task was discussed. However not all 18 students completed every section of the diagnostic tool. Therefore, the sampling size in the evaluation of the use of the diagnostic tool varies between 15 and 18 students depending on how many students have completed the respective section of the diagnostic tool.

Evaluation of the Learning Module

After working with the learning module, the students evaluated it. They answered closed questions using a Likert scale. In addition, comments could be made in a free text field. Based on Arnold et al. (2018, p. 398), criteria for the evaluation could be defined. Acceptance was measured, i.e. the extent to which the learning unit was considered interesting and useful by the students. Another criterion was the comprehensibility of the materials and the learning objectives. In addition, the usability was examined, i.e. the extent to which the learning unit is user-friendly. The results were analysed descriptively and statistically.

Evaluation of the Use of the Diagnostic Tool

In the application task, students were asked to evaluate a pupil's text using the diagnostic tool. The extent to which the students completed the diagnostic tool and the extent to which they actually recognised the weak points of the pupil's argumentation was investigated. In order to be able to assess this, a horizon of expectation was created by the coauthor and a student assistant, for each diagnostic criterion. If a student completely correctly evaluated a diagnostic criterion, (by evaluating the pupil's text) two points (2 P.) were assigned to the student. If the student evaluated a diagnostic criterion partially correctly, one point (1 P.) was assigned to the student. If the students did not correctly evaluate a diagnostic criterion, zero points (0 P.) were assigned to the student. For each criterion, each student could be assigned up to two points. These points were assigned by the coauthor and a student assistant, who came to similar ratings. The scores give an overview of how well the students can diagnose using the questionnaire. The results were analysed descriptively and inferentially.

Results

In the following, the results of the evaluation of the learning module and the evaluation of the use of the diagnostic tool by students are presented.

Evaluation

After using the OER in the seminar, we asked ourselves how interesting, useful, appropriate, understandable and relevant the students found the work with the OER. On average, the students rated the OER as "good" (2.2). The positive overall result is also supported by the evaluation of the individual criteria for evaluating the learning unit (see Fig. 4). Of the 18 students that participated in this study, 11 filled out the sheet to evaluate the learning module. A large proportion of these 11 respondents consider the content of the learning module to be relevant to their studies and future teaching practice. To the question of what they particularly liked, one person wrote in this context:

> The topic of the learning module, as this is important for later everyday working life.

In addition, all of the respondents completely or partly agreed, that it would be useful to apply the diagnostic questionnaire they learned about in the OER in their teaching practice in the future. And the majority (10) completely or partly feel competent to do so (see Fig. 4). One person wrote in this context:

> The diagnostic questionnaire helps me a lot and I can use it again and again now.

This is encouraging as only one of the students had experience with diagnostic tools for argumentative texts before working with the unit. These results show that the students were able to understand the great importance of the pedagogical diagnosis of argumentative competencies through the learning unit and that the diagnostic tool is also evaluated very positively as a means of implementation. The acceptance of the OER is therefore very high.

In addition, the comparison of the answers in the pre-test and post-test to the question of which criteria the students would use to assess argumentative pupil texts shows that before working with the learning unit, the students would have corrected primarily according to the order of argumentation and spelling, and after working with the learning unit, they would also include, among other things, the reference to the material and the naming of actors and their argumentations. Here, the learning effect through the OER is evident, as the students got to know these criteria and applied them with the diagnostic tool.

The students also consider the level of difficulty of the learning unit and the time required to complete it to be appropriate for the most part (see Fig. 5). In addition, they find the learning unit easy to use and only very few technical problems were communicated. Thus, the usability seems to be given. The structure of the learning unit and the interactive elements were rated predominantly positively, but somewhat more negatively compared to the other criteria mentioned. It also fits in with this point that the design of the learning module was not rated as very interesting by everyone (see Fig. 5). In contrast, the authentic pupil's text in the application task was rated very positively. The pupil's text was mentioned by six people in the open question about what they liked:

> I particularly liked the fact that we could read and correct real texts written by pupils with their own handwriting.

Students' Competencies to Use the Diagnostic Tool

In the diagnostic tool, the 17 students[9] who worked on the application task[10] indicated the number of arguments in the pupil's text they analysed. The results varied between

---

[9] As mentioned before, not all of the 18 geography student teachers completed every section of the diagnostic sheet. The sampling size therefore varies between 15 and 18 students.

[10] The number of students varies in the individual categories because students sometimes formally misapplied the diagnostic questionnaire or filled it out incompletely.

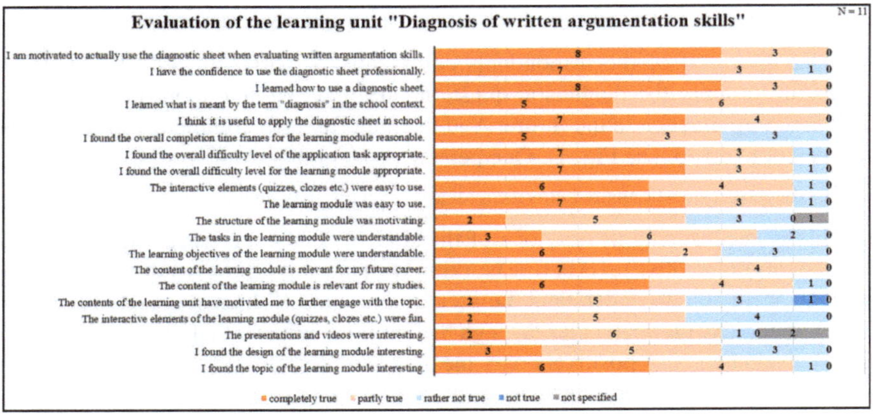

Fig. 5 Evaluation results (own presentation)

three and eleven arguments, which shows that the students still have difficulties in recognising arguments even after working on the OER. The 10 arguments in the student text were fully recognised by only one student. Most students gave a much lower number of arguments (see Fig. 6).

Using the diagnostic questionnaire, students also assessed the general structure of the pupil's text. In this category, the students identified the pupil's competencies completely correctly or partially correctly, with one exception (see Fig. 7).

In addition, the majority of the students assessed the pupil's competencies with regard to his representation of the conflict in the argumentation also either completely or partially correctly (see Fig. 8). For the most part, the students found it easy to recognise that the pupil describes the conflict in his text and reflects the spatial conditions. Furthermore, they were able to recognise that the pupil presents his own position and justifies it in a meaningful way. On the other hand, it was more difficult for the students to recognise that the pupil did not mention the temporal conditions of the conflict. Although the students were able to recognise that the pupil partly weighs

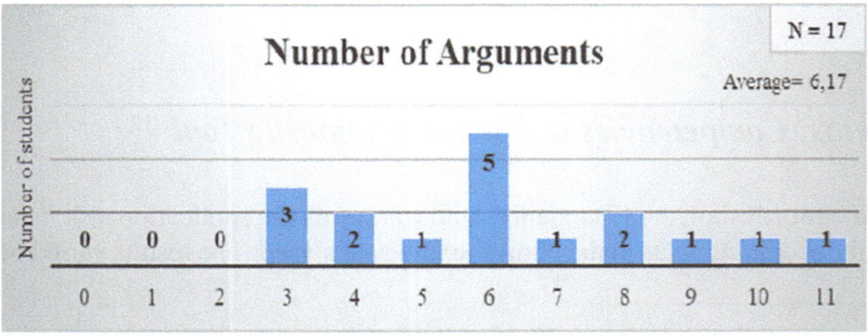

Fig. 6 Evaluation of the students' abilities to determine the number of arguments (own representation)

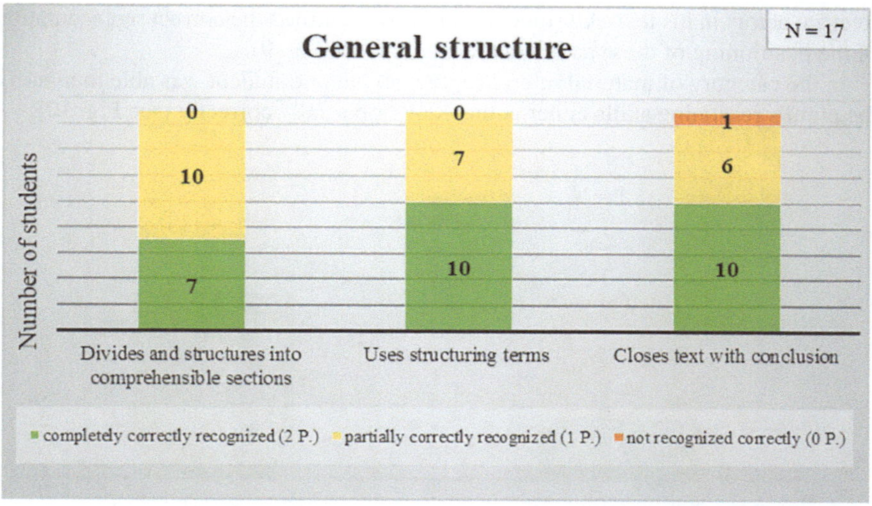

Fig. 7 Evaluation of students' skills in analysing the general structure of the pupil's text (own presentation)

up or invalidates counter-arguments, it was more difficult for them to recognise that the pupil formulates counter-arguments in his text.

Only one-third of the students were completely correct and the other two-thirds were partially correct in their assessment of the fact that the pupil partially names the

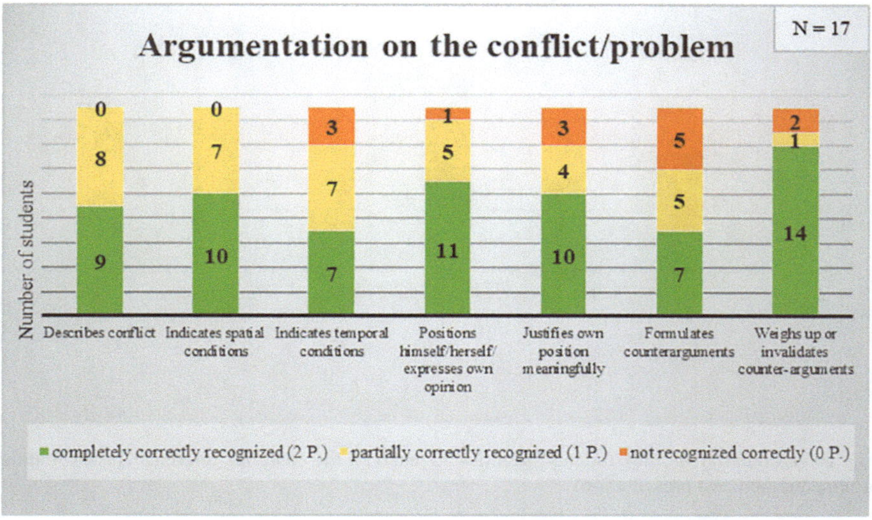

Fig. 8 Evaluation of the students' skills in analysing the presentation of the conflict in the argumentation (own presentation)

relevant actors in his text. All students correctly identified the correct representation of the positioning of these actors in the conflict (see Fig. 9).

In the category of material reference, too, all but one student was able to identify the pupil's reasoning skills either completely or partially correctly (see Fig. 10).

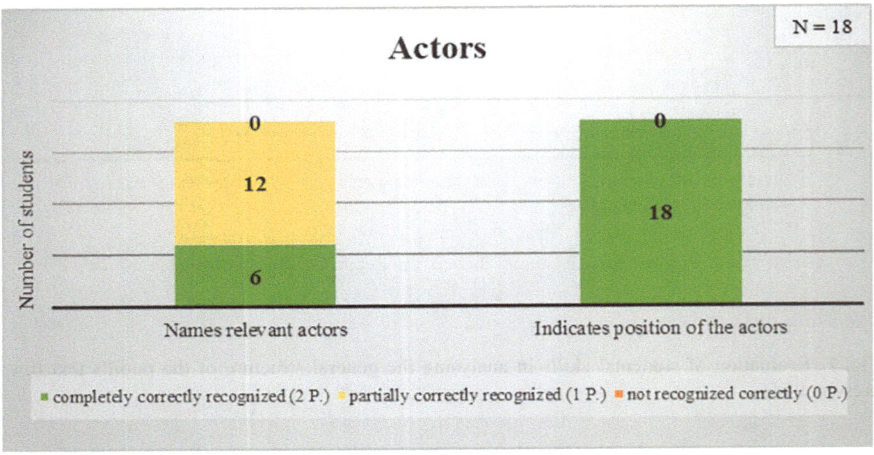

Fig. 9 Evaluation of the students' skills in analysing the representation of actors in argumentations (own presentation)

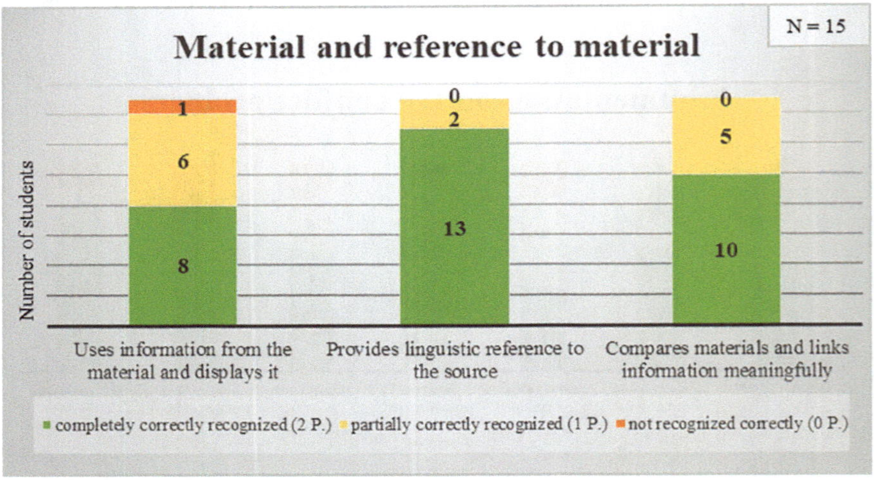

Fig. 10 Evaluation of the students' abilities to analyse the material reference in the pupils' argumentation (own presentation)

Summary and Conclusion

The study showed that after working with the OER, the students were for the most part able to use the diagnostic tool for the diagnosis of written argumentation skills correctly and to apply it competently. In the vast majority of cases (93.7%), they assessed the quality of the pupil's text either completely correctly or partially correctly. The evaluation also showed a positive result, as the OER and the tool were considered highly relevant by the majority, the acceptance was high, as was the comprehensibility of the materials and the usability. It can thus be concluded that the OER created is suitable for the students of the Institute for Didactics of Geography at the University of Cologne and represents enriching material in their studies. The extent to which it produces equally positive results at other locations cannot be answered on the basis of the sample. The diagnostic tool, which is at the centre of the OER, presumably has a particularly great transfer potential, since it was developed on the basis of existing argumentation-theoretical and empirical results on argumentation (see section "Theoretical Background").

Acknowledgements We would like to thank Lena Romeijn, Sebastian Seidel, Sarah Schwerdtfeger, Laura Nieß and Laura Maßmann for their help in creating and evaluating the OER.

Funding This research was funded by the German Federal Ministry of Education and Research (BMBF) as part of the project "Generalisability and transferability of digital subject concepts using the example of responsible digital geomedia use in teacher education" (DiGeo), funding reference 16DHB3003.

References

Arnold P, Kilian L, Thillosen A, Zimmer GM (2018) Handbuch e-learning Lehren und Lernen mit digitalen Medien. W. Bertelsmann Verlag, Bielefeld

Benetos K, Bétrancourt M (2020) Digital authoring support for argumentative writing: what does it change? Journal of Writing Research 12(1):263–290

Budke A, Kuckuck M (2020) Kartenbasierte Argumentationen in der Geographielehrerinnenbildung – eine empirische Untersuchung zur Effizienz von Strukturierungs- und Formulierungshilfen. In: Hemmer M, Lindau A-K, Peter C, Rawohl M, Schrüfer G (eds) Lehrerprofessionalität und Lehrerbildung im Fach Geographie im Fokus von Theorie, Empirie und Praxis. Ausgewählte Tagungsbeiträge zum HGD-Symposium 2018 in Münster (= Geographiedidaktische Forschungen 71). Readbox Unipress, Nürnberg, pp 177–190

Budke A, Creyaufmüller A, Kuckuck M, Meyer M, Schlüter K, Weiss G (2015) Argumentationsrezeptionskompetenzen im Vergleich der Fächer Geographie, Biologie und Mathematik. In: Budke A, Kuckuck M, Meyer M, Schäbitz F, Schlüter K, and Weiss G (eds) Fachlich argumentieren lernen. Didaktische Forschungen zur Argumentation in den Unterrichtsfächern. Waxmann Verlag, Münster, pp 273–297

Budke A, Gebele D, Königs P, Schwerdtfeger S, Zepter AL (2020) Student texts produced in the context of material-based argumentative writing: interdisciplinary research-related conception of an evaluation tool. RISTAL 3:108–125

Budke A (2016) Potentiale der Politischen Bildung im Geographieunterricht. In: Budke A, Kuckuck M (eds) Politische Bildung im Geographieunterricht. Franz Steiner Verlag, Stuttgart, pp 11–23

Budke A (2021) Bedeutung des argumentativen Schreibens im Geographieunterricht im Kontext aller Schreibtätigkeiten—eine empirische Studie auf der Grundlage von Unterrichtsbeobachtungen. In: Budke A, Schäbitz F (eds) Argumentieren und Vergleichen. Beiträge aus der Perspektive verschiedener Fachdidaktiken. Berlin, pp 57–76

Chase BJ (2011) An analysis of the argumentative writing skills of academically underprepared college students. Dissertation, Columbia University

Deutsche Gesellschaft für Geographie (DGfG) (2014) Bildungsstandards im Fach Geographie für den Mittleren Schulabschluss mit Aufgabenbeispielen. Selbstverlag DGfG, Bonn

Feilke H (2014) Argumente für eine Didaktik der Textprozeduren. In Bachmann T, Feilke H (eds) Werkzeuge des Schreibens. Beiträge zur Didaktik der Textprozeduren. Fillibach Verlag, Stuttgart, pp 11–34

Gronostay D (2017) Enhancing the quality of controversial discussions via argumentation training—a quasi-experimental study in civic education classrooms. Bildung Und Erziehung 70(1):75–89. https://doi.org/10.7788/bue-2017-0107

Kopperschmidt J (2016) Argumentationstheorie zur Einführung. Junius Verlag, Hamburg

Lam YW, Hew KF, Chiu TKF (2018) Improving argumentative writing: Effects of a blended learning approach and gamification. Lang Learn Technol 22(1):97–118. https://dx.doi.org/10125/44583

Lytzerinou E, Iordanou K (2020) Teachers' ability to construct arguments, but not their perceived self-efficacy of teaching, predicts their ability to evaluate arguments. Int J Sci Educ 42(4):617–634

Maier V, Budke A (2018) Wie planen Schüler/innen? Die Bedeutung der Argumentation bei der Lösung von räumlichen Planungsaufgaben. GW-Unterricht 149:16–29

Stapleton P, Wu YA (2015) Assessing the quality of arguments in students' persuasive writing: a case study analyzing the relationship between surface structure and substance. J Engl Acad Purp 17:12–23. https://doi.org/10.1016/j.jeap.2014.11.006

Steinseifer M (2018) Materialgestütztes Schreiben und schreibintensiver Deutschunterricht—Reflexionen zum Stand der Debatte. In: Feilke H, Lehnen K, Rezat S, Steinmetz M (eds) Materialgestütztes Schreiben: Erfahrungen aus der Praxis und Perspektiven der Forschung. Fillibach Verlag, Stuttgart, pp 13–26

Toulmin S (1996) Der Gebrauch von Argumenten. Beltz Athenäum Verlag, Weinheim

Uhlenwinkel A (2015). Geographisches Wissen und geographische Argumentation. In: Budke A, Kuckuck M, Meyer M, Schäbitz F, Schlüter K, Weiss G (eds) Fachlich argumentieren lernen. Didaktische Forschungen zur Argumentation in den Unterrichtsfächern. Waxmann Verlag, Münster, pp 46–61

von Aufschnaiter C, Erduran S, Osborne J, Simon S (2008) Arguing to learn and learning to argue: Case studies of how students' argumentation relates to their scientific knowledge. J Res Sci Teach 45(1):101–131

Witt S (2015) Pädagogische Diagnostik. www.die-bonn.de/wb/2015-paedagogische-diagnostik-01.pdf. Accessed 17 Dec 2021

Development of the Online Geospatial Problem-Solving Instrument: Investigating Elementary Students' Perceptual Processes in Geospatial Problem-Solving

Christos Vonapartis Kosmidis and Nikos Lambrinos

Abstract Maps are unique in making symbolic representations accessible even for very young students. The reason for this might be the affinity of the interaction with geospatial representations to direct spatial experiences with physical objects in the physical environment. Could this mean that these two types of interactions share additional common elements? Thus, it would be useful to find out if research findings and theories on the perception of physical objects could be used in explaining children's interaction with maps. One approach explaining the interaction with physical objects that could provide useful insights in this direction is the action-specific account of perception. In our chapter, we will describe the development of an online instrument with which we will attempt to identify if the action-specific account of perception can be applied in elementary students' problem-solving using geospatial representations.

Keywords Geospatial problem-solving · Action-specific account of perception · Geography education · Online instrument

Introduction

Abstract scientific representations can be used successfully by learners when they have reached the formal operational stage (Moore and Slisko 2017; Dickerson et al. 2007; Goodstein and Howe 1978). On the other hand, preschool and early elementary schoolchildren can use maps to complete simple tasks (Uttal et al. 2006; Huttenlocher et al. 2008; Yuan et al. 2017; Salsa et al. 2019). This early accessibility of maps for

C. V. Kosmidis (✉) · N. Lambrinos
Faculty of Education, School of Primary Education, Aristotle University of Thessaloniki, Thessaloniki, Greece
e-mail: ckosmidi@eled.auth.gr

N. Lambrinos
e-mail: labrinos@eled.auth.gr

A. Klonari et al. (eds.), *Re-visioning Geography*, Key Challenges in Geography,
https://doi.org/10.1007/978-3-031-40747-5_2

younger learners may be a result of the primacy of the spatial experience, the fact that children make sense of the world spatially and through their personal geographies, and the familiarity with representational means used in maps, (i.e., the schematization of an aerial view) (Davies and Uttal 2007; Kosmidis and Lambrinos 2018; Catling 2020).

Could this uniqueness of maps as a symbolic representation and the association of maps with direct physical perceptual experiences mean that there are greater similarities between perceptual processes performed when interacting with maps and when interacting with physical objects? Could we then apply research findings and theories on the perception of physical objects in the interaction with maps? If this is the case, we could use this knowledge to enhance and support problem-solving using maps in school education.

To examine this possibility, we developed an online geospatial problem-solving instrument. The instrument is based on the action-specific perception account that is mostly applied in interacting with physical objects. In our instrument, we apply the action-specific perception account in problem-solving using geospatial representations.

The instrument is online and uses a dynamic website. This allows the dynamic and automatic switching of images and tasks and the effective collection of data. An additional advantage of the online nature of the instrument is to face the multiple challenges that the scientific community faces to conduct research during periods like the COVID-19 pandemic, in which access to schools and other institutions was restricted.

In our chapter, we first describe the action-specific perception account and its experimental applications. We outline how we used the processes and assumptions of the action-specific account in our instrument to test its application in elementary students' geospatial problem-solving. We describe the selection of the spatial thinking components addressed in the instrument and how they were addressed in the instrument's tasks. We continue with the technical analysis of the instruments and explain how we used various programming tools for user interface, backend, and frontend development and how graphic design tools were used for the creation of the representations used in the instrument. Finally, we will describe the pilot testing of the instrument and how we will proceed with its application.

The Action-Specific Perception Account

The action-specific perception account proposes that the perception of the environment is affected by the perceiver's ability to act (Philbeck and Witt 2015). For example, wearing a heavy backpack affects the perceived steepness of a hill (Bhalla and Proffitt 1999; Latin et al. 2019), a batter's hitting skills in softball affect the perceived size of a softball (Witt and Proffitt 2005) and the perceived speed of a tennis ball is affected by the player's ability to return the ball (Witt and Sugovic 2010). In these experiments, the effective completion of the task was accompanied by perceptual processes that facilitated it. In the tennis experiments, the players judged the ball to be moving faster when they hit the ball out-of-bounds than on trials where they successfully hit the ball in-bounds in which they judged the ball to move slower. Obviously, the action-specific perception account shares many common elements with Gibson's Affordance Theory. The affordance theory states that the subject perceives the environment through the opportunities for action (affordances) it provides. The subject and the environment construct a unified dynamic system, in which the mind is not a passive information collector but shapes a dynamic reality through perception (Gibson 1979; Chemero 2011).

These perception theories are based on experiments with physical objects and environments. Based on our previous argumentation, it would be useful to find out if this research and findings could provide useful insights into the interaction with symbolic representations and more specifically maps. If the action-specific perception account could be applied to these types of symbolic representations, we would get useful insights that would help us understand how students of various problem-solving skills perceive them. These insights would help differentiate the resources we use depending on the student's skills making them more accessible and adaptable.

The Instrument

In our chapter, we will describe the development of an online instrument. The online instrument presented attempts to identify if the action-specific account of perception can be applied when elementary students solve problems using geospatial representations. The action-specific perception account proposes that the perception of the environment is affected by the perceiver's ability to act (Witt and Riley 2014). Through the application of our instrument, we will investigate if the level of complexity that 5th and 6th graders perceive geospatial representations is affected by their ability to solve a problem using them.

In the instrument, eight spatial thinking components are represented in 12 problem-solving questions using age-appropriate geospatial representations. Each problem-solving question is accompanied by a task in which we identify the level of complexity the learner perceives relevant representations. As shown in Fig. 1 on the first page of the question, the geospatial representation used is projected for two seconds. On the second page of the question, students are asked to select the representation they believe they saw on the first page which is not projected anymore. Each representation on the second page is based on the one presented on the question's first page, but they are characterized by different levels of complexity. Finally, on the problem's third page, a problem using the representation of the first page is presented, and students select the correct answer.

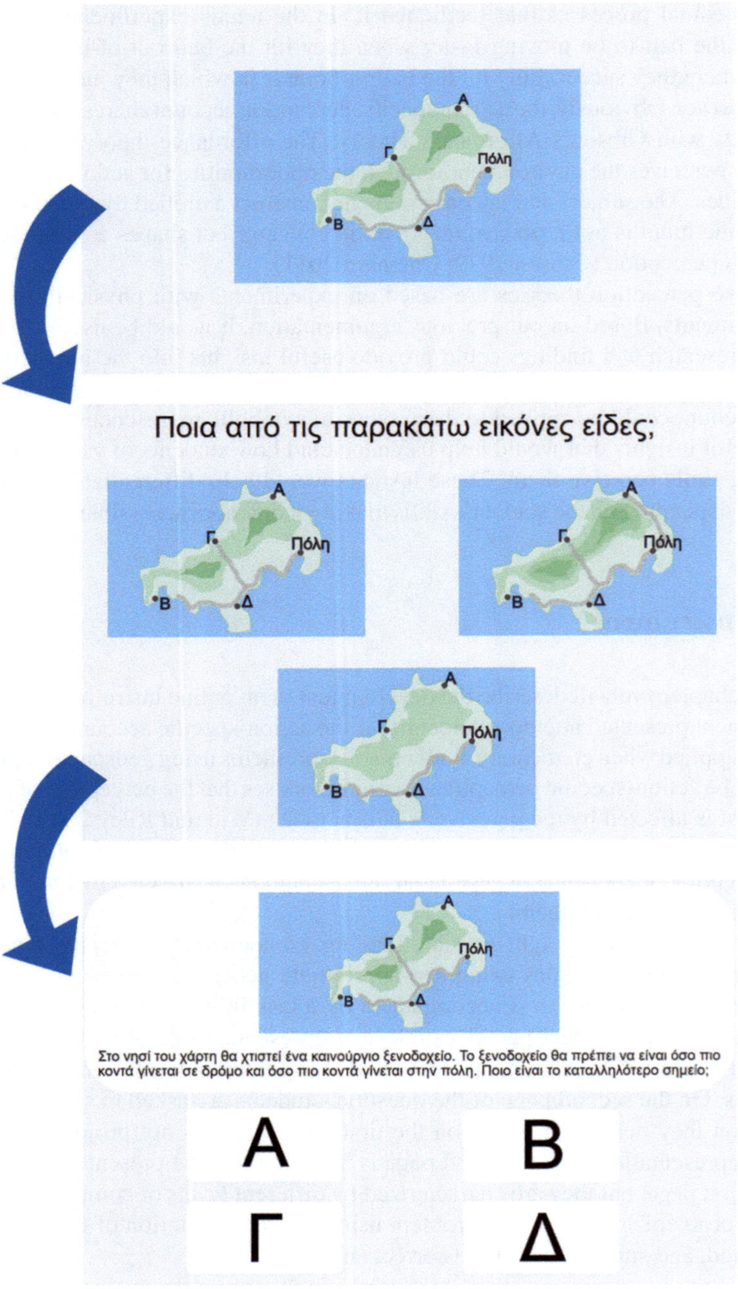

Fig. 1 The sequence followed in each question of the instrument

Spatial Thinking Components

The instrument we were developing was going to assess problem-solving using geospatial representations. Our priority was not to focus on the student's knowledge of maps and mastery of geographical information but on their problem-solving skills in a geographical context. Problem-solving in this case would be an expression of their spatial thinking skills. When discussing spatial thinking assessment, we usually think of the well-known paper and pencil, psychometric, and spatial ability tests (Harle and Towns 2011). These tests usually don't use a geographical context and focus on spatial visualization and spatial orientation.

Problem-solving is a broader activity that uses a larger number of elements and a combination of processes. Freksa et al. (2017) describe spatial problems as a question about a spatial layout or the construction of a spatial layout, with specific properties based on the original. Some actions that may accompany the solution of such problems are:

- Rotation, the circular movement of an object around a given position.
- Movement from one position to another.
- Deformation of objects.
- Separation of objects into parts.
- Assembling objects.
- Combined actions like rotating around a moving position.

These actions are quite general and help us categorize the various applications of spatial problem-solving in various fields. On the other hand, it's challenging to relate them to actual problem-solving activities in a geographical educational context. Geospatial problem-solving is an expression of spatial thinking in a geographic context. The alternating use of terms like spatial intelligence, spatial ability, spatial skills, and spatial thinking can lead to confusion.

The National Research Council & Geographical Sciences Committee in *Learning to think Spatially* (2005) presented a list of spatial thinking components that are typical in geosciences. These components included describing the shape of an object, identifying or classifying an object by its shape, attributing meaning to the shape of a natural object, recognizing a shape among a noisy background, visualizing a three-dimensional object based on one or two-dimensional observations, using a coordinate system to describe the position and orientation of objects in the real world, remembering the location and appearance of an object, envisioning the motion of objects through space in three dimensions, envisioning the shape change of objects, using spatial thinking to think about time and considering multidimensional systems with axes that are not distance. Although these spatial thinking components are typical in spatial thinking activities in geosciences they are usually used combined in problem-solving. This makes their assessment challenging, especially if we intend to use them in a problem-solving instrument.

Gersmehl and Gersmehl (2007) and Golledge et al. (2008) suggest two spatial thinking taxonomies that share common elements and are useful in making connections to actual applications in geographic education in schools. Lee and Bednarz (2009) incorporated these spatial thinking components in the Spatial Thinking Ability test (STAT). They address eight spatial thinking components that are listed below:

- Comprehending orientation and direction
- Recognizing patterns in a map and representing them in a graphic form
- Selecting an ideal location based on the given spatial features
- Visualizing a slope profile based on a topographic map
- Identifying spatial correlations by comparing patterns featured in a set of maps
- Transforming representations and images from one dimension to another
- Understanding types of overlaying
- Comprehending of geographic features represented as points, lines, or polygons

The clear connection between these spatial thinking components to actual educational geographic applications has allowed researchers to use STAT in many countries, age levels and types of interventions. The test has been validated by administering it to junior high, high school, and university students, and although it has been used with Elementary students (Bednarz and Lee 2019), changes and adaptations had to be made to use these spatial thinking components in our instrument.

Development of the Instrument

The instrument includes 12 questions. Each question consists of two sub-questions. The first sub-question identifies the students' perceived level of complexity of a specific geospatial representation and the second sub-question uses the same geospatial representation in a problem-solving activity.

For the first sub-question, the initial representation is projected on the first page for 2 s. After this, three multiple-choice answers are projected. Each one of them is a representation based on the initial one, but they are characterized by different levels of complexity. The student selects which one she or he believes is identical to the one projected on the first page of the question.

For the second sub-question, the user solves a geospatial problem. The STAT spatial thinking components were used for the development of these problems. Considering that the instrument would be used by 5th and 6th graders, we made some adaptations. The eight spatial thinking components were addressed in 12 instead of 16 questions that were used in STAT (Table 1).

One more adaptation was required considering the geospatial representations used in the instrument. To assess problem-solving skills, we had to make sure that geographical knowledge and familiarity with the geospatial representations would not be the crucial factors in answering the questions. The instrument will be used by 5th-grade and 6th-grade students in Thessaloniki, Greece. The types of representations selected were types of representations that are used in textbooks of the Greek

Table 1 Description of spatial thinking components addressed in the instrument's questions

Question number	Spatial thinking components to measure
#1, #2	Comprehending orientation and direction
#3	Recognizing patterns in a map and representing them in a graphic form
#4	Selecting an ideal location based on the given spatial features
#5	Visualizing a slope profile based on a topographic map
#6, #7	Identifying spatial correlations by comparing patterns featured in a set of maps
#8	Transforming representations and images from one dimension to another
#9. #10	Understanding types of overlaying
#11, #12	Comprehending geographic features represented as points, lines, or polygons

Adapted from the Spatial Thinking Ability Test (Lee and Bednarz 2019)

Educational system in various subjects in grades one to four. Additionally, we tried to minimize the effect of geographical knowledge of places by using non-existing geographical areas or existing geographical areas with geographical information not related to the student's personal experience, e.g., the precipitation map of Crete (Fig. 2).

Technical Analysis

Design and User Interface

To make this instrument as accessible as possible in any place at any time, there was the need to host it on a webpage and provide it to students. It was difficult to find the appropriate web tool that would suit our needs. The instrument can be characterized as a quiz with several questions. Each question has two sub-questions: in the first, an image is projected for a specific time and then a multiple-choice question is answered. It became a challenge to find these applications in the available quizzes on the web. We researched resources that were part of Learning Management Systems (LMS) like Moodle or a number of standalone quizzes/surveys but even if there were some interesting approaches, there was no available management system that we could use to customize it to our needs. The applications that were not supported were timed transitions and/or the use of images and charts for multiple-choice answers. Therefore, it was decided to design and develop a custom webpage from scratch that will host our instrument. At the beginning of the design process, it was decided that this webpage should be a dynamic webpage, where questions, answers, timings, etc., could be stored in a database and could be used dynamically. This way, the content could be easily updated.

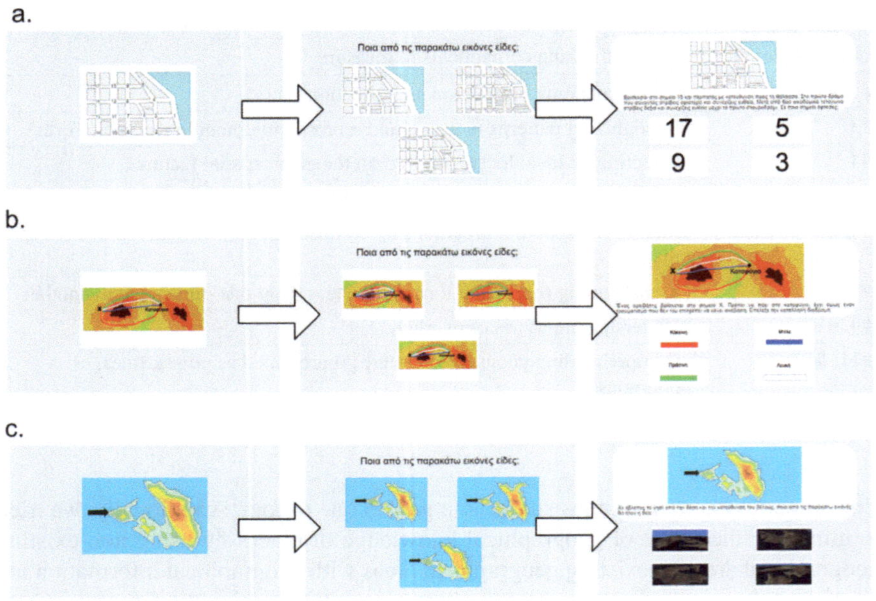

Fig. 2 A sample of three questions from the instrument. The first two slides in each example present the first sub-questions that identify the level of perceived complexity. The third slide presents the problem-solving question. The spatial thinking components addressed in each example are: **a** comprehending orientation and direction; **b** visualizing a slope profile based on a topographic map; and **c** transforming representations and images from one dimension to another

Before δεσψριβινγ the functional details of the instrument, there was a major priority: to design the access process of the webpage and to manage students' data. It was decided that this webpage, and more particularly the participation in the quiz, would not be accessible to anyone via URL. It was going to be a two-level authorization: each school will have a specific password, to access the instrument and each student at this school will have her or his own password in order to start using it. Both levels will be communicated to the relevant school in a secure way. Concerning students' data, we do not store any kind of sensitive data, because this quiz is anonymous and the mentioned students' passwords are the needed information for a particular student to be a unique user of our system. Any other data that are referring to the quiz, like the student's answers, are stored in a database for further statistical analysis.

In the process of designing the instrument's user interface (UI), it was decided that since the instrument's content is mostly colorful images, maps, etc., it was necessary to have a clear, minimalistic environment. The student should not be distracted by any kind of "visual noise" and be fully focused on the provided task. Also, all kinds of animations on this web page (by animations we describe any kind of frame swapping, either from one question to another or from one sub-question to another) will take place instantaneously, without fading modes or delays, to achieve the maximum student's focus on the task.

We will analyze the frontend and backend development and the tools that were used to develop the instrument.

Frontend Development

Frontend development includes any code development that needs to be written, for the web page to have a specific appearance. It indicates the visual structure of the instrument: the background colors and borders, the positions of the dynamic content, the animations of the pages, etc. It was developed in HTML and CSS. Basically it is a single-page website, where through HTML a specific division renders all the content of the quiz. Part of the HTML code that highlights the specific part of the webpage where all content appears (the div with the class named container), is presented below:

```
<div class="container">
   <div class="row">
      <div class="col-md-12">
     <div id="wrapper">
      <div class="jumbotron text-center">
            <a id="start" class="btn btn-primary btn-lg" href="#"
role="button" onclick="">Start!</a>
      </div>
      <div class="row justify-content-md-center" id="subwrapper">
      </div>
     </div>
      </div>
   </div>
</div>
```

All background colors, letter font sizes and color of the buttons, borders and positions of the images are defined through CSS and the code is presented below:

```
body {
   background-color: #CDCEDA;
   font-family: Arial, sans-serif !important;
}

#subwrapper {
   background-color: #eeeeee;
   border-radius: 30px;
   padding: 2% 2% 5% 2%;
}


.answer-button, .answer-button-main-options {
```

```
    cursor: pointer;
}

.answer-button:hover, .answer-button-main:hover {
    border: 1px #000000;
}

.imageContent {
    display: flex;
    justify-content: center;
    padding-top: 1%;
    margin: 0px auto;
}
```

The design of the webpage as a template have been set through HTML and CSS and the transitions of the images and questions are pending implementation. All those kinds of animations were implemented using JavaScript and JQuery. The part of the JavaScript code that is presented below actually shows how a question, image, and answers from the first sub-question load in a specific position and how it transitions to the next sub-question or if there is no question left, to the end of the quiz:

```
loadMainQuestion: function (){
    clearInterval(timer);
    timer = setInterval(game.countdown,1000);
      //$('#subwrapper').html('<div class="col-md-12 progress">
<div id="counter3" class="progress-bar" role="progressbar"
aria-valuemin="0"></div></div>');
      $('#subwrapper').html('<div class="col-md-12 imageCon-
tent" style="max-height: 250px;"><img src="assets/img/
'+game.questions[game.currentQuestion].image+'"/></div>');
        $('#subwrapper').append('<h5 class="col-md-12 text-
center">'+game.questions[game.currentQuestion].question+'</
h5>');
   for(var i=0; i<game.questions[game.currentQuestion].answers.length;
i++){
        $('#subwrapper').append('<div class="col-md-6 imageCon-
tent" style="max-height:120px;"><img class="answer-button"
id="'+game.questions[game.currentQuestion].answersIds[i]+'"
data-name="'+game.questions[game.currentQuestion].questionId+'"
src="assets/img/'+game.questions[game.currentQuestion].answers[i]+'"
/></div>');
    }
},


nextQuestion: function(){
    game.counter=120+1;
    game.counterMain=4+1;
    game.counterMainOptions=4+1;
    game.counterMainQuestion=30+1;
    game.currentQuestion++;
```

```
    if (game.questions[game.currentQuestion].questionId == 27 ||
game.questions[game.currentQuestion].questionId == 31) {
        game.loadMainQuestion();
    } else {
        game.loadIntro();
    }
},
timeUp: function(){
    clearInterval(timer);
    //game.unanswered++;
        $('#subwrapper').html('<h2>Ο χρόνος τελείωσε! Αναμένετε για τον
επόμενο γύρο ερωτήσεων<h2>');
    if(game.currentQuestion === game.questions.length-1){
        setTimeout(game.results,3*1000);
    } else{
        setTimeout(game.nextQuestion,3*1000);
    }
```

Backend Development

The backend development was separated into three tasks: the authorization process, the dynamic data needed to render the instrument, and the students' data storage.

Concerning the authorization, a service was developed that could generate endless random passwords that were then encrypted and stored in the database for each school and each student. A method was created that its core is based on providing a list of available legit characters and choosing as many times as the user wants (this is actually the length of the password). Using a PHP function that generates cryptographically secure pseudo-random integers, we can indicate the position of the characters in a random way. When the length of the password is defined, then the password is encrypted, and it is provided for any use. When the school has its own password, the webpage is accessible to them, and a list of available student slots (with their own individual passwords) is ready for students to use in order to start using the instrument. When a student is logged in via a slot, this slot is locked, and no other student can use this slot.

Concerning the instrument and student data, we designed the following models in the database, and codebase too:

1. **Quiz:** We can create multiple quizzes that can be available to any school. "Quiz" has a unique id and title.
2. **Questions:** Each quiz can have multiple questions. A question has a unique id, the quiz id that is related to, the text, the type that indicates in which sub-question we are, the image URL, and the order it appears in the question.
3. **Answers:** Each question can have multiple answers. An answer has a unique id, the question id that is related to it, and an image URL. All answers in the instrument are provided as images because most of the answers are graphs, maps, and images.

4. **School:** In this entity, we keep the name of the school, an id, and its encrypted password.
5. **User:** The "user" is actually the student slot that we described above, where a school id and an encrypted password are connected.
6. **User Answers**: Finally, in the last entity we keep the most important data for our research, students' answers. A user answer includes a unique id, user id, question id, answer id, and the date that this question was answered.

The codebase was written in PHP using Object Oriented Programming and more particularly MVC (Model View Controller) programming. Data was stored and managed through MYSQL. Part of the code for this section that shows how PHP is connected to MySQL is presented below:

```
private function userExists($userId) {
    try {
        $SQL = "SELECT COUNT(*) FROM user WHERE id =". $userId;
        $result = $this->connect()->prepare($SQL);
        $result->execute();
        return ($result->rowCount() > 0);
    } catch (Exception $e) {
        die('Error QuizConstroller(user) '.$e->getMessage());
    } finally{
            $result = null;
    }
}

public function saveUserAnswers($data) {
        if   ($this->userExists($data['user_id'])   &&   $this-
>answerExistsInQuiz($data['question_id'],    $data['quiz_id']))
{
    return $this->saveUserQuizData($data);
    }
    return false;
}
```

Additional Tools

During the process of designing, developing, and delivering the instrument hosted on a webpage, additional tools were used. For the user interface, all templates of the webpage, as well as all figures, images, and maps that are used as content in the instrument were designed from scratch using Affinity Designer and Blender. The code development that includes frontend and backend development was implemented using NetBeans as an integrated development environment (IDE). For our database, we used PhpMyAdmin. PhpMyAdmin is a user-friendly environment in which, by using MYSQL, we can manage all kinds of data that are stored in the database.

For content management and hosting, we used Parallel Plesk a useful environment that handles the server settings that the webpage is going to be hosted, the code files, the content, and the database (Table 2).

Pilot Testing

The online instrument was administered to fourteen 5th and 6th grade students from Thessaloniki, Greece. Students understood what they had to do when they accessed the instrument. All of them managed to complete the test earlier than the time they had available (45 min). All answers were collected and stored in the database. Two students needed to restart the test. The problem that occurred was that multiple answers were stored for single questions when students clicked on their selected answer multiple times and the instrument crashed. The problem was solved by allowing only one answer for each question.

Comments and Next Steps

The online and interactive nature of the instrument allowed us to define the setting of our investigation in detail, having a timed interchange of tasks, using various types of representations, and making data collection possible.

During the COVID-19 pandemic, researchers faced multiple challenges that led to the postponement of data collection, an adaptation of the research goals, and the following of emergency institutional guidelines (Santana et al. 2021). These compromises can lead to an overall degradation of the research process (Reichertz 2021). Our instrument makes the continuation of our research possible even during challenging periods when restrictions are imposed on free movement and accessibility.

So far, we presented how we developed our instrument and how we overcame the challenges we faced. At the next stage, the online instrument will be completed by approximately 600 students from 20 elementary schools in Thessaloniki, Greece. The schools have been selected through a stratified nested sampling method which ensures the participation of students from all municipalities and diverse socioeconomic backgrounds. The statistical analysis of the results will indicate if there is a dependence between the level of complexity students perceive in geospatial representations and effective problem-solving for specific spatial thinking components when using these representations.

Table 2 The various programming and graphic design tools used for the development of the instrument

Programming languages	**Frontend:** HTML, CSS and JQuery
	Backend: PHP
Graphic design software	Affinity Designer, Blender
Database	MySQL
Database management	PHPMyAdmin
Integrated development environment	Netbeans

References

Bednarz R, Lee J (2019) What improves spatial thinking? Evidence from the spatial thinking abilities test. Int Res Geogr Environ Educ 28(4):262–280. https://doi.org/10.1080/10382046.2019.1626124

Bhalla M, Proffitt DR (1999) Visual–motor recalibration in geographical slant perception. J Exp Psychol: Hum Percept Perform 25(4):1076. https://psycnet.apa.org/doi/10.1037/0096-1523.25.4.1076

Catling S (2020) Reflecting on the purpose of mapwork in primary schooling. Int J Cartogr 6(3):270–283. https://doi.org/10.1080/23729333.2020.1770480

Chemero A (2011) Radical embodied cognitive science. MIT Press.

Coleman JM, McTigue EM, Smolkin LB (2011) Elementary teachers' use of graphical representations in science teaching. J Sci Teacher Educ 22(7):613–643. https://doi.org/10.1007/s10972-010-9204-1

Davies C, Uttal D (2007) Map use and the development of spatial cognition. In: Plumert JM, Spencer JP (eds) The emerging spatial mind. Oxford University Press, New York, pp 219–247. https://psycnet.apa.org/doi/10.1093/acprof:oso/9780195189223.003.0010

Dickerson DL, Penick JE, Dawkins KR, Van Sickle M (2007) Groundwater in science education. J Sci Teacher Educ 18(1):45–61. https://doi.org/10.1007/s10972-006-9019-2

Freksa C, Barkowsky T, Dylla F, Falomir Z, Olteţeanu A-M, van de Ven J (2017) Spatial problem solving and cognition. In: Representations in mind and world. Routledge, New York, pp 56–183. https://doi.org/10.4324/9781315169781

Gersmehl PJ, Gersmehl CA (2007) Spatial thinking by young children: Neurologic evidence for early development and "educability. J Geogr 106(5):181–191. https://doi.org/10.1080/00221340701809108

Gibson J (1979) The ecological Approach to Visual Perception. Houghton-Mifflin, Boston

Golledge R, Marsh M, Battersby S (2008) A conceptual framework for facilitating geospatial thinking. Ann Assoc Am Geogr 98(2):285–308. https://doi.org/10.1080/00045600701851093

Goodstein MP, Howe AC (1978) Application of Piagetian theory to introductory chemistry instruction. J Chem Educ 55(3):171. https://doi.org/10.1021/ed055p171

Harle M, Towns M (2011) A review of spatial ability literature, its connection to chemistry, and implications for instruction. J Chem Educ 88(3):51–360. https://doi.org/10.1021/ed900003n

Huck C, Zhang J (2021) Effects of the COVID-19 pandemic on K-12 education: a systematic literature review. New Waves-Educ Res Dev J 24(1):53–84

Huttenlocher J, Vasilyeva M, Newcombe N, Duffy S (2008) Developing symbolic capacity one step at a time. Cognition 106(1):1–12. https://doi.org/10.1016/j.cognition.2006.12.006

Jirout JJ, Newcombe NS (2014) Mazes and maps: can young children find their way? Mind Brain Educ 8(2):89–96. https://doi.org/10.1111/mbe.12048

Kosmidis C-V, Lambrinos N (2018) Scaffolding spatial problem solving in science: guidelines derived from theory and research. Eur J Geogr 9(4):22–33

Lee J, Bednarz R (2009) Effect of GIS learning on spatial thinking. J Geogr High Educ 33(2):183–198. https://doi.org/10.1080/03098260802276714

Lee J, Bednarz R (2012) Components of spatial thinking: Evidence from a spatial thinking ability test. J Geogr 111(1):15–26. https://doi.org/10.1080/00221341.2011.583262

Lee J, Jo I (2022) Assessing spatial skills/thinking in geography. In: Assessment in geographical education: an international perspective. Springer, Cham, pp 77–97

Laitin EL, Tymoski MJ, Tenhundfeld NL, Witt JK (2019) The uphill battle for action-specific perception. Atten Percept Psychophysics 81:778–793. https://doi.org/10.3758/s13414-018-01652-w

Moore JC, Slisko J (2017) Dynamic visualizations of multi-body physics problems and scientific reasoning ability: a threshold to understanding. Key competences in physics teaching and learning. Springer, Cham, pp 155–164. https://doi.org/10.1007/978-3-319-44887-9_13

National Research Council & Geographical Sciences Committee (2005) Learning to think spatially. National Academies Press.
Philbeck JW, Witt JK (2015) Action-specific influences on perception and postperceptual processes: present controversies and future directions. Psychol Bull 141(6):1120–1144. https://doi.org/10.1037/a0039738
Reichertz J (2021) Die coronabedingte Krise der qualitativen Sozialforschung. Soziologie 50(3):313–315
Salsa A, Gariboldi MB, Vivaldi R, Rodríguez J (2019) Geometric maps as tools for different purposes in early childhood. J Exp Child Psychol 186:33–44. https://doi.org/10.1016/j.jecp.2019.05.004
Santana FN, Wagner CH, Rubin NB, Bloomfield LSP, Bower ER, Fischer S, Santos BS, Smith GE, Muraida CD, Wong-Parodi G (2021) A path forward for qualitative research on sustainability in the COVID-19 pandemic. Sustain Sci 16(3):1061–1067. https://doi.org/10.1007/s11625-020-00894-8
Uttal DH, Fisher JA, Taylor HA (2006) Words and maps: developmental changes in mental models of spatial information acquired from descriptions and depictions. Dev Sci 9(2):221–235. https://doi.org/10.1111/j.1467-7687.2006.00481.x
Witt JK, Riley MA (2014) Discovering your inner Gibson: Reconciling action-specific and ecological approaches to perception–action. Psychon Bull Rev 21(6):1353–1370. https://doi.org/10.3758/s13423-014-0623-4
Witt JK, Proffitt DR (2005) See the ball, hit the ball: Apparent ball size is correlated with batting average. Psychol Sci 16(12):937
Witt JK, Sugovic M (2010) Performance and ease influence perceived speed. Perception 39(10):1341–1353. https://doi.org/10.1068/p6699
Yuan L, Uttal D, Gentner D (2017) Analogical processes in children's understanding of spatial representations. Dev Psychol 53(6):1098. https://doi.org/10.1037/dev0000302

A Spatial Knowledge Infrastructure for the Aegean Archipelago

Vasilis Kopsachilis, Nikos Vachtsavanis, and Michail Vaitis

Abstract The international geospatial data community has long ago recognized the importance of open data and their contribution to SDGs and has a solid tradition in promoting geographic data sharing and interoperability and the operation of Spatial Data Infrastructures. Current technological advances in the field of the semantic web allow the enrichment of data with semantic interpretation capabilities that would lead to increased interoperability among data from different providers and advanced possibilities for the generation of new knowledge. In this fashion, the recent concept of Spatial Knowledge Infrastructures aims to upgrade traditional Spatial Data Infrastructures by the exploitation of semantic web technologies. Accordingly, the University of the Aegean develops services for spatial data semantic annotation and enrichment, integration with data from third-party sources, and semantic data management, querying, and visualization. These efforts aim to the implementation of a public Spatial Knowledge Infrastructure for the Aegean Archipelago, accessible by anyone in order to exploit its content for developing services and exporting spatial knowledge of added value. This paper presents our motives for the adoption of semantic web technologies, the technical components of the Spatial Knowledge Infrastructure, and our so far experience, including examples of the exploitation of the produced spatial knowledge.

Keywords Open data · Geographical data · Semantic web · Linked data · Spatial knowledge infrastructure

V. Kopsachilis (✉) · N. Vachtsavanis · M. Vaitis
Department of Geography, Cartography & GeoInformatics Laboratory, University of Aegean, University Hill, 81100 Mytilene, Greece
e-mail: vkopsachilis@geo.aegean.gr

N. Vachtsavanis
e-mail: nicos.vachtsavanis@aegean.gr

M. Vaitis
e-mail: vaitis@aegean.gr

A. Klonari et al. (eds.), *Re-visioning Geography*, Key Challenges in Geography,
https://doi.org/10.1007/978-3-031-40747-5_3

Introduction

The provision of high-quality open data to professionals and to the public has been recognized as an important factor for promoting, among others, transparency and democratic control, social participation and self-empowerment, innovation and economic growth, and environmental protection.[1] These benefits are aligned with the Sustainable Development Goals (SDGs) defined by the United Nations.[2] The international geospatial data community has recognized the value of open data long ago and related organizations, such as the Open Geospatial Consortium (OGC), have a solid tradition for promoting geographical data sharing. To this end, since the 90s, technical standards have been developed for data interoperability and proposing the implementation of data dissemination information systems, such as Spatial Data Infrastructures (SDI) (Nebert 2004). SDIs are defined as "a coordinated series of arrangements on technology standards, institutional arrangements and policies that enable the discovery and use of geospatial information by users and for purposes other than those it was created for" (Kuhn 2005). During these years, the operation of public SDIs at the local, national, and regional levels of governance and the adoption of related initiatives, such as the INSPIRE directive,[3] has been proven as an important tool for the opening and sharing of previously closed and unknown geographical spatial datasets that eventually allows the exploitation of data by the interested parties and consequently the provision of added-value services and products in accordance with the SDGs. An important contribution of SDIs toward data sharing was the adoption of common technical infrastructures and standards for the representation and dissemination of geographical data over the web, such as the GML representation format and the WMS specification that managed to reduce the syntactic heterogeneity among different data sources (Brodeur 2022).

At the same time, advances in internet technologies, and specifically the emergence of ontologies and the Semantic Web, aim to deal with the semantic heterogeneity of the data on the web. The term Semantic Web originates from the vision of Tim Berners-Lee of a web space where data will be enriched with semantic interpretation so as to be understandable and processable by humans as well as by software (Berners-Lee et al. 2001). The related term Linked Data refers to a set of technical principles that guides the realization of the Semantic Web as a global web of interconnected data or entities (Berners-Lee 2006). The benefits of such developments include increased interoperability among data from different providers, easier data integration from multiple sources, advanced querying capabilities, and possibilities for automated reasoning for the generation of new knowledge.

The characteristics of geographical data and Semantic Web technologies provide the ground for a powerful integration that could unleash new opportunities for the common good and provide benefits for each other. On the one hand, the embedded geometries in geographical objects and their topological relationships can fuel the

[1] Open Knowledge Foundation, Open Data Handbook, https://opendatahandbook.org.

[2] United Nations, The 17 Goals, https://sdgs.un.org/goals.

[3] European Commission, INSPIRE Directive, https://inspire.ec.europa.eu/inspire-directive/2.

derivation of semantic classifications and inferences. Not surprisingly, geographical distance is a factor that is commonly considered to determine whether two objects are semantically related or not (Ngonga Ngomo 2012; Kopsachilis et al. 2020). Establishing such semantic relations between data is the flesh of the semantic web that greatly facilitates data integration from multiple sources and the discovery of new knowledge that could eventually lead to more informed decision-making. On the other hand, the Semantic Web provides the technological background that could offer advanced exploitation capabilities on geographical data and solutions for overcoming the limitations of current SDIs. To this end, the representation of geographical data with shared and common semantic models can reduce semantic heterogeneity and therefore facilitate, or even automate, the integration of geographical data that resides in different databases and SDIs, for the provision of enriched spatial information. Moreover, it will enable the development of services that will operate not only at the dataset level, that currently SDIs offer with regards to documentation, search and view, but also at the "richer" geographical object level. Finally, the adoption of Semantic Web technologies by SDIs will permit the formulation of advanced queries capable of merging results from multiple sources and the application of automated reasoning techniques on geographical data for the generation of new spatial knowledge, which in traditional SDIs requires manual discovery, download, restructure, fusion, and analysis of data. In this fashion, recently it was proposed the transition from the traditional concept of spatial data infrastructures to the emerging concept of spatial knowledge infrastructures (SKI) (Ivánová et al. 2020), where geographical data are integrated with Semantic Web technologies in a manner that constitutes "a network of data analytics, expertise and policies that assist people, whether individually or in collaboration, to integrate real-time spatial knowledge into everyday decision-making and problem-solving" (Duckham et al. 2017). In essence, SKI content is integrated and harmonized into the Semantic Web, which now forms the primary public source of data, on top of which applications and services can be deployed by anyone to offer targeted knowledge to end-users (Arnold et al. 2019).

This paper describes the efforts of the University of the Aegean for the integration of its geographical data with the Semantic Web technologies and the development of a Spatial Knowledge Infrastructure for the Aegean Archipelago. The University of the Aegean produces and gathers an important amount of high-quality spatial data for the Aegean Sea and the wider area of Greece, in the course of its teaching and research activities. During the last years it has developed and maintains web-based geographical information systems, such as the University of the Aegean Spatial Data Infrastructure,[4] the Documentation Systems for Islands and the Greek Islands Atlas,[5] and the Environmental Risk Management Information System—Floods Geoportal

[4] The University of the Aegean Spatial Data Infrastructure, https://sdi.aegean.gr.

[5] Documentation Systems for Islands, http://archipelago.aegean.gr.

ERMIS-F[6] for the provision of regional geographical data to the research community, to public and private organizations, and to the public. Currently, the University of the Aegean researches the opportunities that arise from the integration of SDIs with Semantic Web technologies for the dissemination of its geographical data. To this end, it annotates and enriches its geographical data with semantic information, publishes them on the Semantic Web, and develops tools and services for geo-semantic data querying, searching and viewing. The above actions are designed to be part of a Spatial Knowledge Infrastructure for the Aegean Archipelago that will operate as a public geographical information exploitation and extraction hub and offer upgraded data dissemination services. The integration and management of semantic web geospatial data, and particularly the development of SKIs, are currently open issues with a lot of unexplored aspects and challenges, including data modeling and architectural decisions, and geo-semantic data exploitation potential. This work contributes by presenting the possibilities that arise from the integration of geospatial data and semantic web technologies and by purposing a functional implementation of such integration.

The rest of the chapter is organized as follows: Section "Background" provides a short introduction to Semantic Web technologies and their benefits and presents the related work in the domain of geographical data and semantic web integration with a particular focus on SDIs. Section "Spatial Knowledge Infrastructure for the Aegean Archipelago" presents the components of the University of the Aegean SKI. Section "Discussion" highlights the technical challenges and potential uses of SKIs services and content. The chapter concludes with a summary and pointers for future work.

Background

Semantic Web Technologies

This section provides a short introduction to the basic Semantic Web technologies in order to shed light on the paradigm shift and the new opportunities that it brings. The Semantic Web can be deemed as an extension of the traditional web, in which information is given well-defined meaning so as to be understandable and processable by humans and software as well (Berners-Lee et al. 2001). A major difference between the traditional web and the Semantic Web is that the former is built on linking web pages (web of documents) and the latter on linking resources that describe specific entities (web of data). Semantic entities may correspond to real-world objects (e.g., a specific city, such as Athens, or a specific person, such as Nelson Mandela) or concepts (e.g., the concept of City or Person), uniquely identified by URIs[7] and described by structured and well-defined terms. The technological base

---

[6] Information Risk Management Information System - Floods ERMIS-F, https://ermis-f.eu.

[7] URI (Uniform Resource Identifier) is a way for uniquely identifying web resources. In the traditional web is used for the unique identification of web pages.

Subject	Predicate	Object
http://example.org/Athens	http://example.org/CapitalOf	http://example.org/Greece
http://example.org/Athens	http://example.org/Population	3.167.000

Fig. 1 RDF triples example

of the Semantic Web is the RDF (Resource Description Framework), a graph-based model for describing the semantic resources in the form of triples (subject—predicate—object) (W3C 2014a). For example, the URI http://example.org/Athens may correspond and identify uniquely an RDF resource that refers to the entity of the city of Athens. In the RDF, whoever wants to refer to the city of Athens can use the above URI and conversely, whoever uses the above URI says something about the city of Athens.

An RDF triple makes use of URIs to make a statement. For example, the first triple in Fig. 1 contains in the subject position the URI of the resourcehttp://example.org/Athens, in the predicate position the URI of the resource http://example.org/CapitalOf, and in the object, position the URI of the resource http://example.org/Greece that can be interpreted as a statement that Athens is the capital of Greece. Similarly, the statement of the second triple can be interpreted that the population of Athens is 3.167.000. We note that in RDF the use of literals (i.e., alphanumerical values such as strings, numbers, or dates) is possible but only in the object position of a triple. Usually, RDF resources are dereferenceable, which means that by visiting a URI someone gets more information about the resource that corresponds to this URI or, in other words, gets access to all triples that describe the resource. Therefore, navigation in the Semantic Web is possible by visiting recurrently the URIs that the user discovers. For example, a user may visit the URI of the Athens entity and via this URI navigate to the URI http://example.org/Greece in order to get more information about the entity of Greece.

In the above examples, we interpreted the triples based on our own subjective perception about the possible meaning of the resources. However, in the Semantic Web, the resources are assigned with specific semantics in a structured way through statements that model the world or a domain of interest, that is, through ontologies. Ontologies are developed by (re-) using RDF resources that were assigned with explicit meaning with the use of ontology development languages, such as the RDFS (Resource Description Framework Schema) (W3C 2014b) and the OWL (Web Ontology Language) (W3C 2012). These resources define categories (classes) and properties (predicates) and restrictions on them. Therefore, an ontology may define a resource that corresponds to the class "City", state that the resource "Athens" belongs to (is an instance of) the class "City", define another resource that corresponds to the property "CapitalOf", and restrict the use of the property in triples where the subject resource is an instance of the class "City".

One of the most important RDF resources for the development of ontologies is the predicate http://www.w3.org/1999/02/22-rdf-syntax-ns#type (in short *rdf:type*), which denotes that a resource is an instance of a class. Figure 2 shows some examples

http://example.org/City	rdf:type	http://www.w3.org/2000/01/rdf-schema#Class
http://example.org/Athens	rdf:type	http://example.org/City
http://example.org/CapitalOf	rdf:type	http://www.w3.org/1999/02/22-rdf-syntax-ns#Property

Fig. 2 Usage example of the rdf:type predicate

of the use of the *rdf:type* predicate. The first triple denotes that the resource http://example.org/City is an instance of the RDFS class *Class* (i.e., is a class); the second that the resource http://example.org/Athens is an instance of the class http://example.org/City (i.e., is a city), and the third triple that the resource http://example.org/CapitalOf is an instance of the RDFS class *Property* (i.e., is a predicate).

Two more important terms for the development of ontologies are the RDFS predicates *rdfs:domain* and *rdfs:range*, which apply restrictions on predicates, that is they restrict the use of predicates with instances of specific classes. The *rdfs:domain* defines that only members of specific classes may be used in the subject position of triples with a specific predicate, while the *rdfs:range* defines that only members of specific classes may be used in the object position of triples with a specific predicate. For example, the *rdfs:domain* restriction may be applied for the predicate *CapitalOf* to define that only instances of the *City* class can be used as subjects of the predicate, and the *rdfs:range* may be applied for the same predicate to denote that only members of the *Country* class can be used as objects of the predicate. Such kinds of restrictions are fundamental for the assignment of semantics to resources and for the application of reasoning techniques that infer implicit knowledge based on explicit statements. Therefore, we can not only explicitly describe a resource (e.g., Athens is the capital of Greece, as we stated in Fig. 1), but we can also infer implicit knowledge from existing statements (e.g., since we stated that the object of the *CapitalOf* predicate is a country, then Greece must be a country). The above basic terms for ontology development, that is, *rdf:type*, *rdfs:domain,* and *rdfs:range*, are presented just for indicative reasons and represent only a small subset of the available terms and capabilities provided by the RDFS and OWL.

Regarding geographical information, it can be assigned with explicit semantics commonly understood by everyone, by reusing terms from geographical ontologies, such as the WGS 84 Basic Geo[8] and the GeoSPARQL.[9] These ontologies define classes and predicates for expressing the geometries of geographic objects. For instance, WGS84 Basic Geo is a simple ontology that provides the *long* and *lat* predicates for representing the longitude and the latitude of a point geometry in the WGS84 reference system (Fig. 3). GeoSPARQL is a more powerful ontology that allows the representation of all types of geometries in various Coordinate Reference Systems.

[8] W3C Semantic Web Interest Group, Basic Geo Vocabulary, https://www.w3.org/2003/01/geo/.

[9] Open Geospatial Consortium, GeoSPARQL—A Geographic Query Language for RDF Data, https://www.ogc.org/standards/geosparql.

http://example.org/Athens	http://www.w3.org/2003/01/geo/wgs84_pos#long	23.71
http://example.org/Athens	http://www.w3.org/2003/01/geo/wgs84_pos#lat	37.99

Fig. 3 Usage example the WGS 84 basic geo-ontology

The term Linked Data was introduced by Berners-Lee in 2006 to describe a set of principles for publishing data to the Semantic Web (Berners-Lee 2006). These principles are:

1. Use URIs as names for things.
2. Use HTTP URIs so that people can look up those names.
3. When someone looks up a URI, provide useful information, using the standards (RDF).
4. Include links to other URIs so that they can discover more things.

The adoption of these principles will result in the realization of a *Web of Data*; a global and open data space on the web comprised of commonly understood and interlinked entities. This technological base will increase the semantic interoperability between data and applications, facilitate the integration and enrichment of data from different sources based on their shared semantics, permit the formulation of advanced queries that gather results from multiple sources, and finally, allow the application of reasoning techniques for the generation of new knowledge.

Semantic Web Technologies and Spatial Data Infrastructures

The enrichment of geographical data with explicit conceptual models and of geographical systems with semantic capabilities toward addressing semantic interoperability impediments in data integration and in meaningful information provision is discussed for many years (Fonseca et al. 2002). There are works that recognize that the characteristics of Linked Data, such as easier interlinking and integration of data from multiple sources and the ability to perform reasoning, can benefit SDIs on many levels including increased resource availability, and improved data access and description (Diaz et al. 2012). In the early days of the Semantic Web, many works were researching the use of ontologies in Spatial Data Infrastructures (Lutz 2005; Lutz and Klien 2006; Vaccari et al. 2009; Li et al. 2011). These works were proposing ontologies and concept-matching techniques for improving the geographic web services discovery (Lutz 2005; Lutz and Klien 2006) or achieving harmonization between remote SDIs (Vaccari et al. 2009; Li et al. 2011). Moreover, they were proposing novel architectural models and components for extending SDIs to support these semantic capabilities. Later, the focus shifted from the application of ontologies to the use of Linked Data principles to SDIs (Schade et al. 2010; Janowicz et al. 2010). Many works proposed semantically enabled extensions for the traditional OGC services used in SDIs, such as WFS, CSW, or SOS. For example, some works proposed a semantic enablement layer for SDI that extends OGC services with semantic capabilities (Janowicz et al. 2010), others, proposed solutions to convert WFS formats to Linked Data (Hietanen et al. 2016), and others focused on semantically enabled services that perform spatial data fusion for addressing the data integration from multiple sources problem (Wiemann and Bernard 2016).

Many countries chose a different path for adding semantic capabilities to their datasets by adopting or implementing Linked Data technology stacks, in parallel or on top of Spatial Data Infrastructures. The national mapping agency of Great Britain, OrdanceSurvey, was a pioneer in disseminating administrative boundaries as Linked Data (Goodwin et al. 2008). Nevertheless, it did not rely on standardized means for representing features, spatial relations, and geometries, and as a result, it lacked (re)usability. Other countries that expose national datasets as Linked Data, such as administrative units, geographical names, and statistical datasets, are Ireland (Debruyne et al. 2017), Spain (De León et al. 2010), and Australia (Car et al. 2019). In a recent study (Ronzhin et al. 2019), the efforts of four countries, namely, Finland, Norway, Spain, and Nederlands, are presented toward the annotation and publication of Linked Data as a means to overcome spatial data integration issues across national borders that are present in European national SDIs. The authors claim that exposing national data as Linked Data makes them part of the Semantic Web, which allows shifting focus from collection and dissemination of data to meaningful data consumption. Currently, the Semantic Web is formed by interlinked resources that are residing in several data sources, including geographical data sources; an illustration of which is provided in LOD (Linked Open Data) cloud.[10]

In a position study about the possibility of SDI and Linked Data integration (Schade and Smits 2012), the authors argue for the possible evolution of SDI to applications that incorporate spatial information with the broader knowledge realm and propose the openness of their use to a broader audience and not exclusively to GIS experts. In this fashion, the recent concept of Spatial Knowledge Infrastructure emerged and is associated with the enrichment of spatial data and applications with semantic capabilities to allow the provision of advanced spatial data management and knowledge extraction services to the interested parties (Ivánová 2020; Arnold et al. 2019). The authors propose possible SKI architectures and components that incorporate semantic web technologies and extend traditional SDIs for overcoming the data integration from multiple sources and resource discovery problems and to allow for automatic processing and reasoning capabilities on spatial data. The proposed SKI concept shifts the focus from supply-driven to knowledge on-demand, from data access to knowledge discovery, from dataset-level to feature-level granularity, and from expert to non-expert systems. However, the research on SKIs is still at an early stage. In this work, we present the implementation steps and the challenges that the University of the Aegean faces toward the development of a Spatial Knowledge Infrastructure for the Aegean Archipelago.

[10] The Linked Open Data Cloud, https://lod-cloud.net/.

Spatial Knowledge Infrastructure for the Aegean Archipelago

The Spatial Knowledge Infrastructure (SKI) of the University of the Aegean is formed by a suite of tools and services for: (a) the conversion of spatial datasets to the RDF model (semantic annotation), (b) the publication of RDF data in the Semantic Web, (c) the execution of semantically enabled queries, and (d) the exploration and the map projection of RDF data. The components of the SKI and their relations are depicted in Fig. 4. SKI content mainly consists of geographical data that originate from the geographical information systems, spatial databases, and SDIs that operate the University and are available in traditional GIS formats, such as ESRI Shapefiles and GeoJSON. The conversion of these spatial datasets to the RDF format, the assignment of URIs to geographical entities, and their semantic description using terms from ontologies is achieved through the semantic annotation application (RDF Converter). The generated RDF data are stored in a Triple store, which, on the one hand, makes them available to the Semantic Web and, on the other hand, allows the execution of queries on them. The above technological base makes SKIs content available to anyone and allows the development of applications that exploit its content. Examples of such applications are the RDF explorer for navigating and searching RDF data and the webGIS for projecting RDF data on a map. In the following sections, the components of the SKI are described in detail.

Semantic Annotation Application

The semantic annotation application (RDF Converter) offers a user-friendly web interface for the conversion of vector spatial datasets in traditional GIS formats (e.g., ESRI Shapefile, GeoJSON, GML) to the RDF format. The application implements

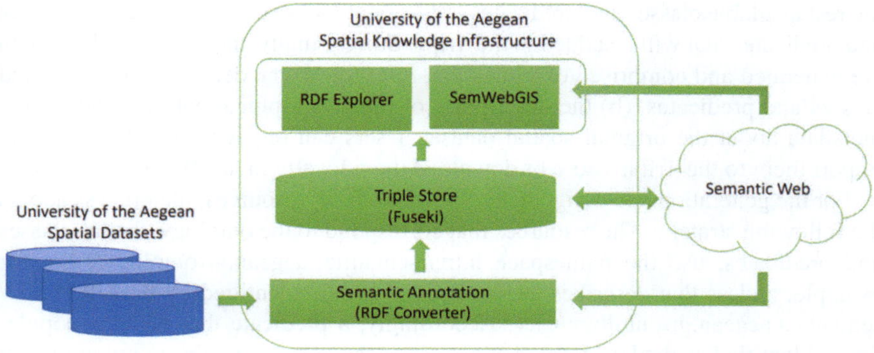

Fig. 4 Overall architecture of the University of the Aegean SKI

a semi-automatic process that analyzes existing semantic knowledge for the recommendation of annotation options but also permits users to intervene in the process by selecting specific classes, predicates, and URIs for the geographical entities. The highlights of the application are: (a) the inclusion of analytical capabilities that produce annotation recommendations based on existing semantic knowledge that minimizes the user involvement during the process, (b) the dynamic and incremental bottom-up building of the underlying ontology based on the spatial datasets at hand, and (c) the ability of the process execution by non-expert users in semantic web technologies.

The semantic annotation process completes in four steps, as depicted in Fig. 5. Initially, the user selects the spatial dataset that wants to convert to RDF and uploads it to the web application (Fig. 5a). Then, the spatial dataset is analyzed, and the application recommends default conversion options that the user can verify or change (Fig. 5b). These options include the selection of: (a) the primary key of the spatial dataset, that is, the column that contains unique values for each geographical entity and will be used for the assignment of URIs to them, (b) the column(s) that contain a short description of the geographical entities and will be used as their labels, and (c) columns that contain "useful" information and therefore will be converted in RDF (e.g., if a column contains only null values or integer that refer to an unknown foreign key, it is not recommended for conversion). Moreover, this step generates and recommends metadata for the original spatial dataset, such as its title, creation date, creator, and source, that are useful for maintaining provenance information for the converted RDF data. In the third step (Fig. 5c), the application recommends an RDF class for the geographical entities of the spatial dataset and RDF predicates for the columns of the spatial dataset that will be converted to RDF. For both class and predicate recommendations, the application analyzes the content of the Triple Store (see section "Triple Store") and recommends available classes and predicates that are suitable for the annotation of the spatial data. The selection of suitable classes and predicates is based on their textual similarity with the spatial dataset name and columns, respectively. In case no suitable recommendations are found or the recommendations do not satisfy the user, there are the options of selecting from the rest available classes and predicates of the Triple Store or of creating new classes and predicates that will be added to the Triple Store. Finally, the annotated RDF data are generated and comprised of triples that refer to: (a) the definition of associated classes and predicates, (b) the description of the geographical entities, and (c) the metadata about the original spatial dataset. Users can preview the RDF data and import them to the Triple Store or download them locally in an RDF file (Fig. 5d).

For the generation and assignment of URIs to RDF resources, the process adopts the following strategy. The resources that correspond to the ontology, that is, classes and predicates, use the namespace http://semantics.aegean.gr/ontology/. So, for example, a class that describes the concept of City is identified by the URI http://semantics.aegean.gr/ontology/City. Accordingly, a predicate that denotes population is identified by the URI http://semantics.aegean.gr/ontology/hasPopulation. The resources that refer to geographical entities use the namespace http://semantics.aegean.gr/resources/ and the URI of an entity is formed by the above namespace, its

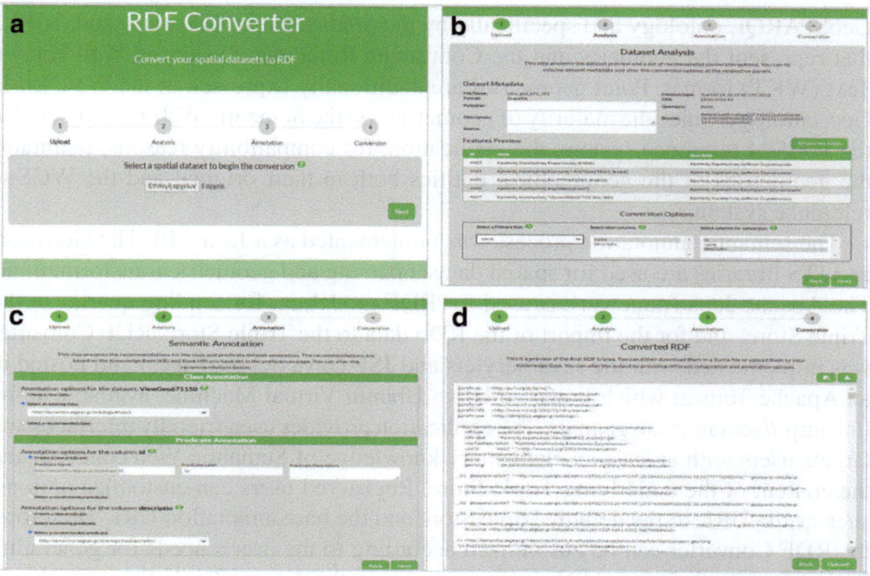

Fig. 5 Steps of semantic annotation in RDF converter: **a** Spatial dataset upload, **b** Dataset analysis, **c** Semantic annotation, and **d** Preview

http://semantics.aegean.gr/resources/City_Athens	rdf:type	http://semantics.aegean.gr/ontology/City
http://semantics.aegean.gr/resources/City_Athens	rdfs:label	"Athens"
http://semantics.aegean.gr/resources/City_Athens	http://semantics.Aegean.gr/resources/City_Athens	3.170.000
http://semantics.aegean.gr/resources/City_Athens	http://www.w3.org/2003/01/geo/wgs84_pos#long	23.71
http://semantics.aegean.gr/resources/City_Athens	http://www.w3.org/2003/01/geo/wgs84_pos#lat	37.99

Fig. 6 Example of semantic annotation in RDF converter

class, and the value of the primary key. For example, the city of Athens is identified by the URI http://semantics.aegean.gr/resources/City_Athens. Finally, the resources that refer to the original spatial datasets use the namespace http://semantics.aegean.gr/metadata/. So, a dataset is identified by the above namespace and a randomly generated unique code e.g., http://semantics.aegean.gr/metadata/c7093d. Figure 6 illustrates an example of generated triples that follow the above-described strategy and concern an entity about the city of Athens that was described in a Shapefile containing Cities.

The semantic annotation application is heavily relying on the re-use of terms from well-known ontologies. The RDFS predicates *rdf:type* and *rdfs:label* are used for the declaration of entities' membership to a class and for the assignment of their labels, respectively. The Dublin Core ontology[11] is used for the description of the original spatial datasets metadata, such as the title, the creation date, and their spatial extent. The geometries of geographical entities are assigned according to the

[11] Dublin Core, Ontology, https://www.dublincore.org/resources/glossary/ontology/.

GeoSPARQL ontology and specifically by the predicates *hasGeometry* and *asWKT* that represent the geometry and the Coordinate Reference System in Well Known Text (WKT) format. Point geometries are additionally expressed in the W3C Basic Geo-ontology. Since the majority of geometries on the Semantic Web are projected in the WGS84 reference system, the application, for compatibility reasons, maintains the geometries of the geographical entities both in their original and the WGS84 reference system.

The semantic annotation process was implemented as a Java API. The GeoTools and JTS libraries are used for spatial dataset parsing and geometric transformations. The Apache Jena framework is used for RDF modeling, for sending queries to the Triple Store, and for the import of the RDF data to the Triple Store. RDF Converter was implemented with the Java Servlets and JSP (Java Server Pages) and hosted in an Apache Tomcat which is served by an Ubuntu Virtual Machine, available at the link http://semantics.aegean.gr. The application provides user-friendly interfaces and targets users with at least an elementary knowledge of Semantic Web concepts and the content of the underlying Triple Store. Registered users can customize options, such as the URL of the default Triple Store and the base annotation URI. Therefore, the RDF Converter can be customized according to the user's needs for generating RDF data with their own URIs and for storage in third-party Triple Stores.

Triple Store

Triple Store is responsible for the storage of the RDF data and their publication to the Semantic Web. In addition, it provides a SPARQL endpoint, i.e., an interface for the execution of SPARQL queries and the return of the responses over the HTTP protocol. SPARQL (SPARQL Protocol and RDF Query Language)[12] is a powerful query language for RDF data. Its syntax is based on graph patterns and acts for RDF like SQL does for relational databases. The query interface of the Triple Store is depicted in Fig. 7, where users can formulate a SPARQL query (above) and get the results (below). In this simple example, the query asks for airports that are stored in the Triple Store and return the airports' URI and their labels. However, the SPARQL protocol allows for more complex queries and provides advanced features, such as the execution of federated queries, that is, the simultaneous query on multiple endpoints, and the setting of inference rules for the generation of implicit knowledge. The software that was selected for the Triple Store is Fuseki, because is a mature and stable software that supports GeoSPARQL and the execution of spatial queries and spatial functions, such as distance, buffer, contains, intersection, etc.

[12] W3C, SPARQL Query Language for RDF, https://www.w3.org/TR/rdf-sparql-query/.

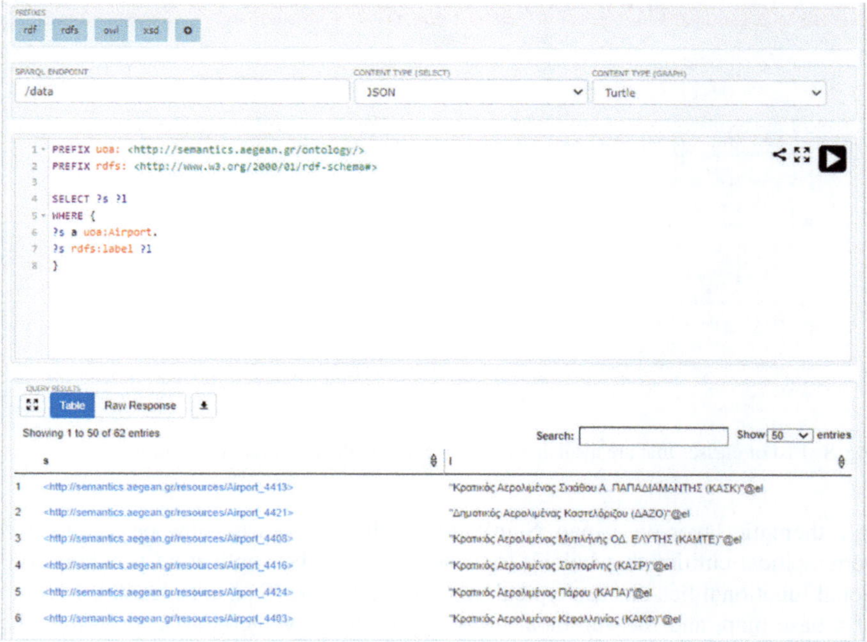

Fig. 7 Query interface of the Triple Store

RDF Explorer

RDF Explorer is a user-friendly web application that allows the navigation and search of RDF data without the need to formulate SPARQL queries. Specifically, users can explore classes, predicates, geographical entities, and original spatial datasets metadata that are available in the Triple Store. The list of classes in the RDF Explorer is illustrated in Fig. 8. Users can enter a text to search for a class and view the instances that belong to a class and the predicates that are related to a class. Similar functions are provided for the predicate, geographical entities, and spatial datasets lists. RDF Explorer is built on simple HTML pages and Javascript code for its communication with the Triple Store through SPARQL queries for the retrieval of the store's content.

SemWebGIS

SemWebGIS application allows the projection of the RDF entities on an interactive map. The application combines the typical functionality provided by traditional webGIS applications with the capabilities of Semantic Web technologies. For instance, like traditional webGIS, where users can view the objects that belong

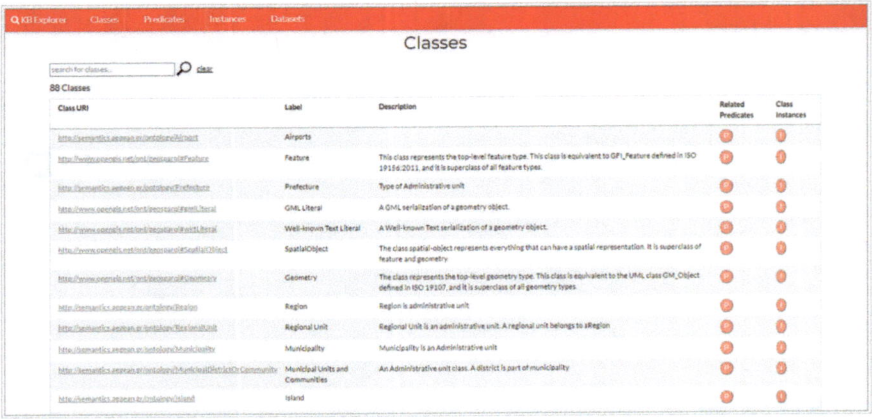

Fig. 8 List of classes that are available in Triple Store in the Explorer Application

to a thematic layer on a map, SemWebGIS allows users to view on a map RDF geographical entities that belong to an RDF class. The application provides additional functionalities that are typical to traditional webGIS such as legend, selection of a base map, and measurement tools. Figure 9 shows geographical entities that belong to the class *Airport*. By clicking on a geographic entity on the map, a popup window appears with more information about the entity, such as its URI and a short label. By visiting the URI of an entity, it appears an HTML page that corresponds to the RDF resource and contains the triples that are associated with the entity and its map projection (Fig. 10). This feature, that is, the ability to get an HTML description of the triples that are associated with a resource by visiting its URI is a capability of the Semantic Web which is called URI dereferencing and is one of the Linked Data principles.

SemWebGIS supports the execution of SPARQL queries and the projection of the results on the map. In this way, SemWebGIS makes use of the SPARQL capability for formulating queries that combine information to generate knowledge. Figure 11 shows an example of this functionality. The SPARQL query on the right combines information from municipalities, NATURA areas, and airports, and specifically ask for municipalities that in their jurisdiction contain NATURA areas within which operate Airports. At the left, the results of the query, that is, the matched municipalities are projected on the map. In future versions, we schedule to incorporate innovative functionality that exploits more Semantic Web capabilities, such as the appearance of semantic relationships between geographical entities on the map and the appearance of relations between geographic entities of the SKI and entities that are available through third-party data sources, such as DBPedia or Geonames.

The development of the SemWebGIS application was based on HTML, CSS, and Javascript. The functionality that is related to the map operation is provided by the Leaflet library and the available base maps are provided by OpenStreetMap, Mapbox και Google Maps. For the URI dereferencing functionality, the Pubby software is used.

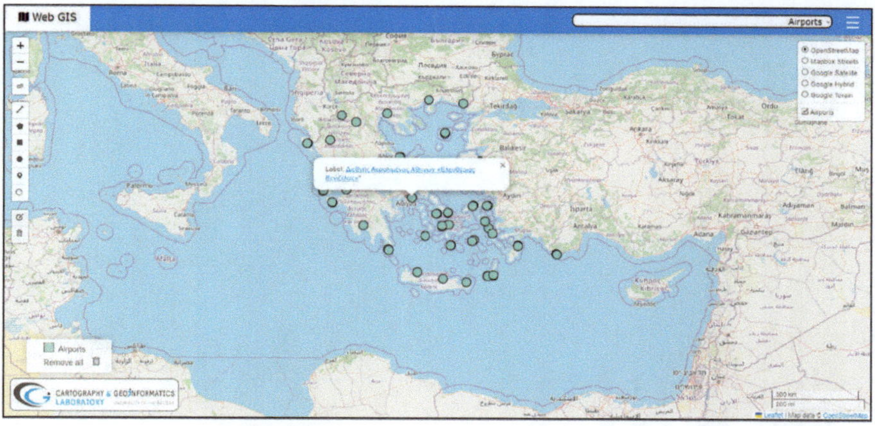

Fig. 9 Map projection of geographical entities of the Airport class

Discussion

In the previous section, we presented the overall architecture of the Spatial Knowledge Infrastructure for the Aegean Archipelago, which consists of components for (a) the generation and publication of RDF data on the Semantic Web and (b) the exploitation of semantically enabled data. The development of the SKI aims to upgrade the data dissemination services of the University of the Aegean currently provided through its Geographical Information Systems and Spatial Data Infrastructures. Specifically, SKI aims to form a public geo-semantic data hub that will be integrated with the Semantic web and offer services for supporting the decision-making process, the development of services and smart applications, and the export of added-value spatial knowledge by any interested party.

SKI contains RDF descriptions of geographical objects that belong to various thematic categories such as administrative boundaries (e.g., municipality and regions boundaries), geomorphology (e.g., rivers, caves, volcanoes), infrastructures (e.g., airports, schools, sports facilities), and other POIs (e.g., museums, temples, banks) and covers the Aegean Archipelago and the wider Greek areas. The original data sources are vector spatial datasets that are already maintained by the University of the Aegean and disseminated through its geographical information systems, and which are semantically annotated by the SKI component, RDF converter. The tool was designed to simplify the actions of converting spatial datasets to syntactically correct RDF and their publication to the Semantic Web, even for users that are not experts in these technologies. Behind the scenes, it re-uses terms from well-known ontologies, such as the RDFS and the GeoSPARQL, to increase semantic interoperability and tighten the integration of SKI's content with the Semantic Web. It also builds a bottom-up ontology based on the spatial datasets at hand. The alignment

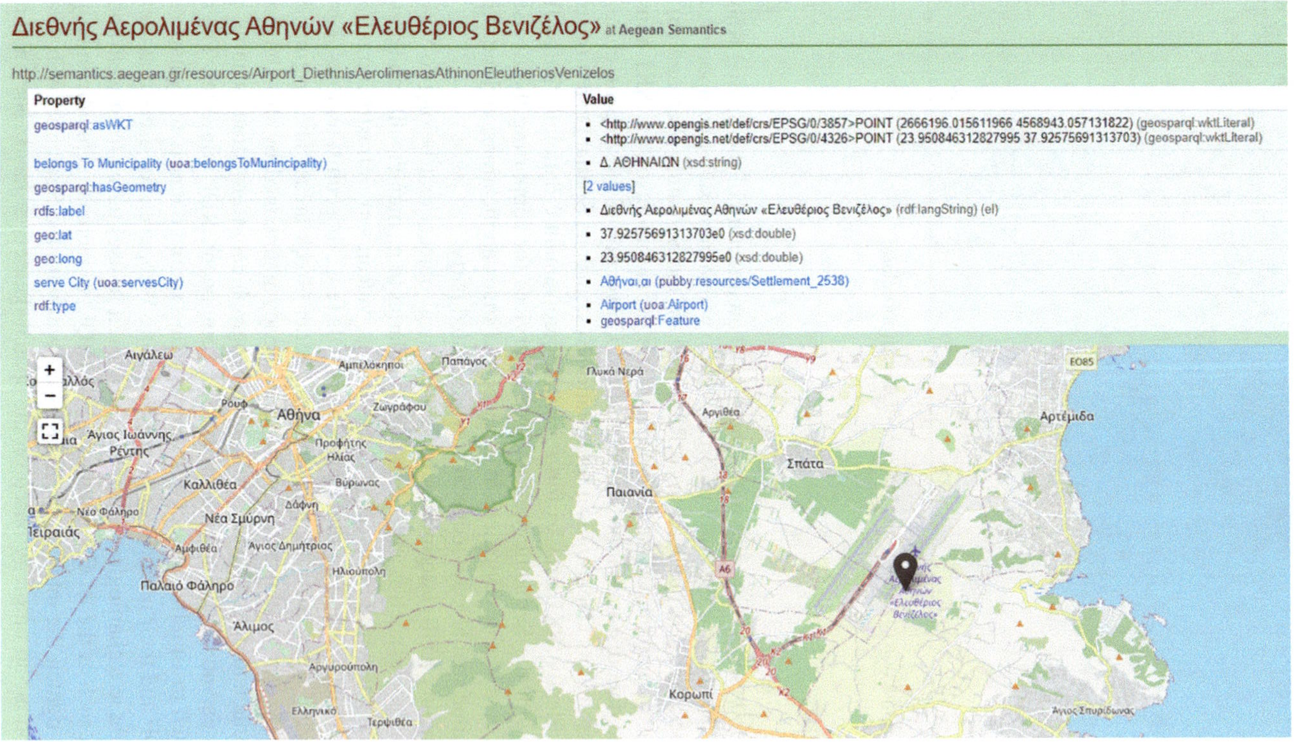

Fig. 10 "Eleutherios Venizelos" Airport derefenced web page

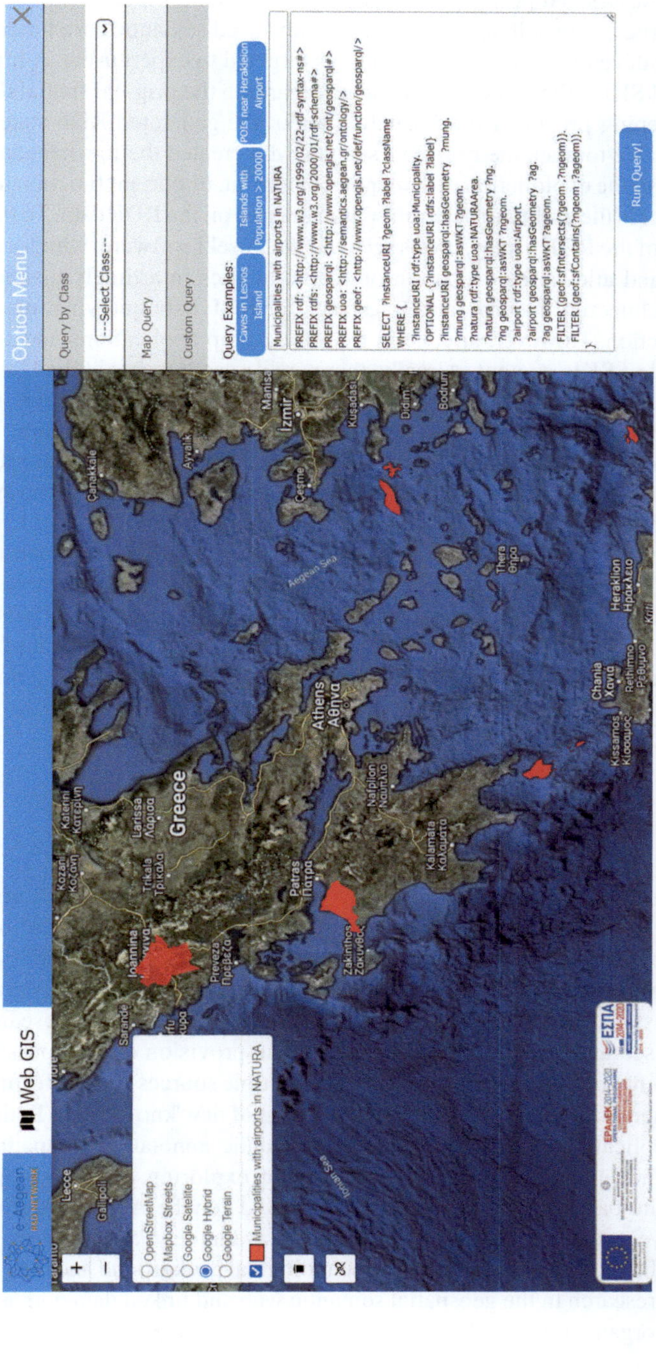

Fig. 11 SPARQL query that shows municipalities that contain NATURA areas within the boundaries of which there are airports

of the SKIs ontology with third-party ontologies, including the definition of relations between classes and properties and the establishment of links between related semantic resources, is a challenge and concerns advanced semantic enrichment tasks that require the development of additional tools targeted to expert users. A highlight of the proposed SKI is the storage of metadata regarding the original spatial datasets that help in keeping provenance information about the generated RDF statements, so as users are able to track their origins, such as who created them, when, and how.

With regard to the exploitation of semantic content, an important function that any SKI should offer is the ability to perform spatial queries on the RDF data. To this end, for the storage of the RDF triples, it was selected the Fuseki software, which supports GeoSPARQL and allows the execution of spatial queries in a timely manner. The current SKI architecture incorporates interfaces for RDF data querying, exploring, and map projection. However, anyone is free to develop applications and services that would exploit SKIs content. Already, there are research teams at the University of the Aegean that provide examples of this potential. Specifically, Moraitou et al. (2022) develop a system for the management of cultural heritage semantic trajectories generated from Unmanned Aerial Vehicles that enrich these trajectories with nearby POIs that are available in the SKI. More interfaces are needed for the automation of knowledge extraction that will implement advanced SPARQL features, such as federated queries and reasoning, through easy-to-use interfaces targeted to non-expert users. After all, the SKI concept refers to public infrastructures whose power lies in the ability that offers to third-party bodies to export spatial knowledge. The challenge is to highlight the new possibilities that the semantic enrichment of spatial data offers in terms of data integration, querying, and reasoning and pinpoint ways of exploitation of SKI services toward knowledge extraction.

Conclusion

This work presented the efforts of the University of the Aegean toward ingesting semantic web technologies for spatial data management and dissemination. The University recognizes the importance of open data in supporting SDGs and is already a major hub in providing spatial data in Greece through web-based geographical information systems and its efforts aim to upgrade its services. The research and application of semantic web technologies target the provision of new possibilities for spatial data integration and enrichment from multiple sources and the formulation of advanced queries and reasoning for the generation of new knowledge. To this end, we develop applications and services for the semantic annotation of spatial data, publication to the Semantic Web, RDF data querying, exploring, and map projection. The above is the basis for the development of a SKI for the Aegean Archipelago that would serve as a public geo-semantic data hub available to anyone to exploit its services. In addition, SKI aims to offer the technological background and the network for supporting research in the geospatial semantic web and linked data domains, and aid third-party organizations and companies that may want to exploit the capabilities of Semantic Web and be part of it (e.g., semantic annotation services). This chapter contributes to the geosciences community by presenting the new possibilities that arise from the integration of geographical data with the semantic web, discussing the evolution from Spatial Data Infrastructures to Spatial Knowledge Infrastructures, and providing directions, with regard to data modeling and architectural decisions, toward a possible implementation of the latter. However, the development of the SKI is in progress and there is much work still to do for the full exploitation of its potential. This includes the interlinking of geographical entities with entities provided by other data sources (e.g., DBpedia, GeoNames), the alignment of the ontology with other ontologies, the evaluation of the RDF data quality and the SKI services, further research on possible ways of exploiting SKI services and content, the development of user-friendly interfaces for the public, and the application of reasoning techniques for the generation of new knowledge.

Funding This research was funded by the Research e-Infrastructure [e-Aegean R&D Network], which is implemented within the framework of the "Regional Excellence" Action of the Operational Program "Competitiveness, Entrepreneurship and Innovation". The action was co-funded by the European Regional Development Fund (ERDF) and the Greek State [Partnership and Cooperation Agreement 2014–2020].

References

Arnold LM, McMeekin DA, Ivánová I, Armstrong K (2019) Knowledge on-demand: a function of the future spatial knowledge infrastructure. J Spat Sci 66(3):365–382. https://doi.org/10.1080/14498596.2019.1654942

Berners-Lee T, Hendler J, Lassila O (2001) The semantic web. A New form of web content that is meaningful to computers will unleash a revolution of new possibilities. Sci Am 284(5):34–43. https://www.lassila.org/publications/2001/SciAm.pdf

Berners-Lee T (2006) Linked data—design issues. https://www.w3.org/DesignIssues/LinkedData.html. Accessed 10 Nov 2022

Brodeur J (2022) Geosemantic interoperability and the geospatial semantic web. In: Kresse W, Danko D (eds) Springer handbook of geographic information. Springer Handbooks. Springer, Cham. https://doi.org/10.1007/978-3-030-53125-6_17

Car NJ, Box PJ, Sommer A (2019) The location index: a semantic web spatial data infrastructure. In: Lecture notes in computer science (including subseries lecture notes in artificial intelligence and lecture notes in bioinformatics) 11503 LNCS. Springer International Publishing. https://doi.org/10.1007/978-3-030-21348-0_35

De León A, Saquicela V, Vilches LM, Villazón-Terrazas B, Priyatna F, Corcho O (2010) Geographical linked data: A Spanish use case. In: ACM international conference proceeding series, pp 1–3. https://doi.org/10.1145/1839707.1839753

Debruyne C, Meehan A, Clinton É, McNerney L, Nautiyal A, Lavin P, O'Sullivan D (2017) Ireland's authoritative geospatial linked data. Lecture Notes in Computer Science (Including subseries lecture notes in artificial intelligence and lecture notes in bioinformatics) 10588. LNCS, pp 66–74. https://doi.org/10.1007/978-3-319-68204-4_6

Diaz L, Remke A, Kauppinen T, Degbelo A, Foerster T, Stasch C, Rieke M, Schaeffer B, Baranski B, Bröring A, Wytzisk A (2012) Future SDI—impulses from geoinformatics research and IT trends. Int J Spat Data Infrastructs Res 7:378–410. https://doi.org/10.2902/1725-0463.2012.07.art18

Duckham M, Arnold L, Armstrong K, McMeekin D, Mottolini D (2017) Towards a spatial knowledge infrastructure, pp 1–19. https://www.crcsi.com.au/assets/Program-3/CRCSI-Towards-Spatial-Knowledge-Whitepaper-web-May2017.pdf

Fonseca FT, Egenhofer MJ, Agouris P, Cmara G (2002) Using ontologies for integrated geographic information systems. Trans GIS 6(3):231–257. https://doi.org/10.1111/1467-9671.00109

Goodwin J, Dolbear C, Hart G (2008) Geographical linked data: The administrative geography of Great Britain on the semantic web. Trans GIS 12(Suppl. 1):19–30. https://doi.org/10.1111/j.1467-9671.2008.01133.x

Hietanen E, Lehto L, Latvala P (2016) Providing geographic datasets as linked data. ISPRS Int Arch Photogramm. Remote Sens Spatial Inf Sci XLI-B2:583–586. https://doi.org/10.5194/isprsarchives-XLI-B2-583-2016

Ivánová I, Siao Him Fa J, McMeekin DA, Arnold LM, Deakin R, Wilson M (2020) From spatial data to spatial knowledge infrastructure: a proposed architecture. Trans GIS 24(6):1526–1558. https://doi.org/10.1111/tgis.12656

Janowicz K, Schade S, Bröring A, Keßler C, Maué P, Stasch C (2010) Semantic enablement for spatial data infrastructures. Transactions in GIS 14(2):111–129. https://doi.org/10.1111/j.1467-9671.2010.01186.x

Kopsachilis V, Vaitis M, Mamoulis N, Kotzinos D (2020) Recommending geo-semantically related classes for link discovery. J Data Semant 9:151–177. https://doi.org/10.1007/s13740-020-00117-4

Kuhn W (2005) Introduction to spatial data infrastructures. Presentation held on 14 Mar. https://slidetodoc.com/introduction-to-spatial-data-infrastructures-werner-kuhn-march-2/

Li W, Yang C, Nebert D, Raskin R, Houser P, Wu H, Li Z (2011) Semantic-based web service discovery and chaining for building an Arctic spatial data infrastructure. Comput Geosci 37(11):1752–1762. https://doi.org/10.1016/j.cageo.2011.06.024

Lutz M (2005) Ontology-based service discovery in spatial data infrastructures. In: International conference on information and knowledge management, proceedings, pp 45–54. https://doi.org/10.1145/1096985.1096997

Lutz M, Klien E (2006) Ontology-based retrieval of geographic information. Int J Geogr Inf Sci 20(3):233–260. https://doi.org/10.1080/13658810500287107

Moraitou E, Angelis S, Kotis K, Caridakis G, Papadopoulou EE, Soulakellis N (2022) Towards engineering drones semantic trajectories as knowledge graphs. In: Proceedings of the 5th international workshop on geospatial linked data (GeoLD 2022), co-located with the 19th European semantic web conference (ESWC 2022). Crete, Greece, 30 May

Nebert D (2004) The SDI cookbook, version 2.0. Global spatial data infrastructure association, technical working group report

Ngonga Ngomo AC (2012) On link discovery using a hybrid approach. J Data Semant 1:203–217. https://doi.org/10.1007/s13740-012-0012-y

Ronzhin S, Folmer E, Lemmens R, Mellum R, Ellett Von Brasch T, Martin E, Romero E, Kytö S, Hietanen E, Latvala P (2019) Next generation of spatial data infrastructure: lessons from linked data implementations across Europe. Ijsdir Sadl Kuleuven Be 14:83–107. https://doi.org/10.2902/1725-0463.2019.14.art4

Schade S, Granell C, Díaz L (2010) Augmenting SDI with linked data. CEUR In: Workshop on linked spatiotemporal data 2010 (LSTD-2010). Zurich, Switzerland, 14 Sept 2010. CEUR workshop proceedings 691. https://ceur-ws.org/Vol-691/paper3.pdf

Schade S, Smits P (2012) Why linked data should not lead to next generation. IEEE Int Geosci Remote Sens Symp 2012:2894–2897. https://doi.org/10.1109/IGARSS.2012.6350721

Vaccari L, Shvaiko P, Marchese M (2009) A geo-service semantic integration in spatial data infrastructures. Int J Spat Data Infrastructs Res 4:24–51. https://doi.org/10.2902/1725-0463.2009.04.art2

W3C (2012) OWL 2 web ontology language. https://www.w3.org/TR/2012/REC-owl2-overview-20121211/. Accessed 10 Nov 2022
W3C (2014a) RDF 1.1 Concepts and Abstract Syntax. https://www.w3.org/TR/rdf11-concepts/. Accessed 10 Nov 2022
W3C (2014b) RDF Schema 1.1. https://www.w3.org/TR/rdf-schema/. Accessed 10 Nov 2022
Wiemann S, Bernard L (2016) Spatial data fusion in spatial data infrastructures using linked data. Int J Geogr Inf Sci 30(4):613–636. https://doi.org/10.1080/13658816.2015.1084420

Using Story Maps for SDG Visibility and Education: A Scoping Review

Juan José Pons Izquierdo

Abstract Story maps are one of the most common and useful ways to conduct virtual field trips or virtual tours, in places both near and far. Though new, they are growing rapidly and cover many themes, including the 2030 Agenda and Sustainable Development Goals (SDGs). The ease with which story maps can bring environmental case studies into the classroom in a visually attractive way, while combining different materials (maps, texts, photos, videos, etc.) on one webpage and promoting user-interaction, make them an excellent teaching tool. They can also be employed in many non-academic contexts, such as the dissemination or awareness of values among the population as a whole or among specific groups. Although there are thousands of environmental story maps, related research and scientific articles have been scarce. Through a review of the main bibliographic databases (in Science and Scopus), the importance of story maps in support of the SDGs is analysed and their main characteristics described: approaches, tools used, methodologies, training levels, etc. The study finds growing interest in story maps relating to environmental issues among researchers. There is a particular focus on education: half deal with various aspects of story maps as a pedagogical tool. Finally, this paper highlights that the vast majority of articles address specific case studies, rather than more general issues.

Keywords Story map · Web map · SDG · Environment · Scientific production

Introduction

The term 'story map' encompasses a set of web tools that allow the combination of interactive maps with other types of explanatory content, such as texts, graphics, photos, videos, podcasts and hypertext links (Cope et al. 2018). This integration of different digital materials around cartography facilitates the implementation of

J. J. Pons Izquierdo (✉)
Department of History, History of Art and Geography, University of Navarra, Pamplona, Spain
e-mail: jpons@unav.es

National University of Education at Distance (UNED), Pamplona, Spain

A. Klonari et al. (eds.), *Re-visioning Geography*, Key Challenges in Geography,
https://doi.org/10.1007/978-3-031-40747-5_4

narratives of complex spatial phenomena in a more accessible and creative way (Lo Presti 2022), making them easier to understand.

These characteristics of story maps make them very useful tools for teaching (Baker 2015; Hong 2014), due to their dynamic and visual nature and their capacity for user-interaction. They can also be used in many non-academic contexts, such as disseminating or raising awareness of values among the population as a whole or among specific groups (De Lázaro et al. 2020).

This versatility means that, although the use of story maps is very recent—the creation of geographic information systems (GIS) in the cloud was a novelty only a decade ago (Caquard and Dimitrovas 2017)—their use has exploded in recent years, especially in education (Antoniou et al. 2018). Environmental Systems Research Institute (ESRI), the company that has most driven the development of this technology and has the largest number of users, reports that there are currently more than 2.2 million ArcGIS story maps hosted on its cloud service (Fig. 1), up from 120 a decade ago (ESRI n.d.).

This rapid and massive development has led to story maps being used in all kinds of fields and for very different purposes, although they always focus on a spatial theme and its narrative character (storytelling). One of these themes is the 2030 Agenda and Sustainable Development Goals (SDGs), which is an increasingly popular topic of research and publication (De Lázaro et al. 2020; Pons 2021; Puertas-Aguilar et al. 2021; Quinn et al. 2019).

However, to our knowledge, no scientific studies have analysed the literature on story maps in the field of SDG dissemination. For this reason, the aim of this study is to delineate and assess this literature, and particularly to address questions such as: what kind of studies have been conducted on story maps? how have they been developed? where are they published?, and what is their content and methodology?

In sum, this work represents a first review, which allows us to know the state-of-the-art on a subject of undoubted current interest and future projection.

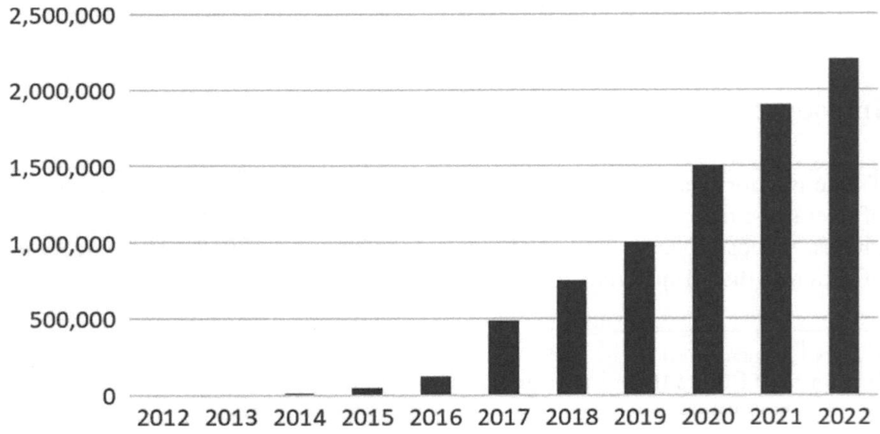

Fig. 1 Number of story maps hosted on ESRI cloud service by year (ESRI n.d.)

Methodology

A review of scientific story map articles in leading bibliographic databases was carried out, with criteria included in the *Preferred Reporting Items for Systematic Reviews and Meta-Analyses* (PRISMA)—in their 2020 revision (Page et al. 2021)—used as an illustrative framework. Specifically, the study follows the extension of scoping reviews (PRISMA-ScR), detailed by Tricco et al. (2018), and is based on this topic's foundational contribution by Arksey and O'Malley (2005).

Therefore, our methodology includes a literature review protocol based on the PRISMA-ScR checklist (http://www.prisma-statement.org/Extensions/ScopingReviews), which ensures the systematisation and traceability of the analysis performed. Unfortunately, some basic data cannot be reproduced in this chapter due to its length (in particular, the complete list of articles reviewed).

Several web applications to support researchers in this task (Stefanovic et al. 2021) exist, including some specific software, such as *SESRA*, *ReLis* and *Parsifal*, but in this case they were not used, since all the information was provided in a shared *Google Sheets* document (https://bit.ly/SDG_storymap).

The review consisted of eight consecutive stages: (1) definition of the framework and e-research question; (2) identification of relevant articles in bibliographic databases; (3) filtering of the selected files to purify the results (elimination of duplicates and articles not available in full text); (4) thematic review of the articles to select those appropriate for the study; (5) in-depth content analysis; (6) charting of the data; (7) presentation and discussion of the results; and finally, (8) drawing of conclusions.

The following methodological sections explain the sources used and process followed to collect, filter and analyse the relevant information (steps 2 to 5).

Data Sources and Search Strategy

The electronic bibliographic databases used to search for scientific papers were Clarivate Analytics' Web of Science (WOS) and Elsevier's Scopus. These are traditionally the most widely employed in bibliometric analysis as they are more selective in their content (Pranckutė 2021; Singh et al. 2021). The final search of both databases was carried out on 3 December 2022.

The selection process was as follows: firstly, a number of subject-related terms were identified and searched using logical operators. This was done in three specific database areas: "title", "abstract" and "keywords". In WOS, this triple search was carried out in the subject area. The query was therefore the same in both databases, but with different search routines:

- WOS: "story map" (Subject) or "storymaps" (Subject) or "story mapping" (Subject) or "storymapping" (Subject).

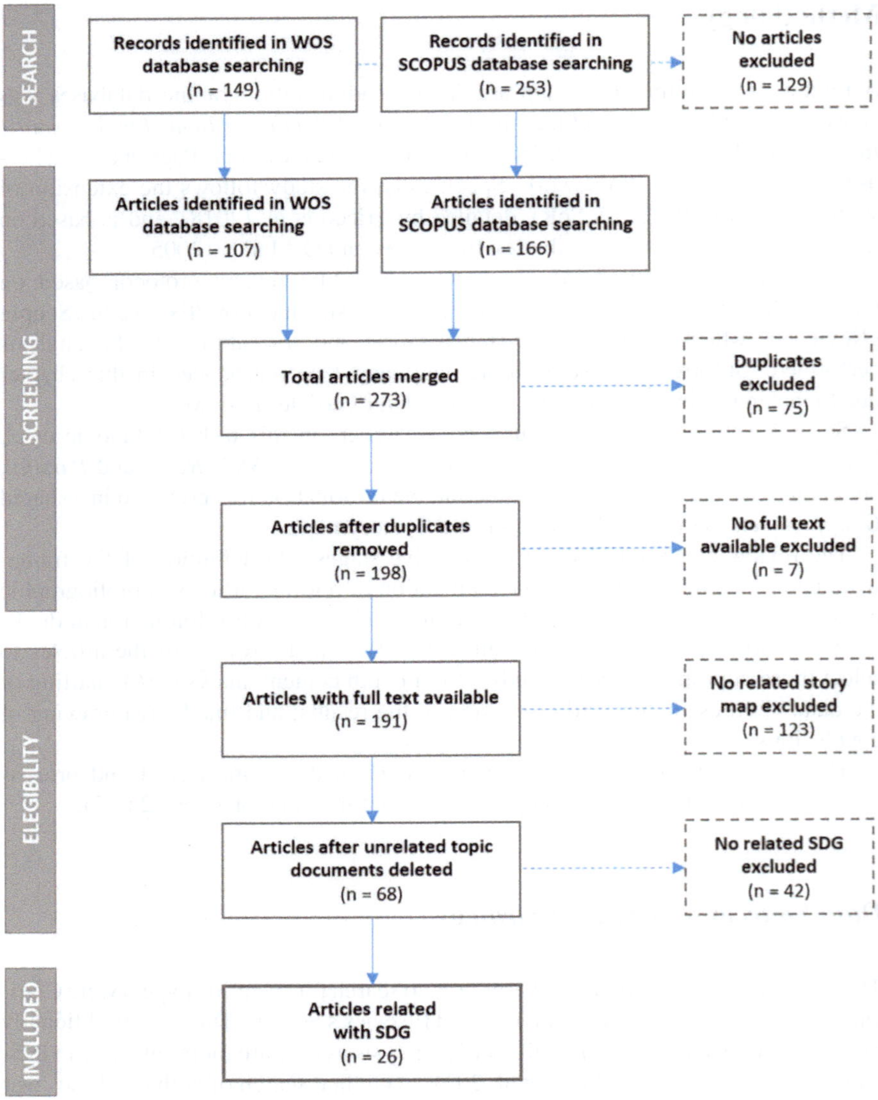

Fig. 2 Selection process flowchart, following the PRISMA Statement

- Scopus: (TITLE-ABS-KEY ("story map") OR TITLE-ABS-KEY ("storymap") OR TITLE-ABS-KEY ("story mapping") OR TITLE-ABS-KEY ("storymapping").

In the case of WOS, only references from the Core Collection were retrieved.

Consequently, a total of 402 documents were collected: 149 in WOS and 253 in Scopus (see Fig. 2).

This represented an unmanageable number of resources for this study. A preliminary examination of the records revealed that many contained the term "story map" (or one of its derivatives), but this did not correspond to the meaning relevant here.

In order to make a shortlist, the possibility of adding a new term to the description search that included a reference to geography was tested.

However, this idea was rejected because the search options in WOS are less numerous than in Scopus and would therefore have distorted information retrieval.

The second step was to filter both searches by document type, leaving only 273 articles in peer-reviewed scientific journals (107 in WOS and 166 in Scopus) and rejecting the others. Although this decision significantly reduced the number of documents, it was taken because of the heterogeneity of many of those retrieved, which were not relevant for this review. Consequently, a significant reduction in the number of references available for analysis (129 in total) was assumed, which improved the accuracy and quality of the remaining references.

Data Management and Screening

Once this task was completed, the next step was to prepare and filter the information for content analysis. Each search was exported to a spreadsheet (Microsoft Excel 2019) and merged into a single database containing the 273 articles. The 75 duplicates were then removed from the database, first by an automatic function and then by a manual check, leaving 198 articles to work with.

Records for which the full text was unavailable (only 7) were also excluded, as further analysis was not possible.

Eligibility Criteria

A two-stage analysis process was used to assess the thematic fit and consistency of the studies identified in the search. First, the remaining 191 records were examined for content, retaining only those whose semantic content was directly related to the topic discussed in this article. For this purpose, three of the database fields were analysed: article title, journal title and abstract. In about 25% of the cases, it was also necessary to check the full text, as the cited fields did not give a clear picture of article content.

A large proportion of these scientific publications deal with issues in the field of pedagogy, where story maps are conceived as graphical tools to facilitate the learning of people with autism and other difficulties (Idol and Croll 1987; Kamps et al. 1995). The remaining articles rejected at this stage dealt with story maps in a very peripheral way, with little relevance to the overall content of the document.

Table 1 List of selected articles on story maps related to the SDGs (with link to the full text)

Authors	Title	Year
Aahlaad, M. et al.	An Object-Based Image Analysis of WorldView-3 Image for Urban Flood Vulnerability Assessment and Dissemination Through ESRI Story Maps	2021
Aldinger, J. M. M.	Story Maps and Disability Studies: A Digital Blueprint for Teaching Community Engagement	2018
Budowle, R.	Narratives of Place: Integrated Digital Storytelling and Story-Mapping for Sustainable Recreation Management	2022
Du, J. T. et al.	A Framework for Co-designing and Developing Participatory Storymaps with Aboriginal Communities	2022
Gillespie, N., and Encinas, J.	An Aquatic Organism Passage Story Map	2017
Hunter, J. et al.	Learning on Harare's Streets Under COVID-19 Lockdown: Making a Story Map with Street Youth	2021
Kirby, E. et al.	Queering the Map: Stories of Love, Loss and (Be)Longing Within a Digital Cartographic Archive	2021
Johansson, T. et al.	Web Map Application to Support Spatial Planning, Decision-Making and Transition Toward Climate-Smart Landscapes in the Taita Hills, Kenya	2019
Li, J. et al.	Web GIS for Sustainable Education: Towards Natural Disaster Education for High School Students	2022
Lung-Amam, W. S., and Dawkins, C.	The Power of Participatory Story Mapping: Advancing Equitable Development in Disadvantaged Neighbourhoods	2020
Malkowski, J. A., and Klenke, C. M.	Rhetorical Cartographic Story Maps as Public Work	2020
Mychajliw, A. M. et al.	Using the Anthropocene as a Teaching, Communication and Community Engagement Opportunity	2015
Northmore, L., and Hudson, M. D.	Digital Environmental Impact Assessment: An Exploration of Emerging Digital Approaches for Non-technical Reports	2022
Oubennaceur, K. et al.	Flood Risk Communication Using ArcGIS StoryMaps	2021
Park, Y. M.	A GPS-Enabled Portable Air Pollution Sensor and Web-Mapping Technologies for Field-Based Learning in Health Geography	2021

(continued)

Table 1 (continued)

Authors	Title	Year
Pietsch, M. et al.	Analyzing Ecosystem Services in Armenia Using ArcGIS Online: A Case Study of the Geghard Monastery and the Kotayk Province	2019
Riggsbee, K. A. et al.	More Than Fast Food: Development of a Story Map to Compare Adolescent Perceptions and Observations of Their Food Environments and Related Food Behaviors	2019
Samuels, K. L., and Platts, E. J.	An Ecolabel for the World Heritage Brand? Developing a Climate Communication Recognition Scheme for Heritage Sites	2020
Schnitzler, C.	Telling Human Stories of Climate Change With ArcGIS Story Maps. Geography Teacher	2020
Schroth, O., and Mertelmeyer, L.	Telling the Story of a Landscape Plan Online	2020
Taylor, J. et al.	Participatory and Spatial Analyses of Environmental Justice Communities' Concerns About a Proposed Storm Surge and Flood Protection Seawall	2022
Thürkow, D. et al.	Using Interactive Story Maps Enriched by Direct Knowledge Queries for the Development of E-Learning Modules on Climate Change	2019
Thürkow, D. et al.	Conception of Online Learning Formats for Climate Adaptation in Central Germany Using Geoinformation Technologies—A Contribution to Education for Sustainable Development	2020
van Blerk, L. et al.	Creating Stories for Impact: Co-producing Knowledge with Young People Through Story Mapping	2022
Vollstedt, B. et al.	Co-production of Climate Services: A Story Map for Future Coastal Flooding for the City of Flensburg	2021
Walther, S. C. et al.	Using GIS and Remote Sensing to Map Grassroots Sustainable Development for a Small NGO in Nepal	2019

Finally, certain documents were excluded because of errors in the previous filters. For example, some book chapters were found at this stage that should have been removed earlier, as well as documents with insufficient relevance to the study.

Additionally, two articles in different languages (Malay and German) were identified, which prevented their analysis.

Following this systematic review, the number of studies removed from the database was 124, compared to the remaining 68 articles.

The second step involved closely working with these datasets, to identify the extent to which their content was aligned to one or more of the SDGs (26 articles, Table 1). This required analysis of not only the title and abstract, but also the full text. A new column in the database was created for this purpose.

Results

From the articles catalogued as part of the corpus of scientific literature on story maps, an analysis of the most relevant quantitative and qualitative aspects was carried out. Whenever possible, the presence of articles dealing with the SDGs is distinguished.

Number and year of Publication

As said, academic literature on story maps is very new (ESRI n.d.), almost as recent as the development of cloud-based GIS itself. Even more current is the United Nations Agenda 2030 (UNGA 2015), adopted by the General Assembly on 25 September 2015. Therefore, scientific articles related to this topic can be traced from this date and not before.

Figure 3 shows the publication sequence of the 68 articles selected in the search on story maps. Those that are related to one or more of the SDGs are highlighted in a different colour.

This graph shows, primarily, that the number of scientific articles dealing with story maps increased rapidly from 2015 to 2022, with the exception of 2021, which can probably be considered an anomaly due to the impact of the COVID-19 lockdown on scientific production (Myers et al. 2020). It is expected that by the end of 2022;

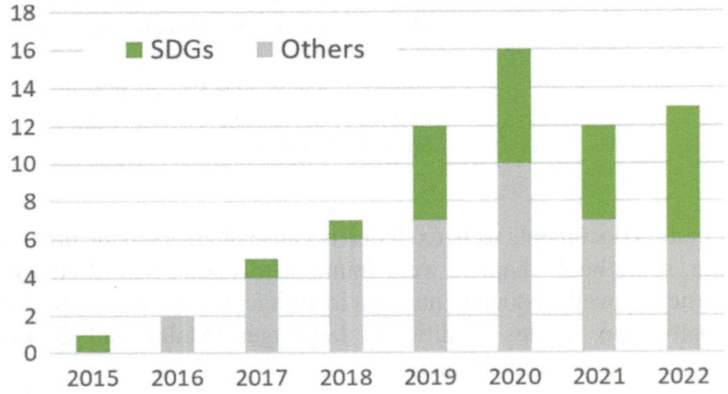

Fig. 3 Frequency of story map articles per year published in scientific journals, with an indication of those with SDG content

the number of articles published on this specific topic will be equal to or higher than in 2020.

On the other hand, there is also growing interest in studies related to the SDGs (Straza and Schneegans 2021). For example, a Scopus search revealed more than 1,500 scientific articles citing them in the title, keywords or abstract in 2022. This increase is also reflected in story map articles. Over the last four years, as shown in Fig. 3, the average number of articles addressing SDGs represents almost 40% of the total, and reached 53% in 2022.

Sources

To date, articles on story maps have been published in 56 different journals, indicating that scientific literature in this field is highly dispersed. Figure 4 shows the journals in which two or more papers have been published. The Journal of Geography in Higher Education stands out, with six articles on the topic. As seen below, this is in line with the predominant thematic orientation of the corpus of articles.

Articles on the SDGs are also very scattered. Only two journals (International Journal of Environmental Research and Public Health and Journal of Digital Landscape Architecture) have published more than one article on this topic.

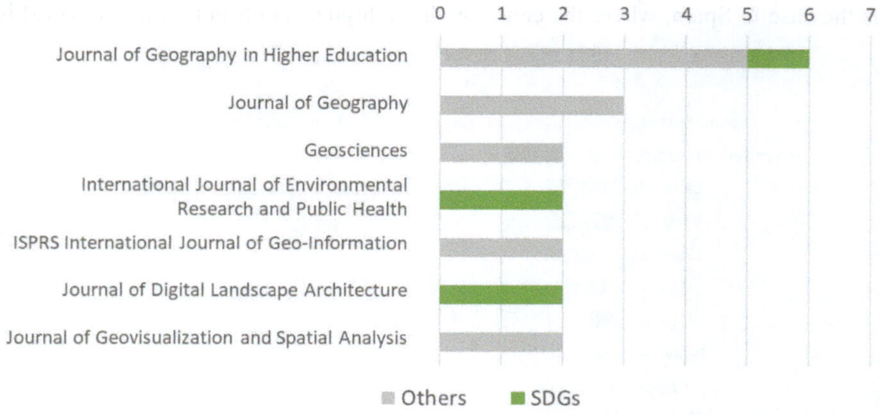

Fig. 4 Frequency of story map articles published by source, with an indication of those with SDG content

Origin and Language

The articles selected for this review are from 14 different countries. This figure does not include the total of 216 participating authors, as only the corresponding or first author of each paper was considered in order to assign the article to a country, city or work centre. However, collaborations between researchers from different institutions and even international partnership are commonplace. This is possible because the average number of authors per paper is higher than three, with 13 single-author papers and several with a very high number of authors: six with five, three with six and one each with seven, eight and 12 authors.

The distribution of papers by country is very uneven (Fig. 5). The United States (US) alone accounts for almost half of all contributions (31 papers). This is followed by the United Kingdom (UK; eight), a group of three countries (Canada, Germany and Greece) with five papers each and Spain with four.

Within this country analysis, only half contributed articles relating to the SDGs. Numerically, the US stands out (12), but the most relevant case is that of Germany, whose five articles fall into this group, and the UK, with 50%, well above the overall 38%.

The map (Fig. 6) shows the distribution of articles by institution of the main author. There is a clear correspondence with the countries mentioned above, but there are some nuances. For example, in the US, despite the large number of scientific articles, production is spread over many institutions (28 out of a total of 31). The opposite is the case in Spain, where the centre with the highest contribution in the world is

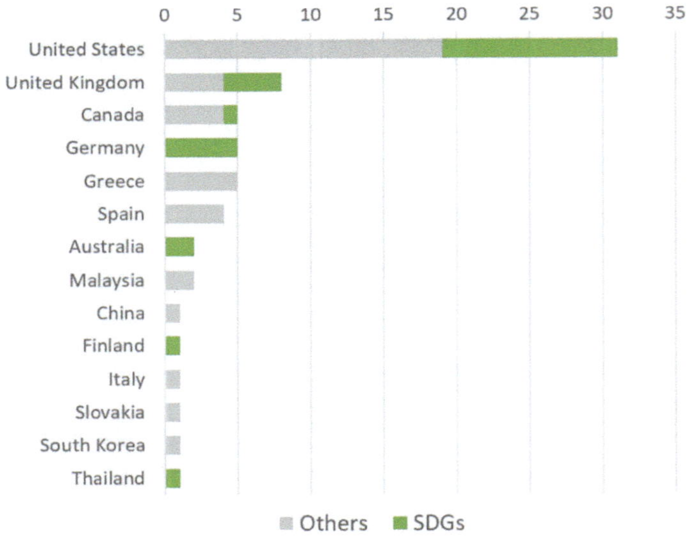

Fig. 5 Frequency of story map articles published by the main author's country, and SDG content

Fig. 6 Frequency of story map articles published by work place of the main author (in the range from one to three, depending on the diameter), with an indication of those with SDG content (green)

located (Universidad Complutense de Madrid, three studies), accounting for 75% of the country's total production.

In addition to this academic centre, there are other institutions with outstanding contributions: in the US, Clemson University, the University of Maryland and the University of Wyoming; in Germany, Martin Luther University Halle-Wittenberg; in Greece, the National and Kapodistrian University of Athens; in Malaysia, the National University of Malaysia; and finally in the UK, the University of Dundee and University College London. Each institution contributes two articles, while the remainder total 58 articles.

Most articles were published in English (96%), although this is the native language of only two-thirds of the origin countries. The exceptions were three articles written in Spanish, German and Malay, respectively.

Topics

Probably the most interesting analysis that can be carried out on the selected articles is that of their subject matter and content.

In this sense, the first analytical approach reveals a strong pedagogical orientation in the scientific literature on this subject (Fig. 7), since half (34) relate to teaching experiences developed at different educational levels. All involved groups of students at different levels. Most of the projects described were carried out at university level (26), while a further seven were at secondary and four at primary level (Fig. 8). The sum of these experiences exceeds the total number, as one took place at primary and secondary level simultaneously, and two others at secondary and university level.

One fifth of the teaching experiences had a theme related to the SDGs.

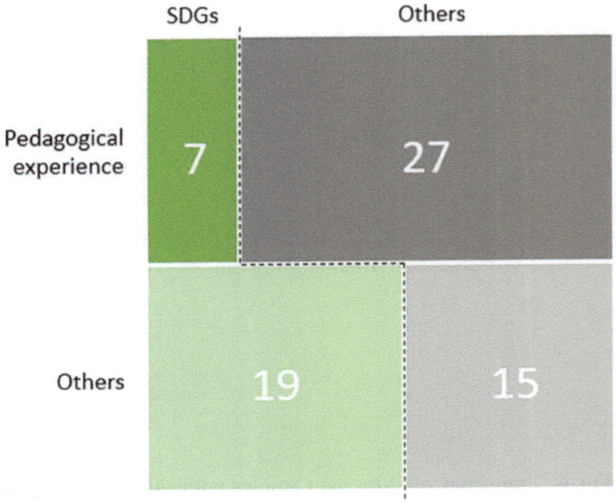

Fig. 7 Distribution of published story map articles, indicating whether they are pedagogical experiences and include SDGs content

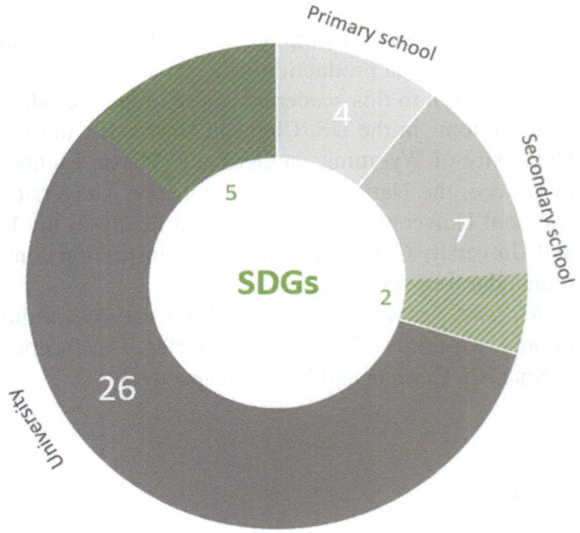

Fig. 8 Distribution of published story map articles based on pedagogical experiences according to educational level, with an indication of those with SDG content

The second type of work that stands out, in terms of number, are methodological studies (11 in total). These articles present the story map tool and analyse its narrative possibilities, the techniques for its development, the software that can be used and even the specific steps—instructions—for constructing a story map.

In some cases, in addition to the more focused aspect of story maps as a tool, these papers also have a specific theme: so in addition to methodology, they are counted as having a social theme (2 articles) and a collaborative experience theme (1 article).

The remaining articles focus on a specific theme. The most common is environmental sustainability (11), followed by social issues (7), history and cultural heritage

(3) and tourism (2) and volcanology (2). Among the articles on environmental sustainability, four articles on the threat posed by climate change to increased flooding in urban areas stand out numerically.

Most of these articles, and of those analysed in general—64 out of 68—focus on a case study. In other words, the content explains or reflects a specific experience.

The distribution of the 26 articles that deal with the topic of sustainability, through the SDGs, is shown in Fig. 9. The largest number of articles relates to Goal 6 (12 articles). Goals 10 and 13 (6 articles) and 15 (articles) are far behind. There are several articles that have been allocated to two objectives because they were strongly linked to both, and in one case to three.

Finally, the content of the articles was examined by means of a co-occurrence mapping analysis of the terms appearing in the titles and abstracts of the 68 articles (Fig. 10). This was undertaken using the VOSviewer software, version 1.6.18.

The results show a 'cloud of terms' with their corresponding relationships. The size of the circles represents occurrence of the terms, and the thickness of the lines, as well as the distance between the nodes, indicates the relationship between them (the number of times they are cited together).

The organisation of the words into three clusters, represented by colours, is clearly visible. The first (green) refers to the educational aspect of the story maps, with the word 'student' occupying a prominent place in the analysis and being closely related to 'geography', 'education' and 'teaching'.

The second cluster (red) contains the most terms and is the most semantically diverse. The dominant word is 'development' and it is noticeable that there are several thematic terms next to it ('history', 'climate change', 'sustainable development' and others), which gives an idea of the large number of 'case studies'—another of the terms that appear—dealt with in the articles.

Finally, the blue cluster is dominated by two terms, "story" and "narrative", which are very close and connected to the red cluster, which tells us about the communicative aspect of story maps.

This analysis of terms allows a better global understanding of a large and very diverse corpus of articles by means of a tool designed to automatically establish links between the words used. Although it is insufficient in itself to draw conclusions, it can help reveal connections and outline explanations.

Discussion and Conclusions

Using PRISMA as a tool to ensure transparency and accuracy throughout, this review identifies as many studies as possible that relate to story maps.

Some of them analyse the story mapping tool and one or more sustainable development goals simultaneously (26 articles in total). These articles represent only a part of the academic literature on story maps, which serve different purposes.

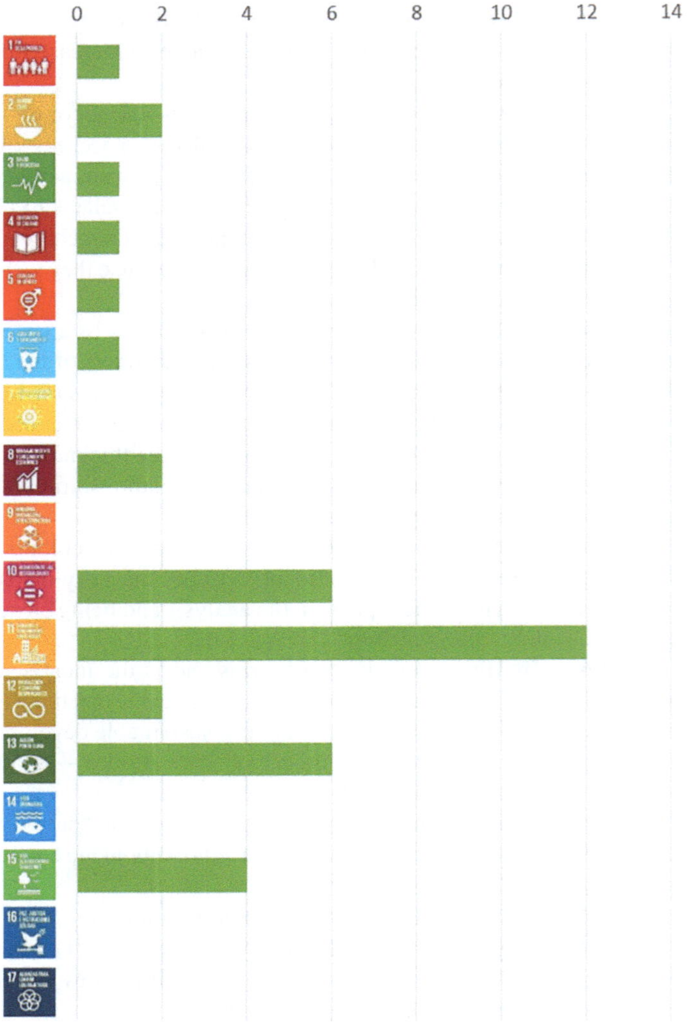

Fig. 9 Frequency of published story map articles by SDG

To date, there have been no published studies that have systematically addressed the development and characterisation of articles written on such a specific topic. This in itself is the greatest strength of this work.

However, it is also important to note this study's limitations. Firstly, choices were made regarding the selection of documents to be analysed, which significantly reduced the base of available publications. The decision to use Web of Science and Scopus, while guaranteeing a certain level of quality of the papers, meant that many were excluded. Systematic literature reviews based on these databases have an

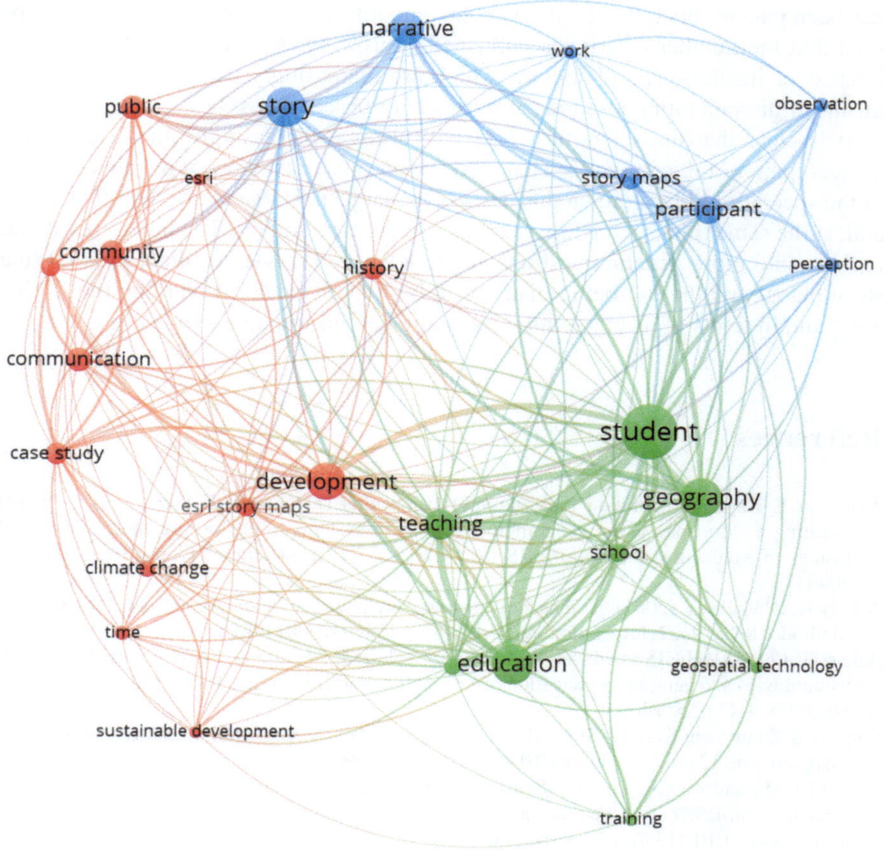

Fig. 10 Co-occurrence map of clustered terms based on the title and abstracts of 68 selected articles

inherent bias: both platforms underrepresent, in terms of scientific output, languages other than English and the fields of Social Sciences and Humanities (Tennant 2020). Searches in other sources may have led to the discovery of additional publications and, of course, to changes in the results found.

The terms used to describe the topic ('story map' and other derived terms) are broad enough to capture the majority of studies, according to some tests in the process, but may exclude scientific literature in other languages.

In addition, due to the large number of documents retrieved, it was decided that all those that were not journal articles, such as book chapters, conference papers, reviews, journal editorials, notes, etc., were not included. The reason for this was that these documents implied a reduction in the sample. As a consequence, the resulting surplus was found to be more homogeneous.

This study affirms that, despite the enormous number of story maps created over the last decade (over two million using ESRI software alone), little scientific attention has been paid to this type of online geographic information tool. However, it can be seen that the number of articles published is growing rapidly, in parallel with the interest of institutions and society, which are increasingly using story maps as a means of disseminating all kinds of territorially-based issues.

It is clear that the scientific world will become increasingly interested in the subject matter of this research. This work represents a first view of the development of the specialised literature on this issue in the largest databases of scientific articles and, at the same time, an attempt to transcend the case studies. The latter is essential to enter into more general questions and to advance knowledge about the tool (the story maps) and the use being made of it, both thematically (regarding the SDGs, e.g.), and in terms of its elaboration (pedagogical, informative tool, etc.).

References

Antoniou V, Ragia L, Nomikou P, Bardouli P, Lampridou D, Ioannou T, Stentoumis C (2018) Creating a story map using geographic information systems to explore geomorphology and history of Methana peninsula. ISPRS Int J Geo Inf 7(12):484. https://doi.org/10.3390/ijgi7120484

Arksey H, O'Malley L (2005) Scoping studies: towards a methodological framework. Int J Soc Res Methodol 8(1):19–32. https://doi.org/10.1080/1364557032000119616

Baker TR (2015) WebGIS in education. In: Muñiz Solari O, Demirci A, Schee J (eds) Geospatial technologies and geography education in a changing world. Springer, Tokyo. https://doi.org/10.1007/978-4-431-55519-3_9

Caquard S, Dimitrovas S (2017) Story Maps & Co. The state of the art of online narrative cartography. Mappemonde 121. https://doi.org/10.4000/mappemonde.3386

Cope MP, Mikhailova EA, Post CJ, Schlautman MA, Carbajales-Dale P (2018) Developing and evaluating an ESRI story map as an educational tool. Natural Sciences Education 47(1):1–9. https://doi.org/10.4195/nse2018.04.0008

De Lázaro ML, Borderías P, Morales FJ (2020) Citizen and educational initiatives to support sustainable development goal 6: clean water and sanitation for all. Sustainability 12(5):2073. https://doi.org/10.3390/su12052073

ESRI (n.d.) What are ArcGIS StoryMaps? Retrieved November 28, 2022, from https://arcg.is/15L5aH

Hong JE (2014) Promoting teacher adoption of GIS using teacher-centered and teacher-friendly design. J Geogr 113(4):139–150. https://doi.org/10.1080/00221341.2013.872171

Idol L, Croll VJ (1987) Story-mapping training as a means of improving reading comprehension. Learn Disabil Q 10(3):214–229. https://doi.org/10.2307/1510494

Kamps DM, Leonard B, Potucek J, Garrison-Harrell L (1995) Cooperative learning groups in reading: an integration strategy for students with autism and general classroom peers. Behav Disord 21(1):89–109

Lo Presti L (2022) Leaving or rescuing the (story) map?—commentary to Saxinger, Sancho Reinoso and Wentzel. Fennia 200(1):68–71. https://doi.org/10.11143/fennia.120599

Myers KR, Tham WY, Yin Y, Cohodes N, Thursby JG, Thursby MC, Schiffer P, Walsh JT, Lakhani KR, Wang D (2020) Unequal effects of the COVID-19 pandemic on scientists. Nat Hum Behav 4:880–883. https://doi.org/10.1038/s41562-020-0921-y

Page MJ, McKenzie JE, Bossuyt PM, Boutron I, Hoffmann TC, Mulrow CD, Shamseer L, Tetzlaff JM, Akl EA, Brennan SE, Chou R, Glanville J, Grimshaw JM, Hróbjartsson A, Lalu MM, Li T,

Loder EW, Mayo-Wilson E, McDonald S et al (2021) The PRISMA 2020 statement: an updated guideline for reporting systematic reviews. BMJ 372:n71. https://doi.org/10.1136/bmj.n71

Pons JJ (2021) Acercar el territorio al aula: la utilización de Storymap Webs como recurso docente. In: Martín MA, Soria C (coords) Cuestiones transversales en la innovación de la docencia y la investigación de las ciencias sociales y jurídicas: especial referencia al impacto del COVID-19, las nuevas tecnologías y metodologías, las perspectivas de género y la diversidad. Dykinson, pp 757–775. https://dialnet.unirioja.es/servlet/libro?codigo=850952&orden=0&info=open_link_libro

Pranckutė R (2021) Web of science (WoS) and scopus: the titans of bibliographic information in today's academic world. Publications 9(1):12. https://doi.org/10.3390/publications9010012

Puertas-Aguilar MÁ, Álvarez-Otero J, De Lázaro-Torres ML (2021) The challenge of teacher training in the 2030 agenda framework using geotechnologies. Education Sciences 11(8):381. https://doi.org/10.3390/educsci11080381

Quinn C, Lippert C, Mclean S (2019) A place-based investigation of the United Nations SDGs: case report on a course assignment for undergraduate sustainability science majors. Sustainability 12(2):124–128. https://doi.org/10.1089/sus.2018.0030

Singh VK, Singh P, Karmakar M, Leta J, Mayr P (2021) The journal coverage of web of science, scopus and dimensions: a comparative analysis. Scientometrics 126:5113–5142. https://doi.org/10.1007/s11192-021-03948-5

Stefanovic D, Havzi S, Nikolic D, Dakic D, Lolic T (2021) Analysis of the tools to support systematic literature review in software engineering. IOP Conference Series: Materials Science and Engineering 1163(1):012013. https://doi.org/10.1088/1757-899X/1163/1/012013

Straza T, Schneegans S (2021) Are we using science for smarter development? In: Schneegans S, Straza T, Lewis J (eds) UNESCO science report: the race against time for smarter development. UNESCO Publishing. https://unesdoc.unesco.org/ark:/48223/pf0000377433

Tennant JP (2020) Web of science and scopus are not global databases of knowledge. Eur Sci Ed 46:e51987. https://doi.org/10.3897/ese.2020.e51987

Tricco AC, Lillie E, Zarin W, O'Brien KK, Colquhoun H, Levac D, Moher D, Peters M, Horsley T, Weeks L, Hempel S, Akl EA, Chang C, McGowan J, Stewart L, Hartling L, Aldcroft A, Wilson MG, Garritty C et al (2018) PRISMA extension for scoping reviews (PRISMA-ScR): checklist and explanation. Ann Intern Med 169(7):467–473. https://doi.org/10.7326/M18-0850

UN General Assembly [UNGA] (2015) Transforming our world: the 2030 Agenda for Sustainable Development. A/RES/70/1. October 21. https://www.refworld.org/docid/57b6e3e44.html

Education for Sustainable Development in Teacher Training Through Multinational Cooperation: Goals, Opportunities, and Challenges

Tal Yaar-Waisel, Sandra Sprenger, and Caroline Leininger-Frézal

Abstract The study of global challenges, such as climate change or water scarcity, requires specific educational concepts. This chapter talks about an international project that conducted parallel courses on geography education for teacher training in Germany, France, and Israel. This international virtual academic collaboration was facilitated using different long-distance communication options. The goal of the project was to enable students to have experiential and meaningful distance learning while implementing ESD in teacher training seminars in geography. The main content of the collaboration was called "Water is Life" which focused on the thematic area of water and was connected to the Erasmus Plus V-Global project. During the seminar, the students developed virtual teaching concepts for geography lessons. The feedback from many of the students indicated that it was an extraordinary opportunity to meet and work together. The project allowed them to discover the issues of water from different perspectives. Multi-linguistic challenges, cultural differences, and online learning difficulties were an integral part of this project and constituted its challenges. Although many obstacles emerged, this project enabled students to use the skills they acquired in their work as future teachers. Finally, the article outlines the conceptual basis and results of the seminar.

Keywords Educational Sustainable Development (ESD) · Teacher training · Distance learning · International collaborations · Intercultural learning

T. Yaar-Waisel (✉)
Oranim College of Education, Kiryat Tiv'on, Israel
e-mail: yaar.tal@gmail.com

S. Sprenger
Universität Hamburg, Hamburg, Germany
e-mail: sandra.sprenger@uni-hamburg.de

C. Leininger-Frézal
Université Paris Cité, Paris, France
e-mail: caroline.leininger-frezal@u-paris.fr

A. Klonari et al. (eds.), *Re-visioning Geography*, Key Challenges in Geography,
https://doi.org/10.1007/978-3-031-40747-5_5

Introduction

The COVID-19 crisis brought significant changes to university teaching and learning (Day et al. 2021). Within a very short time, face-to-face courses had to be converted into digital formats, confronting both teachers and students with considerable challenges. For example, geographical working methods, such as fieldwork, can no longer be carried out (Day et al. 2021). At the same time, this change also gave way to several opportunities. For example, transferring teaching to the digital space offered great potential for international exchange and the internationalization of teaching. Expanding on a bi-national project (Leininger-Frézal and Sprenger 2022; Yaar-Waisel and O'Reilly 2022), a third country, Israel, was included in the international cooperation under the present project. The content of the collaboration focused on the thematic area of water, and the project called "Water is Life" is connected to the Erasmus Plus V-Global project. It is an international geography teaching project that took place from 2021 to 2022 through cooperation among the Universität Hamburg (Germany), Université Paris Cité (France), and Oranim College of Education (Israel). The focus of the content is on global challenges, such as climate change and water scarcity, which require specific educational concepts (Fensham 2012). One of these is the Educational for Sustainable Development (ESD; Leicht et al. 2018), and within this context, "water" is considered one issue. Global change and rising water consumption all over the world have led us to investigate this issue and make them the focus of the project, which was completed via parallel courses on geography education for teacher training. International virtual academic collaboration facilitated the completion of this project through the use of different long-distance communications options. Due to the current pandemic situation, different digital teaching opportunities were also used. The basis of our work was the perception that international collaboration allowed students to get to know the subject better from different perspectives.

Contextual and Theoretical Backgrounds

Teacher Training in Three Different Countries

The education systems, including teacher education, varied significantly in the three countries involved in the project. The teacher training curriculum is under the responsibility of the ministries of education in each of the three countries.

In France, teacher training begins after a bachelor's degree in one of the disciplines taught. Students then enter the *Institut National Supérieur de Professorat et de l'Education* (INSPE) to prepare for obtaining a master's degree dedicated to teaching (Master MEEF). At the end of their master's degree, they must pass a teaching exam that is quite selective. Students who do not pass the exam can be hired as temporary teachers.

In Germany, teacher training is divided into two stages (Kultusministerkonferenz 2021). The first stage is a degree program in higher education at a university. In the majority of states, a consecutive structure of study with bachelor's and master's degrees (BA/MA) exists, including periods of practical training. The university stage is followed by a second phase, which is a practical training course in a school setting called *Vorbereitungsdienst*. This takes place in teacher training institutes (*studienseminar*) or other comparable institutions.

Meanwhile, teacher training in Israel is highly diverse. The accepted route is to study for a first degree (BEd) while also obtaining a teaching certificate. Completing a degree lasts four years, but many studies for a teaching certificate only after completing their first degree (for one or two years). There are unique tracks for master's degree studies and a teaching certificate requirement. In fact, many teachers study for master's degrees separately from their teacher training. Teacher training in Israel also includes theoretical studies in colleges and universities, in addition to actual teaching experience in schools. After completing their academic studies and obtaining a teaching certificate, students must also undergo an internship year, which will allow them to obtain a teaching license from the country's Ministry of Education.

Implementation of ESD in Teacher Training

Education for sustainable development (ESD) "aims at developing competencies that empower individuals to reflect on their own actions, taking into account their current and future social, cultural, economic, and environmental impacts, from a local and a global perspective. Individuals should also be empowered to act in complex situations in a sustainable manner, which may require them to strike out in new directions; and to participate in socio-political processes, moving their societies towards sustainable development" (UNESCO 2017, p. 7). Within ESD, Wiek et al. (2011) developed a framework that includes five key sustainability competencies: (1) systems thinking, (2) future thinking (or anticipatory), (3) values thinking (or normative), (4) strategic (or action-oriented), and (5) collaboration (or interpersonal). In the context of teacher training, further models have been developed, including the RSP Competence Framework (Vare et al. 2019), which consists of a matrix of 12 competencies.

Given that teacher training itself varies significantly in the three countries, the conditions for implementation also vary. ESD was first introduced in France in 2004 through a transversal circular. Then, at the end of the 2000s, sustainable development was introduced into the curriculum, particularly in geography and science. A label was set up at the same time for establishments that engaged in educational projects related to sustainable development, which were called E3D schools (establishments in the sustainable development process). In teacher training in Germany, the implementation of ESD at the formal and informal levels in the curricula has progressed in recent years. Sprenger and Nienaber (2017) demonstrated that the formats applied at universities are highly diverse. Depending on the university, a wide range of formats has been used—from individual seminars based on the initiative of individual teachers to the use of modules and whole-institution approaches. In Israel, the ESD forms a significant part of the geography study program. The curriculum in Israel from Grades 2–12 mainly deals with the subject of "education for sustainability." The values that lead to the teaching of the subject as they appear in the curriculum include "A person's responsibility for the environment and for development through sustainability" and "Environmental justice." The curriculum states that "a person has a responsibility towards the natural environment, and he also develops sustainable processes" (Ministry of Education 2022). In this project, sustainable development is addressed in a disciplinary framework through the lens of geographical education.

Interculturality

Another central theoretical aspect of our project seminar was intercultural learning or intercultural competence (e.g., Barrett 2018). Barrett (2018: p. 94) defines intercultural competence "as the set of values, attitudes, skills, knowledge, and understanding that are needed for understanding and respecting people who are perceived to be culturally different from oneself, for interacting and communicating effectively and appropriately with such people, and for establishing positive and constructive relationships with such people." This includes a whole framework of 14 components that are to be assigned to the following areas: (a) values, (b) attitudes, (c) skills, and (d) knowledge and critical understanding (Barrett 2018). For the school context, Barrett (2018) suggested various activities to promote intercultural competence, which can also be applied to the university context. These include Internet-based intercultural contact and pedagogical approaches, such as cooperative learning (Higgitt et al. 2008) or project-based learning.

Concept of the Seminar

Goals of the Project

The main goal of the project was to enable students to realize ESD and to have experiential and meaningful distance learning. It also aimed to help them identify the challenges of ESD and prepare them for implementing it in their classrooms in the future. To this end, the seminar aimed to accompany students in the development of the five skills listed above (Wiek et al. 2011). The first three competencies (systems thinking, future thinking, and values thinking) are developed through the contents of the training and are also implemented in the seminar and the framework of the

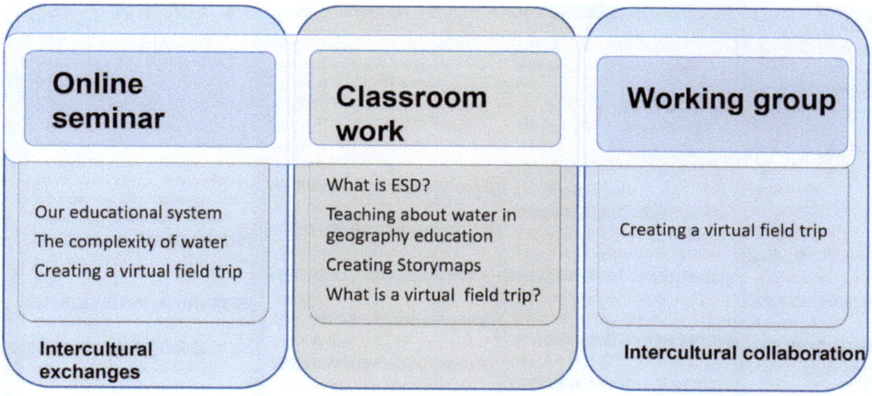

Fig. 1 Organization of the seminar (own illustration)

courses in which they are included. Indeed, as explained in detail, the international seminar was carried out in each of the universities involved as part of a geography education course. The last two competencies (strategic and collaboration) constituted the pillars on which the seminar was built, as shown in Fig. 1.

Organization of the Seminar

The seminar was organized into three parts: online seminar, face-to-face course, and working group.

Online Seminar

The first part consisted of four online meetings among students from the three countries and was focused on the educational systems in Germany, Israel, and France, as well as the water-related issues in each national context. In the beginning, the online seminar required the students to get to know one another and create trust-building steps, perform fun icebreaker activities, and share their experiences. Such meetings among people who were in different places but partners in the same profession allowed fruitful exchanges, although they faced linguistic and online learning challenges, which were integral parts of this project. Overall, this project enabled students to use various skills they will acquire in their work as future teachers. The online seminar was a collaboration-based learning activity using participative tools participatory tools are those that allow collaboration between students or between students and their teachers. There is a wide range of participatory tools that allow students to write a text collaboratively, and collect data, such as Padlet and Survey123. On each of the topics, the students identified similarities and differences between the three countries. They then conducted research and formalized presentations on their country's educational systems, which they eventually shared in groups. On the theme

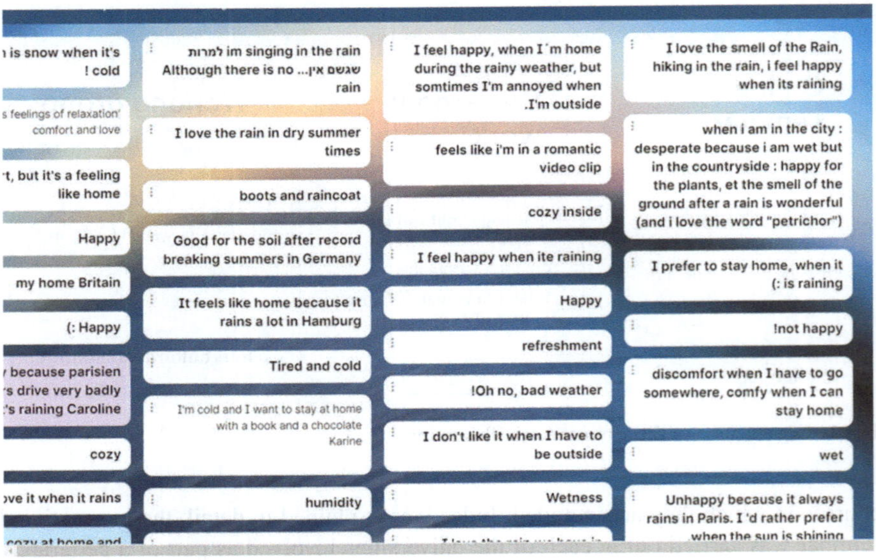

Fig. 2 "Padlet" board with students' perceptions of rain (permission to publish)

of water, the students had the opportunity to learn from others regarding the current state of water in their respective countries, the present challenges, and existing solutions. For instance, the *Padlet* screenshot in Fig. 2 highlights variations in students' reactions to rain. While some French and German students did not appreciate it and preferred to stay at home, the Israeli students welcomed it with enthusiasm. The online seminar was articulated in a regular face-to-face course and in working group activities.

Classroom Work

Between the online seminars, the university teachers developed some relevant content for ESD and for the project in their own classrooms. The following four topics were taught:

- What is education for sustainable development?
- Teaching about water in geography education.
- What is a virtual field trip?
- Creating a Story map.

Water that enables life and is one aspect of sustainability. These topics were relevant to the German, Israeli, and French curricula. Therefore, it was interesting to develop them in a national context. These topics were discussed from a comparative perspective, both in the webinar and in the group work.

Working Group

The students worked for several weeks in international groups with the following instructions:

- Choose and explore a specific water issue that is relevant to the three countries.
- Raise questions and collect data to explore these questions.
- Create a virtual field trip based on several case studies or data from the three countries. The students were required to use Story Map.

The student groups worked independently, while the teachers intervened to guide and advise them when necessary. The students and teachers encountered various difficulties, which will be presented in Sect. 4. The students presented their virtual field trips in the last session of the seminar, after which they were allowed to ask questions about their respective productions. They were also asked to discuss in groups the water issues they identified in their countries and how such issues were taught in the subject of geography. This led to the next step of group work ("*Story Maps*"). The groups collected or produced data on a chosen theme related to a space and processed it. On the basis of this data, they created a Story Map to lead other students to explore the space and identify the global change issues related to a theme.

Evaluation—Methods

Academics describe action research as part of their teaching and research activities. Action research matches the educational goals of the project and provides benefits for both teachers and students. Action research is a model of professional development that promotes collaborative inquiry, reflection, and dialogue (Lesha 2014). The action research part of the current project was organized in a successive feedback loop aimed at adjusting the content of the seminar and its organization to our objectives, the student's experiences, and their learning. After each online session, a review was conducted with the students in the classroom. This feedback was then discussed among the partner universities to debrief the last seminar session and anticipate the next one. Necessary adjustments and clarifications were made throughout the project. Discussions at the end of the year were intended to take stock and look forward to the following year. The diagram in Fig. 3 illustrates the feedback system that guided the project and the educational actions behind it.

The evaluation methodology consisted of two qualitative document types (Creswell and Creswell 2018): (1) analysis of the students' logbooks and (2) analysis of the lecturers' notes.

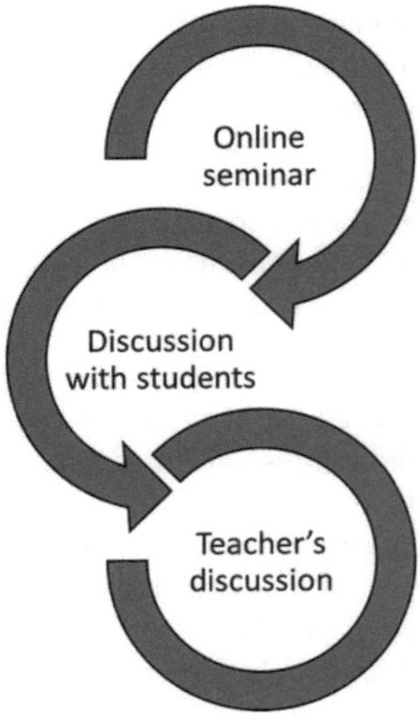

Fig. 3 The research action plan (own illustration)

Analysis of Students' Logbooks

The logbook is a reflective tool in which students can record various types of data throughout the project, including their observations, feelings, reflections, and questions. The students were asked to write in a logbook what they learned, what surprised them, their questions, and the strategies they planned to deploy to answer these questions. Below are the instructions given at the beginning of the logbooks distributed to the students.

> This document is a tool to help you learn. After each international seminar, please write here what you have learned, what wonders you, your questions (regarding the point on ESD, on intercultural, language skills, digital skills, etc.), and how you will answer your questions. You will also complete this document during the working group.

The evaluation of the seminar was independent of their assessment, as it was decided that they should be given freedom in the official assessment of the course. In each university, different evaluation procedures and measurements were accepted. The use of a logbook is not common in any of the countries involved in this seminar; yet to encourage students to express their ideas, we asked them to write at least after each webinar and during the group work. For this reason, the dates of each meeting were entered into the logbook. The contents of the logbooks were not shared among the students, although they were regularly discussed in class. At the end of the project, the teachers collected the logbooks from the students who agreed to share them. We analyzed these using a content analysis methodology (Mayring 2014). However, this analysis was not conducted in a comparative manner because the students from the different countries involved had profiles that were too disparate to be comparable. The data were also not collected in the same way: the Israeli students were asked

Table 1 Profiles of students involved in the project

University, Country	France	Germany	Israel
Number	12	30	22
Number of logbooks collected	8	–	21
Profiles	In-service teachers (10) Teacher trainers (2) France Senegal (2) Ivory Coast (2)	Pre-service teachers, MA degree program	Geography education students. Half of them already obtained their BA, while the others are currently studying for their bachelor's degree (in their 3rd year) The students—all Israelis—were Jewish, Muslim, Christian, and Druze

to complete an online logbook after each lesson as part of their summary reflection, while the French students did it on a Word document. For organizational reasons, no data have been collected in Germany.

Students from the University Paris City were all experienced teachers. Some were teacher-educators (2), others were PhDs (1) or post-docs (2). They came from all over the French-speaking world: France; Senegal, Ivory Coast (see Table 1). During the years 2021–2022, three African students who participated in the project faced very different conditions from the other students (French, Germans, and Israelis). Classes in Ivory Coast and Senegal had between 60 and 120 students. Furthermore, the African teachers had limited resources: no or few textbooks for the students, random access to the Internet, and erratic electricity supply. Meanwhile, the Israeli and German students were in pre-service teacher training. In Israel, some students were Arabs, while others were Jews. The cultural diversity of the project came not only from the dialogue and collaboration among students from different universities but also from the highly diverse backgrounds of the participants.

Analysis of the Lecturers' Notes

Before and after each international meeting, the notes taken by the three lecturers were compiled and then analyzed using qualitative content analysis (Mayring 2014). During this method of content analysis, the texts (in this case the students' logbooks) are systematically examined. It is used to identify patterns or structures in the texts studied, based on this, categories are then formed. The categories formed are reflected in the results. In this case, they were formed inductively.

Evaluation—Results

The theoretical and conceptual principles of the seminar (teacher education, intercultural learning) form the basic structure of the chapter, and the results are presented accordingly in each case.

Students' Learning Outputs—French Results

Intercultural Dimension

The Université Paris Cité students discovered other educational systems compared to the French one, and this did not fail to generate debates and questions, as demonstrated by the following quotes:

> Both Germans and Israelis were upset by the length of the school days for our pupils! (Student UPC 7)
>
> Thus, after the presentation of these educational systems, we were able to note that the Israeli educational system is based on religion and relies on three types of schools: secular public schools, religious public schools, and orthodox schools. As for the German educational system, we noted that unlike in Senegal, where school programs are applied throughout the country, in Germany, each state is responsible for the content of the school program in effect in the schools located there. (Student UPC 6)

The French students were unaware that their relationship with water was culturally anchored. The relationship of Hamburg students was marked by risks, that of Israeli and African students by its scarcity, and that of Parisian students by the unfavorable weather. As some of them stated:

> For European countries, it seemed to me that it was sadness and cold, negative, that was associated with rain; for Israel, [it was a] joy. (Student UPC 1)
>
> It seems that the question of mental/social representations is very present without this being explicitly formulated for the students. I am beginning to understand that a cultural project is built on the basis of national representations, perhaps in response to the initial hypothesis I formulated for myself at the last meeting: Are there national specificities in the teaching of the issue of water? It is clear today that there are! The next question is: Which ones? (Student UPC 2)

The French students also noted differences in their understanding of ESD. The French curriculum places greater emphasis on weak sustainability (Aubertin and Vivien 2006) than on strong sustainability, unlike the German and Israeli curricula.

> I think I have also become aware of some of the limitations of education for sustainable development, as it is organized in the French school system, despite my goodwill to make pupils think about current issues. In the curricula, as well as in the textbooks and in my lessons, we have too much of a tendency to look for examples far from France, which reduces, I think, the ability of pupils to project themselves into the reality of the issues. (Student UPC 2)

Furthermore, the students mentioned that there was no physical geography taught in secondary schools in France, unlike in the other two countries:

> A student sent me his work. I realize that physical geography is very important for Germans [...]. His work focuses on the physical causes of the 2021 flood in a large German region. (Student UPC 4)
>
> The German students were really surprised that we don't teach physical geography in France (or so few issues) because many of them are going to teach biology and geology together with geography. In Israel, physical geography is also important—maybe to avoid social or political issues? (French student 7)

Finally, the students noted that they had different relationships with burning social issues.

> I explained that in France, the issue of water does not seem to be debated and that French programs rather present diffuse issues, often on an international or other countries' scale, but rarely on the French scale. The German students explained that ecological issues are being discussed a lot by teachers in Germany at the moment, rather than through the entry: How to reduce your carbon footprint. Israeli students presented their material on water management in Israel: How the country has successfully met the water needs of its population. Water management is presented as a model, but they do not address sensitive issues (e.g., resource sharing with the Palestinian territories). (Student UPC 2)

Learning Outputs in Terms of Digital Skills

Beyond intercultural learning, the Université Paris Cité students seemed to have developed digital skills. However, it was difficult to find evidence of these skills in the students' logbooks. Most of the time, they pointed out the difficulties they added with StoryMaps and other tools they were asked to use:

> So I started working on the StoryMap; the tool is interesting, and not so complicated to use (except for the ArcGIS maps: impossible to load shape files to make layers... I pull out my hair for many hours). (Student Journal 5)

The students' virtual field trips showed that they were able to appropriate the StoryMap. The content of the virtual field trips highlighted their ability to search for or create data, analyze them, and produce maps or images relevant to the topic. The StoryMap implements different resources, such as maps, photography, and 360° photos. However, we decided not to analyze the final productions of the students in detail. Nevertheless, we feel it is important to highlight what the logbooks do not say. Students' digital skills were translated into action. All groups managed to produce virtual terrains on the topic of water. We had the same difficulties showing the collaborative skills developed by the students. The logbooks addressed the difficulties previously highlighted in the sessions: language differences as well as different conceptions of ESD and what was to be done:

> We had a discussion in our group because some members didn't want to speak in English in front of everyone and wanted one person to represent the group. It's not easy for anyone to speak a foreign language! Another difficulty was the question of interactivity in the case studies. The case studies correspond rather to a juxtaposition of examples without any real

> complementarity; even if there are common themes, there is no crossing of local issues. (Student UPC 2)

These virtual field trips were conducted in a concerted manner in groups of students from each of the partner universities.

Professionalization of Students Undergoing In-Service Teacher Training

Finally, the particularity of the French students involved in the project was that they were training to become teacher trainers. Thus, they were the only experienced students in the groups. However, this tended to put them in a difficult position. The pre-service teachers were sometimes more focused on the content than on how to teach it, giving priority to a lecture rather than the design of the learning activities.

> The second dimension of this experience is professional. It has indeed allowed me to reflect on my professional practice, in a reflective approach (Perrenoud 2010), and in particular on teaches training that I am a stakeholder (initial training). (Student UPC 2).

Students' Learning Outputs—Israeli Results

Intercultural Dimension

Many of the Israeli students addressed the language barrier and the difficulties related to the Internet. These difficulties obscure the creation of communication and learning. In some groups, it was crucial and too difficult to learn. After overcoming the language and communication barriers, the enthusiasm to meet people from "overseas" was viewed positively:

> I got to know really nice, warm, and smart people from different cultures.
>
> I learned that it is good to get to know more people and their perceptions of geography and learn from them.
>
> It surprised me that there were a lot of things in common between us, from a geographical point of view.
>
> It surprised me that the students were really enthusiastic about Israel and the entire composition of the country.
>
> "It was very interesting to get to know a different culture and hear about the different education system in each country."

The element of "surprise" for the Israelis can be argued: the participants listened to them and were interested in what was happening in Israel. This was a recurring motif in what they had written. Israeli students thought, although the Israeli students encountered diverse cultures in their academic studies, the distance created a special kind of excitement. It seemed that the project experience and the multicultural meeting opportunities created more excitement than the content itself. Overall, the Israeli students liked taking part in multicultural mixed groups.

Learning Outputs

Some of the Israeli students suggested how to improve teaching and learning methods, such as enlarging the groups due to the lack of presence of participants, which could serve as a "backup" to support the language difficulties experienced by some of the participants. They also recommended the preparation of visual learning materials:

> It surprised me that the conversation was successful and there were no awkward silences; on the contrary, the time was not enough for us, and we wanted to keep talking and listening.
>
> How can all participants be encouraged to take an active part in the conversation? In some parts, I felt that there were few people who were active in the exercise and thinking about the questions, while some were silent.

Cultural differences, learning habits, and teamwork stood out in the cooperative learning process. These were actually parts of the overall goal of the project, and the students had opportunities to adapt to these differences.

Professionalization of Students Undergoing In-Service Teacher Training

The students studying teaching saw the project as a demonstration of a project-based learning method. As such, they had to deal with both the teaching method and the chosen content. Both will serve them in the future as geography teachers. In particular, the students expressed interest in both the education system and the water situation. They were generally happy to learn what was happening elsewhere. In addition, they were happy to share their knowledge with people whom they felt were interested in their topics. In terms of content, when learning was enabled, it seemed to be successful. Furthermore, the water subject proved to be interesting, especially the differences in attitudes resulting from climate differences. As one student noted, this was "*an interesting topic that may be relevant to each country but of the same topic.*" Another student mentioned that "*the issue of water economy management, various dealings with floods*" was the most interesting topic.

As future teachers and those who experienced teaching, they appreciated the efforts required in constructing such a structure, which "*make a kind of competition not between countries but between groups.*" Other students shared the following opinions:

> [It] was very interesting and experiential. I learned a lot of new tools and enjoyed collaborating with students from Germany and France.
>
> I am so delighted to participate in this program.
>
> Thank you for the opportunity.

Apart from the positive evaluations, the Israeli students were not afraid to criticize and offer suggestions for improvement. First was the issue of time: they thought the lessons were too short, and they lacked time for group assessment. They also raised issues related to differences in culture, knowledge, preparation for the meetings, and even in software experience:

> Mixed feelings, I was excited, scared, but it really [developed] me a lot.
>
> Very exciting and stressful [at the same time].

For the Israeli students, this was part of their preparation for teaching in schools, which was mostly done in homogeneous groups. The success of the project may lead future teachers to bring different classes together.

Analysis of the Lecturers' Notes

Based on the notes taken by the lecturers, four areas of multinational collaboration in teacher education were identified as being associated with challenges and opportunities (Fig. 4): (1) organizational aspects, (2) content, (3) language and communication, and (4) the digital teaching and learning environment.

Organization
The collaborative work required a very precise organization of the work plan ahead of time so that it would be clear to the lecturers as well as the students. Each meeting was well-planned in advance and in detail. It was agreed in advance how much time would be given to each part and which of the lecturers would be responsible. However, in advance of the planning, the different semester times at the three universities proved to be a challenge. For example, the semesters began and ended at different times, and there were also interruptions due to vacations at different times so the (relatively small) overtime had to be identified. Another opportunity—and also a challenge—was the structure of the educational system at the school level as well as the level of teacher education. This meant that a solution had to be identified for the simultaneous seminars, which could not be held on the same days of the week for various reasons.

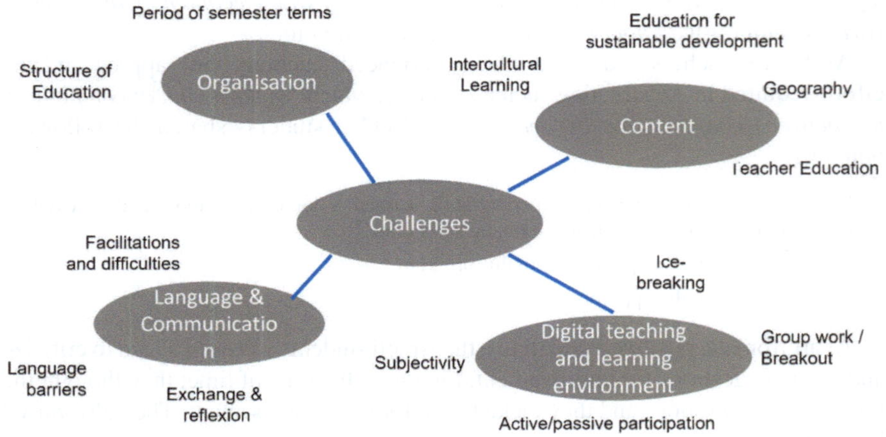

Fig. 4 Challenges to and opportunities for multinational cooperation (own illustration)

Content

The content of the seminar was determined by the educational concept of ESD. Other points of reference were intercultural learning and teacher training in geography. In the course of the seminar, it became clear that although we were talking about the same topic, the background, and framework conditions were very different (see Sects. 2.1. and 2.2). At this point, it was very important to have a regular exchanges with the other lecturers about this topic. Collaborative tasks were also challenging, both in terms of the content and the level of thinking required of the students, especially in terms of the differences in mentality, language, and schedules of the students.

Language and Communication

Regarding this aspect, the biggest challenge was the language barrier experienced by the students. During the seminar, it became apparent that there were very different language levels among the students; while some were able to engage in the conversation almost effortlessly, other students had greater difficulties. Three things proved helpful in advance or during the seminars: scaffolding in the sense of linguistic formulations or vocabulary lists, a written preparation of the task, and/or working in small groups/breakout rooms. In particular, the work in smaller group rooms and not in the whole plenum led to a larger group of students being active.

Digital Teaching and Learning Environments

In the area of digital teaching and learning environments, we started with icebreaker activities at the beginning of the seminar to first introduce the international students to one another and get them talking to each other. This included, for example, a digital map activity in which everyone was asked to locate themselves. Another activity was a collaborative digital board (Padlet) to introduce selected aspects of their own countries. These activities were also conducted regularly at the beginning of each international session.

A central challenge of collaboration in a digital learning environment is the sometimes very different collaboration of students. On the one hand, some were highly engaged and active and had advanced the content of the seminar. Others, on the other hand, were more passive and, often, only the black tile was visible to some students during the videoconference. This was resolved during the seminar by having group work take place in smaller rooms/breakouts. Despite some strengths, some limitations also became apparent. These lay primarily in the fact that fieldwork is directly associated with geography, which is not possible or only possible to a limited extent in digital space.

The sentences summarizing the students' words, "*Mixed feelings, I was excited, scared but it really promoted me a lot*" and "*Very exciting and stressful together*" can also express well the feelings of the lecturers: it is hard work that requires an investment of time and thought mental flexibility, and the ability to adapt to the requirements and limitations of the other partners. The initial nature of the project obliged the lecturers to preparatory meetings, which took place before the beginning and during the semester. Checking the student feedback after each lesson and adjusting the following lessons accordingly. It taught the double time of teaching: 12 h of preparation. The collaboration between the lecturers was made possible thanks to the videoconference tool. Notably, the shared presence excited and aroused curiosity. This was also a new teaching method, and they had to adapt to it.

Discussion

The purpose of this chapter was to present the concept and evaluate tri-nation cooperation in geography education as part of teacher training. In this chapter, a case study in the context of ESD in teacher education was presented to show the goals, opportunities, and challenges involved. Based on the results, it became clear which opportunities and challenges exist. On the one hand, this development toward international cooperation was favored by the COVID-19 crisis, where teaching was converted to digital formats within a very short time (Day et al. 2021). This benefited the present seminar, as both lecturers and students were already used to digital videoconferencing.

In general, the present study presents the potential for international collaboration in geography and teacher education in geography. With regard to this general aspect, the results are consistent with those of other studies (e.g., Higgitt et al. 2008; Leininger-Frézal and Sprenger 2022) that have also demonstrated international cooperation in geography. The virtual field trips served as relevant tools for fostering international collaborations, especially in the context of ESD. The students were able to explore each other's territories from a collaborative perspective; the objective being to design a virtual excursion together. Thus, the virtual field trips were not only learning objects but also links between students and between distant territories. This is one of the innovative aspects of this chapter.

Research has shown the benefits and limitations of using virtual field trips in different disciplines, such as geography (Friess et al. 2016), biology (Haris and Osman 2015), and geosciences (Qiu and Hubble 2002), as well as from the perspective of environmental education or climate change education (Markowitz et al. 2018; Petersen et al. 2020). Those studies are focused on the learning achieved or on the curriculum implemented (Detyna and Kadiri 2020). In the current study, the virtual field trips served as tools for linking teachers in pre-service or in-service teacher training from the perspective of their professionalization. Furthermore, although experiments have been carried out in secondary education, little work has been done to introduce virtual field trips in teacher training. This is the second innovative aspect of this project.

Overall, our findings highlight the challenges and opportunities in transnational collaboration in teacher training. An interesting opportunity is that students from different places found a common interest in the chosen subject (i.e., water). Although the perceptions of the subject and everyday life created very different experiences, the results of this research show initial findings of cooperative international teaching, encouraging the motivation of young students to get to know one another, participate, and learn in this way. The common goals of teaching also led to overcoming difficulties and challenges, such as language and mentality differences.

In the literature, Barrett (2018) has suggested various activities to promote intercultural competence, which can also be applied to the university context, including Internet-based intercultural contact. To implement this, the activity we proposed encouraged personal, cultural, and national representations, for example, getting to know the students' favorite foods, favorite music, etc. The message to future teachers is that teachers have to accommodate all of their students' cultures. From a teacher's perspective, these are categorized into four areas: organization, content, language and communication, and digital teaching and learning environments. Cooperative learning requires flexibility in guidelines and acceptance of differences, both among lecturers and students.

The present study represents the first attempt to address issues relating to an international teacher training seminar. Nevertheless, the study also has limitations. First, the research is a case study that represents an example of collaborative work; however, it has no claim to general validity. Thus, a quantitative approach to the different dimensions of ESD could be an ideal topic for future research. Second, the research is limited by the differences in various guidelines and regulations for conducting studies among countries (e.g., for data collection).

This project is an example of the internationalization of secondary school teacher training. It is part of a larger V-Global project for university students and teachers with the objective of developing ESD at universities. Furthermore, the virtual field trips served as tools to introduce ESD into the university curriculum through the development of blended learning.

References

Aubertin C, Vivien F-D (2006) Le développement durable enjeux politiques économiques et sociaux. Documentation Française, Paris

Barrett M (2018) How schools can promote the intercultural competence of young people. Eur Psychol 23(1):93–104. https://doi.org/10.1027/1016-9040/a000308

Creswell JW, Creswell DJ (2018) Research design qualitative, quantitative, and mixed method approaches. Sage, Thousand Oaks

Day T, Chang I-CC, Chung CC, Chung KLC, Doolittle EW, Housel J, McDaniel NP (2021) The immediate impact of COVID-19 on postsecondary teaching and learning. Prof Geogr 73(1):1–13. https://doi.org/10.1080/00330124.2020.1823864

Detyna M, Kadiri M (2020) Virtual reality in the HE classroom: feasibility, and the potential to embed in the curriculum. J Geogr High Educ 44(3):474–485. https://doi.org/10.1080/03098265.2019.1700486

Echart T (ed), Kultusministerkonferenz [Standing Conference of the Ministers of Education and Cultural Affairs] (2021) The education system in the Federal Republic of Germany 2018/2019. KMC, Bonn. https://www.kmk.org/dokumentation-statistik/informationen-zum-deutschen-bildungssystem/dossier-englisch.html

Fensham PJ (2012) Preparing citizens for a complex world: the grand challenge of teaching socio-scientific issues in science education. In: Zeyer A, Jyburz-Graber R (eds), Science| environment| health. Springer, Dordrecht, Heidelberg, New York, London, pp 7–29. https://doi.org/10.1007/978-90-481-3949-1

Friess DA, Grahame OJH, Quak MSY, Lau AYA (2016) Incorporating «virtual» and «real world» field trips into introductory geography modules. J Geogr High Educ 40(4):546–564. https://doi.org/10.1080/03098265.2016.1174818

Haris N, Osman K 2015 The effectiveness of a virtual field trip (VFT) module in learning biology. Turk Online J Distance Educ 16 (3):102–117.

Higgitt DL, Donert K, Healey M, Klein P, Solem M, Vajoczki S (2008) Developing and enhancing international collaborative learning. J Geogr High Educ 32(1):121–133. https://doi.org/10.1080/03098260701731629

Leich A, Heiss J, Byun WJ (eds) (2018) Issues and trends in education for sustainable development. UNESCO Publishing. https://unesdoc.unesco.org/ark:/48223/pf0000261954

Leininger-Frézal C, Sprenger S (2022) Virtual field trips in binational collaborative teacher training: opportunities and challenges in the context of education for sustainable development. Sustainability 14(19):12933. https://doi.org/10.3390/su141912933

Lesha J (2014) Action research in education. Eur Sci J 10(13):379–386. Accessed from https://core.ac.uk/download/pdf/328024278pdf

Mayring P (2014) Qualitative content analysis—theoretical foundation, basic procedures and software solution. https://www.ssoar.info/ssoar/handle/document/39517

Markowitz DM, Laha R, Perone BP, Pea RD, Bailenson NJ (2018) Immersive virtual reality field trips facilitate learning about climate change. Front Psychol 9:2364. https://www.frontiersin.org/articles/10.3389/fpsyg.2018.02364

Ministry of Education, Israel (2022) Geography, Adam Sviva [Geography, man and environment. Geography curriculum]. https://pop.education.gov.il/tchumey_daat/geography_adam_sviva/

Nobaizura H, Kamisah O (2015). The effectiveness of a virtual field trip (VFT) module in learning biology. Turk Online J Distance Educ 16:102–117. https://doi.org/10.17718/tojde.13063

Petersen G, Klingenberg S, Mayer R, Makransky G (2020) The virtual field trip: investigating how to optimize immersive virtual learning in climate change education. Br J Edu Technol 51(6):2099–2115. https://doi.org/10.1111/bjet.12991

Qiu W, Hubble T (2002) The advantages and disadvantages of virtual field trips in geoscience education. China Pap: Tert Sci Math Teach 21st Century 1(1):75–79. https://www.researchgate.net/publication/253691199_The_Advantages_and_Disadvantages_of_Virtual_Field_Trips_in_Geoscience_Education

Sprenger S, Nienaber B (2017) (Education for) sustainable development in geography education: review and outlook from a perspective of Germany. J Geogr High Educ 42(2):157–173. https://doi.org/10.1080/03098265.2017.1379057

UNESCO (2017) Education for sustainable development goals. Learning Objectives. https://www.unesco.de/sites/default/files/2018-08unesco_education_for_sustainable_development_goals.pdf

Vare PA, de Hamer G, Del Gobbo A, de Vries G, Farioli G, Kadji-Beltran F, Kangur C, Mayer M, Millican M, Nijdam R, Réti CM, Zachariou A (2019) Devising a competence-based training program for educators of sustainable development: lessons learned. Sustainability 11(7):1890. https://doi.org/10.3390/su11071890

Wiek A, Withycombe L, Redman LC (2011) Key competencies in sustainability: a reference framework for academic program development. Sustainability 6(2):203–218. https://doi.org/10.1007/s11625-011-0132-6

Yaar-Waisel T, O'Reilly G (2022) Small but smart: international shared virtual class link-up during the pandemic 2020–2021—third-level students in Ireland and Israel. Eur J Geogr 13(5):58–74. https://doi.org/10.48088/ejg.t.yaa.13.5.58.7

Connecting Literature and Web Maps: Hungarian Writers and Poets Online

José Jesús Reyes Nunez and Krisztina Irás

Abstract The Erasmus+ KA201 project, "Biographical map library of European authors" (BIO-MAPS) started in December 2020 with the participation of three bilingual secondary schools and three universities from Spain, Portugal, and Hungary, as well as a regional organization (EUROGEO). The project has three main aims: first, the making of an online library of writers and poets from the participating countries, which is written in Spanish and developed with a GIS-based technology freely accessible on the web. Second, an evaluation of the role that Geoinformatics plays in Geography teaching in each participant country. Finally, the mobility of secondary school pupils to test the literary routes designed according to the StoryMaps. In this chapter, authors describe in more detail the research related to the first two aims developed in Hungary, while focusing on their relationship with the Sustainable Development Goals adopted by the United Nations Member States in 2015.

Keywords Literature · Map · StoryMap · Secondary school · Web · BIO-MAPS

Introduction: Organizing the BIO-MAPS Project

The organization of the current Erasmus+ KA201 project was developed in late 2019 and early 2020. The San Roque Secondary School (Badajoz, Spain) contacted secondary schools in Portugal and Hungary, which were partner institutions in previous Erasmus projects, as well as a higher educational institution in each country and an international organization (Table 1). The project started on 31 December 2020 and will be finished on 30 June 2023. Participant institutions decided to propose a project that combines school mobility with research activities to directly apply the results in secondary education.

J. J. Reyes Nunez (✉) · K. Irás
Institute of Cartography and Geoinformatics, Eötvös Loránd University, Budapest, Hungary
e-mail: jesusreyes@ik.elte.hu

K. Irás
e-mail: iras@map.elte.hu

A. Klonari et al. (eds.), *Re-visioning Geography*, Key Challenges in Geography,
https://doi.org/10.1007/978-3-031-40747-5_6

Table 1 Institutions and organization participating in the project

Spain	Instituto de Educación Secundaria San Roque (coordinator) Universidad Nacional de Educación a Distancia
Portugal	Escola Secundária Quinta das Palmeiras Instituto Politecnico do Porto
Hungary	Kispesti Károlyi Mihály Magyar–Spanyol Tannyelvű Gimnázium Eötvös Loránd Tudományegyetem
Organization	European Association of Geographers (EUROGEO)

As the official language of the project is Spanish, the online library is also being developed in this language. This means that the online library can be used not only by any bilingual secondary school teaching Spanish in the participant countries or in Europe, but it can also be a helpful interactive material for secondary schools in Latin America.

The main theme selected for the project was the online presentation of writers and poets from the three countries who met two main principles: they must be included in the national curriculum and at least one of their works has already been translated into Spanish. Consequently, the official title of the project is "Biographical Map Library of European authors (BIO-MAPS)". However, this is not the only task planned by and for the participants: we also studied the current use or teaching of Geoinformatics in our secondary schools in order to make proposals for improving it as well as for organized school mobilities (literary routes) to test the online StoryMaps in practice created during the project. This chapter describes the research and activities carried out in Hungary.

From the outset, Hungarian institutions involved in the project realized that results to be obtained during this research would significantly help to meet the majority of the Sustainable Development Goals (SDGs) adopted by the country as a member of the United Nations (UN) in 2015 (United Nations 2023). Among the 17 SDGs to be met by member nations, this project almost inherently makes a significant contribution to one of them: number 4, dedicated to Quality Education. However, during our working sessions, we were able to see that our results also contribute to other goals that are no less important, which provided further motivation to continue developing our initial ideas.

Geography, Cartography and Geoinformatics in the Hungarian Secondary Schools

One of the aims of the current project is to develop studies and proposals for the possible use of Geoinformatics (GIS)-based solutions as effective and trans-disciplinary teaching tools in schools, and to determine their current presence in the curricula of each participating country. In order to achieve this objective, the

Table 2 Geography and Literature in the Hungarian secondary education

Secondary level					
Name	Grades	Ages	Subjects	Geography	Literature
Technical Education	9 to 10	15 to 16 years	Subjects of the selected technical profession	Not taught	Not taught
Vocational training and baccalaureate Secondary school	9 to 12	15 to 18 years	– Subjects specific to the profession and general knowledge subjects (Humanities and Sciences) – General knowledge subjects (Humanities and Sciences)	Grades 9 and 10 For the university entrance, it is optional (depending on the targeted university degree) to take the baccalaureate exam in Geography	Grades 9 to 12 For the university entrance, it is compulsory to take the baccalaureate exam in Literature

Hungarian team developed research to work out to what extent Geoinformatics is present in the curricula and used in classrooms, and what are the possibilities for its introduction in Hungarian secondary schools.

In Hungary, the education is compulsory until the age of 16. The structure of the Hungarian educational system is as follows: elementary level from grades 1 to 8 (ages 6 to 14) and secondary level from grades 9 to 12 (ages 15 to 18). Table 2 summarizes the place of Geography and Literature subjects within the structure and type of secondary education.

Elementary Level

The Hungarian pupils begin to learn topics on Geography and Geography-related sciences in the Environmental Studies subject from the 2nd to 5th grades in the Elementary School. The subject is renamed to Studies on Nature in grade 6. Both subjects include knowledge related to Natural and Earth Sciences including not only Geography but also Biology, Astronomy, Cartography, Geology, etc. The teaching is helped with the use of a School Atlas specially designed for this level according to the guidelines determined in the national and framework curricula.

Geography is taught as an independent subject in the 7th and 8th grades of Elementary School. In these two grades, the pupils can review some of the general concepts previously learned (e.g., those about maps), learn on Geography of Hungary, Europe, and the rest of continents, as well as general knowledge mainly related to international economic and social processes. Teachers and pupils find that textbooks and workbooks use different types of graphic and cartographic solutions more systematically, e.g., visualizing data with complex charts (population pyramids and climatic diagrams among others) and diverse types of thematic maps (population, vegetation, precipitation, economic and social themes, etc.).

Secondary Level

Relationship Between Geography Competences, Geoinformatics and SDGs

The competences and skills to be developed by the Geography teaching at secondary level are determined by the methodology proposed in the Hungarian National and Framework Curricula (Educational Office 2020). As mentioned in the introduction, this project makes a remarkable contribution to the fulfilment of the SDGs, especially number 4 (Quality Education), but also to other no less important goals. More specifically, the contribution of Geography and Geoinformatics as unreplaceable tools for the economic and social development of a country is indisputable, and so they are also present in the solutions offered for the fulfilment of the SDGs. Nor should we forget the leading role that Literature and History play in this project, which broadens its social and cultural vision and thus its contributions to other SDGs too.

Geoinformatics (regardless of whether it is taught or used as a teaching tool) can play a relevant role in the development of the following basic competences and skills, while laying the foundations needed to meet the SDGs:

Learning competences: Geography (and Geoinformatics) contributes to the development of the ability to acquire and process information as well as to open-minded thinking essential for understanding and comprehending our changing world. Its most important aim is making pupils able to identify and collect geographical, economic, social and environmental information in the real world (e.g., in the field) and from different (printed and digital) information sources (e.g., newspaper articles, graphs, maps, news, reference texts, pictures). Geography teaching (can be helped or completed by Geoinformatics) intends to promote the use of the acquired knowledge in different areas of everyday life, to support the development of self-directed learning skills and self-development according to individual. Open-minded thinking is required to fulfil goals as numbers 1 (no poverty), 5 (gender equality), 10 (reduced inequalities), 12 (responsible consumption and production), 13 (climate action), 16 (peace, justice, and strong institutions) and 17 (partnerships for goals), as well as data collection and processing are present in the achievement of practically all goals.

Communication skills: The development of reasoned discussion based on the understanding of geographical information is a relevant skill. This development is supported by written and oral tasks requiring the interpretation of geographical information, including presentation tasks. In the last years, Geoinformatics has been developing interactive and online solutions for a multifaceted presentation of results obtained during its use not only for specialists, but also for the public. Communication skills are also needed for the successful presentation (not only for specialists, but for the public too) of results obtained in each of the SDGs outlined by the UN in 2015.

Digital competences: The learning-teaching process is based on the use of digital maps, GIS-based systems and/or applications, as one of the most important and indispensable elements in Geoinformatics. The individual or group research projects develop the ability to share knowledge online. The presentation skills can be also developed by encouraging the use of digital tools for the displaying of geographical processes. The use of digital tools is also a determining factor to achieve the SDGs, e.g., numbers 3 (good health and well-being), 6 (clean water and sanitation), 7 (affordable and clean energy), 9 (industry, innovation, and infrastructure) and 11 (sustainable cities and communities).

Mathematical and thinking skills: The development of thinking skills (analysis, systematization, reasoning, and problem solving) is especially important in groups and then individually. Other important aim is to develop analogical, analytical, reflective and synthesis thinking, as well as developing creative thinking together with the pupils' decision-making, reflection on alternatives, risk-taking, evaluation, reasoning and selection of the best solutions. All these options are implicitly included in the teaching and/or use of Geoinformatics and are needed for future specialists to be able to solve problems and tasks that arise in their working life, which is a prerequisite for the aims set by the SDGs to continue to be met beyond the year 2030.

Personal and interpersonal competences: Geography (and Geoinformatics) contributes to the understanding and development of interest in the socio-cultural diversity of the world. The use of cooperative methods enables pupils to develop social skills too. These competences are vital in those goals directly related to social development, e.g., numbers 1 (no poverty), 2 (zero hunger) and 5 (gender equality).

Skills for creativity, creative work, self-expression and cultural awareness: The exercises that require the elaboration (creation) of a product (e.g., a model, a presentation), either independently or in a group, contribute to the development of creative work and self-expression. Similar to mathematical and thinking skills, all these skills are fundamental elements in the training of those future specialists and in general of those members of society who make the achievements to be obtained in the SDGs a reality.

Work, innovation and entrepreneurial skills: The learning about current socio-economic and environmental processes through the modern geographical education (and currently the modern geographical education cannot be conceived without the inclusion of Geoinformatics) is a determinant contribution to the pupils' citizenship formation. By introducing the successful actors of modern economic life, education contributes to understanding the role of innovation and the needs of the labour market, which in turn has an impact on the development of the skills of employees and employers. Sustainable environmental processes and development are main targets within the SDGs, making necessary the continuing education of the young generations in the importance of living in a world that is developing and simultaneously capable of being ecologically sustainable. SDGs as numbers 12 (responsible consumption and production), 13 (climate action), 14 (life below water) and 15 (life on land) are typical examples of areas related to these skills.

Geoinformatics in the Textbooks

The Geography teaching is continued in the first two years (grades 9 and 10) of Secondary School. Grade 9 is essential concerning the general knowledge related to Cartography and Geoinformatics. The first chapter of the Geography 9 textbook, "Orientation in cosmic space and time" has 16 pages (Arday et al. 2020). The themes on Cartography and Geoinformatics are presented in the first sub-chapter, "Modern technique in cartography" from pages 8 to 11 (Fig. 1). This sub-chapter includes the following topics: remote sensing, types of satellite orbits, remote sensing in the practice (how it is used in meteorology, agriculture and protection of environment), our digitalized environment, GIS definition, use of GIS in mining, studies on ground water, urban development, navigation systems and business tasks. Some of these concepts are not explained in detail, which can be considered an invitation extended to the pupils (and teachers too) to collect more information about them, e.g., satellite, geostationary orbit, satellite image, GPS, Geoinformatics, GIS and remote sensing. All these topics are closely related to the achievement of SDGs too, as shown in Fig. 1. The text included in the figure is entitled "Our digitized environment" and briefly shows how Geoinformatics can provide solutions to different tasks not only related to the environment, but also included in many of the SDGs determined by the UN. Some of the specific examples are listed in Fig. 1: mining, underground water resources, region and settlement development.

A digitalizált környezetünk

Hazánkban a földrajzi információs rendszert (GIS) inkább geoinformatika néven ismerik. Lényege, hogy az adott területre vonatkozó térbeli adatokat (elhelyezkedés, alak, kapcsolat más objektumokkal) számítástechnikai eszközök használatával gyűjtik, tárolják, kezelik, feldolgozzák a különböző információszinteket és megjelenítik azokat.

Nézzünk néhány példát a geoinformatika alkalmazására!
Bányászat: számítógépes programokkal modellezték például a Pannon-medencét létrehozó folyamatokat, és jelölték ki, hogy hol érdemes még szénhidrogén után kutatni.
Felszín alatti vízkészletek: segítségükkel modellezhetőek a várható változások (a talajvízáramlás és az esetleges szennyezés terjedési iránya).
Térség- és településfejlesztés: közműnyilvántartás, iskolai körzetek kialakítása, városüzemeltetés.
Navigációs rendszerek: a járművek helyzetének folyamatos érzékelése és irányítása, a forgalmat befolyásoló tényezők minél teljesebb körű figyelembevételével tervezik a járatokat, módosítják a járatsűrűséget, tájékoztatják az utasokat.
Üzleti élet: telephely kiválasztása, a konkurencia figyelése, a házhozszállítás szervezése, a legjobb eladási helyek meghatározása.

Szemléletesen jelenik meg a térképeken a város területrendezési adatainak térbeli eloszlása, összefüggései. A különböző típusú térképi elemeket a számítógép rétegenként tudja kezelni, így egy-egy réteg adatainak frissítésével a megjelenítés automatikusan igazodik az új helyzethez (1.8.).

1.8. A város területrendezési adatainak térbeli eloszlása

Fogalmak
műhold | geostacionárius pálya | műholdfelvétel | távérzékelés | geoinformatika (térinformatika) | földrajzi információs rendszer (GIS) | GPS

Fig. 1 Explanation related to Geoinformatics on page 11 of the "Modern techniques in cartography" sub-chapter (Arday et al. 2020)

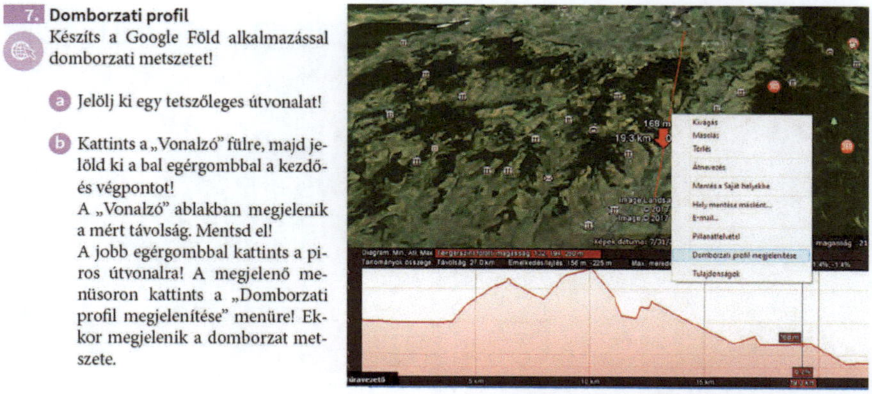

Fig. 2 Exercise using Google Earth on page 9 of the Geography 9–10 workbook (Arday 2021)

The Geography 9–10 workbook follows the chapters of the textbook with the first four pages dedicated to the "Modern technique in cartography" sub-chapter (Arday 2021). The exercises include questions related to Geoinformatics and GIS, its use in different branches of sciences, comparison between printed thematic maps and GIS maps, and maps with satellite images. Three exercises highlight the use of Google Maps and Google Earth to create simple maps and topographical cross sections, as well as to determine the geographic coordinates, the altitude and the distance between the pupils' house and other places (Fig. 2).

Teachers and pupils can also use in the classes school atlases created specifically for the secondary level, which follow the guidelines given in the national and framework curricula. The current official school atlas, School Atlas for secondary pupils (Balassa et al. 2021), is characterized by the use of more thematic maps of each continent, a more detailed presentation of the country using different types of maps (physical, political and thematic maps), as well as the replacement of the six introductory pages made for elementary pupils by twelve satellite images of different regions of the world at the end of the atlas, e.g., making a comparison of changes that occurred over a period of time (Fig. 3). In this way, the school atlas can also be considered an important contribution to understanding why the defence of the environment (together with other topics) can be considered a common denominator of the targets set by the United Nations with the SDGs.

Geoinformatics in the Current Secondary Teaching

Based on the above description of the presence of Cartography and GIS in the Geography textbooks and workbooks, readers can suppose that the basic conditions for teaching or using GIS and GIS-based solutions are ensured in the Hungarian educational system. The real situation is more complicated because two factors exert a negative influence on its implementation. The first one is a negative tendency observed in

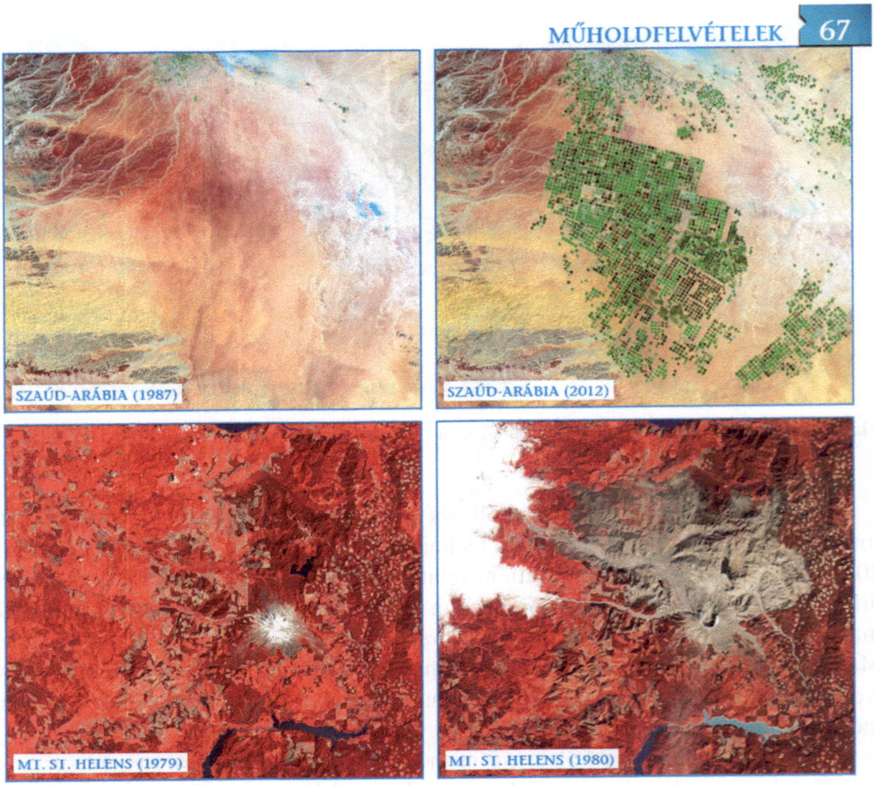

Fig. 3 Comparison of satellite images in the School Atlas for secondary pupils (Balassa et al. 2021)

the majority of European countries: the decrease in the number of hours dedicated to Geography teaching. This fact seriously hampers any attempt to incorporate any new content (e.g., Geoinformatics) into the teaching activities. The second factor is related to the difficulties with the GIS formation of teachers: the lack of time to develop lessons and the lack of practice of the teachers to use GIS in the classrooms were identified as determinant factors in different international projects, e.g., by Wheeler et al. (2010), Höhnle et al. (2011) and Singh et al. (2012).

Therefore, it has become an urgent need in today's geographical education to seek for other Geoinformatics-based (GIS-based) solutions that are easier for both teachers and pupils to assimilate. The current BIO-MAPS project aims to be an easily understandable and feasible option for teachers and pupils in all grades of secondary education (and in upper primary education). The selected authors and the content in general, as well as the online solutions that are applied for the visualization of the data, wish to go beyond the boundaries of traditional subjects to become a bridge that links them to each other. In this way, the StoryMaps can be used in subjects such as Digital literacy (grades 9–10), Geography (grades 9–10), History (grades 9–12),

Literature (grades 9–12), Foreign languages (grades 9–12), History of Art (grades 10–11), Civilization (grades 9–11) and Socio-cultural subjects taught in bilingual schools.

Of course, teachers should pass a training course, but it will be much easier and less time-consuming than a "more traditional" GIS course. The training course can include basic digital literacy training, basic geo-computing skills and basics of the use of the recommended online platform, in this specific case ArcGIS Online.

Creating a Biographical Map Library of Hungarian Authors

Hungarian literature, looking back to a c. 800 year-long History, has always been one of the greatest pillars of Hungarian culture by preserving and developing the language. Therefore, school subject of Literature has an outstanding role in our primary and secondary education.

For the "Biographical map library of European authors" (BIO-MAPS) project, organizers of the participant countries chose fifteen of their most important and internationally-nationally recognized writers and poets. Several aspects had to be considered in the selection process; for example, the authors' life and literary work must be included in the national curricula. At the same time, the selected authors should represent both classical and contemporary era, they should represent male and female authors, and a special aspect in the selection of the Hungarian authors was to have translations into Spanish in high quality (Reyes Nunez et al. 2022). The following authors (listed in alphabetical order by surname) meet all the above-mentioned criteria: Endre Ady (1877–1919), Éva Bánki (1966), György Faludy (1910–2006), Géza Gárdonyi (1863–1922), Mór Jókai (1825–1904), Imre Kertész (1929–2016), Dezső Kosztolányi (1885–1936), Imre Madách (1823–1864), Sándor Márai (1900–1989), Ferenc Molnár (1878–1952), István Örkény (1912–1979), Sándor Petőfi (1823–1849), Miklós Radnóti (1909–1944), György Spiró (1946) and Magda Szabó (1917–2007). During the selection of authors, Hungarian colleagues considered important to also include women writers and poets with a significant literary work, thus fulfilling the task set by SDG number 5 (gender equality).

StoryMaps in Education

A StoryMap is a complex, text-based form of online publication. It is created with the story authoring cloud-based application of ArcGIS called StoryMaps, which is free to access via internet after registration (ESRI 2022). Within the application, there are several in-built features to illustrate the text, e.g., linking videos and other multimedia elements, creating a timeline, photo gallery. The most important feature is the possibility to create or insert static and interactive web maps. All these features

Fig. 4 Covers of the Hungarian authors in BIO-MAPS project

provide information on a multi-channelled way about the topic. Our aim was to use these elements in the widest possible range to show all the ways we can approach literary topics through History and Geography. We planned to determine a theoretical-conceptual standard for the structure of the StoryMaps in BIO-MAPS projects that would result in a certain structural commonality; but due to the diversity of life stories, of the nature of the works and of the material that is possible (and valuable) to share, it turned out that personalizing the StoryMaps to each author is more effective. The differences in content types and in quantity of available source material are the main reason of the diversity in the visual components and in the content components in the BIO-MAPS series (Fig. 4).

A StoryMap is also considered one of the most powerful online tools for the multimedia-based communication of scientific results. Therefore, its use to publicize the results obtained in projects related to the implementation of the SDGs is also highly recommended, because it allows publicizing these results in a very didactic and easy-to-understand graphic language, which can easily attract the attention of the public.

This series of StoryMaps is a new initiative to expand the collection of visual/interactive aids of high school level of literary studies (De Lázaro-Torres et al. 2022). It works on two ways: StoryMaps can be prepared by teachers to illustrate and complement the lessons, or by students who are also able to create them as homework or other presentations. StoryMap editing and web mapping in ArcGIS online are easy for everyone, but they need certain self-education through tutorials, which are provided on the arcgis.com website.

Data Collection

A StoryMap is built up in three large work phases such as data collection, map making and StoryMap editing. Collecting data is the most important and the most time-consuming part of StoryMap making, where research needs to cover literary, cultural (even pop-cultural) and geographic areas (De Lázaro-Torres et al. 2021). It includes biographical data in the first place but with an emphasis on those geographical (georeferenced) data that connect an author to specific places where he lived and worked. Depending on the content of the author's work in question, further important geographical data are those that appear in the discussed works (mostly in novels) because these allow us to create interactive maps on the content of these books. Research also targets visual material as well, such as original photos of people, places, books, and other objects (e.g., buildings, sculptures, paintings, manuscripts) included in the main text. Monotony of text reading can be broken with inserts of relevant video or audio material. A video file is also a great opportunity to present complete works with subtitles as for example the wonderful, animated adaptation of the Tragedy of Man (in the StoryMap of Imre Madách) by Marcell Jankovics. We can use videos to give only a tiny glimpse on the sound of a poem in its original Hungarian language, as a short poem recorded in a TV studio for Vers mindenkinek (Poetry for all) cultural series and professionally performed by an actor, as it can be seen in Endre Ady's StoryMap. In this case, it was important to find a poem that has Spanish translation (placed right next to the video block), that is short enough for non-Hungarian speakers (not to make them feel bored or confused), and that was recorded in perfect quality (Fig. 5). Another example of using multimedia elements is a radio show broadcasted by the Spanish Radio Sefarad where György Faludy's Spanish translator Alfonso Martínez Galilea is interviewed about Faludy's world-famous novel My Happy Days in Hell.

Phase of data collection also covers a secondary task on searching for visual material to illustrate the cartographic material. StoryMap components like *Map tour* and *Interactive maps* allow the users to add photos, multimedia files and links to each map element. These details provide possibilities for the reader to connect the topic to the modern era and to place the events shown in the map into the modern environment (Fig. 6).

Map Making

Maps per se are visual tools for telling stories. Effective and professional map making is based on deep research, because fundamental decisions on content, on scale, on background map, on graphic methods used in the web map and even on the number of maps can be made only by knowing the represented story in detail. Considering the communicative potential implied in the maps, their use is self-evident for the representation of social and economic problems raised in the SDGs and results obtained in

Fig. 5 Actor Attila Menszátor Héresz performing a short poem (*Őrizem a szemed* /I guard your eyes/) by Endre Ady with Spanish translation on the left

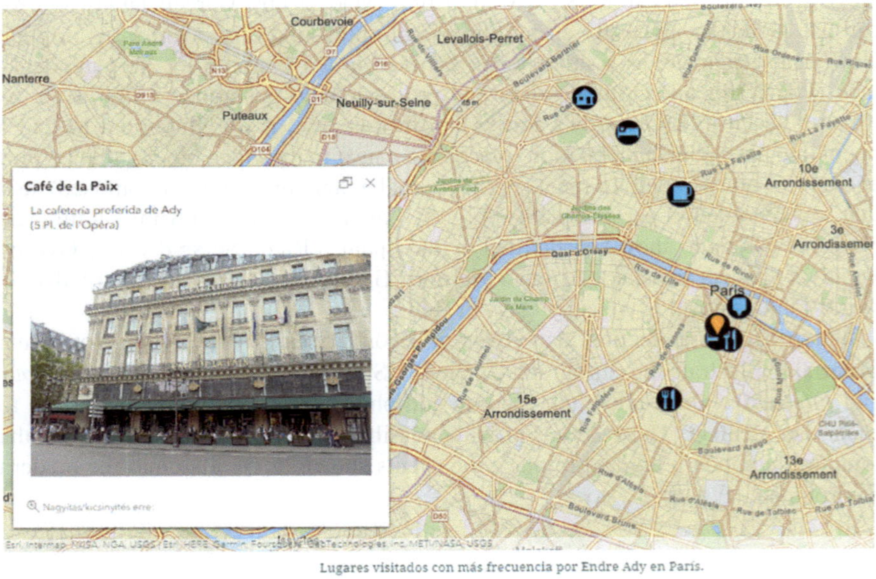

Fig. 6 Pop-up blocks in large-scaled web maps provide information and visual material that cannot be discussed in the main text. Current map includes name, address, and photo of a café in Paris mentioned only in the main text

projects developed to find solutions to those problems. Specialists can geographically locate the regions affected, e.g., by a natural catastrophe, pandemic or an economic crisis. They can use general maps and thematic maps in their analyses to find a solution for those situations and can also represent on maps the planned results of the proposed solutions.

In the BIO-MAPS project, all the geographic data with their additional information are placed in in-built web maps of the Map Viewer tool on arcgis.com website. Different perspectives within an author's "story" demand different maps. According to the structure of storylines, at least three types of maps are recommended. These are:

- *An overview map of Europe or of the world*: A small-scale map is useful when the author moved to other countries, or his/her travels were important for some aspects concerning life or work activities. This map only indicates the author's movements and the countries he/she visited or lived in, as shown in Fig. 7. Other details are included in larger scale maps.
- *A country map* in larger scale than the overview map is precisely prepared with all the data related to the author's life. This is the map type where the typical biographical data are discussed with additional visual material (Fig. 8). Figure 6 also shows this map type in an even larger scale to locate tiny details of the author's life, like cafés and hotels where the author was a regular guest. This scale also allows us to place the locations of memorial elements, e.g., plaques, statues, grave site, streets named after the author and others; thus, we can show how the author's

Fig. 7 Overview map about György Faludy's movements in the World

memory is kept alive. Besides all these, different extracts of maps in this scale range can be used as static figures in the StoryMap.

- *A map illustrating one of the books* written by the author if mapping the story is possible (Fig. 9). In the case of poets, inserting a large-scale map of the town where the author lived and got inspired is recommended.
- Additionally, thematic maps or combinations of modern and historic maps could be also created, but it is important to keep focus on the author's life and works.

Fig. 8 Country map showing the most important locations of novelist Mór Jókai's life

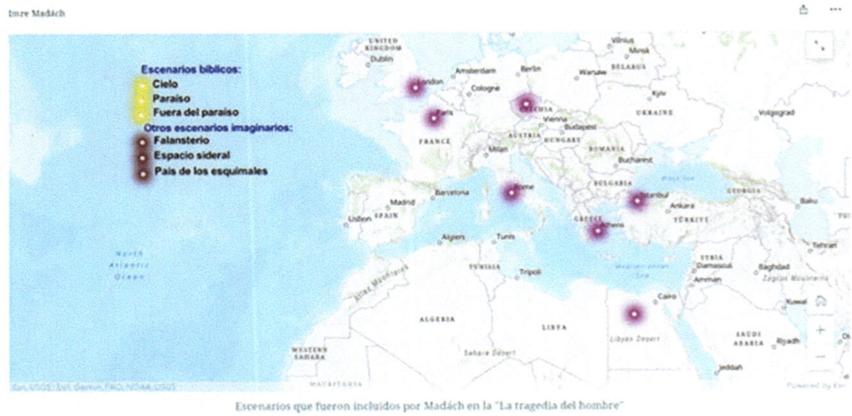

Fig. 9 Locations of the earthly (purple), biblical (yellow) and imaginary (brown) scenes in the Tragedy of Man by Imre Madách

Considering that future creators and users of maps in the BIO-MAPS project are not GIS experts or professional cartographers, but high school professors and pupils without cartographic experience, map design follows general-purpose maps and basic thematic maps. Background maps vary according to the scale: it is a general-purpose map (called National Geographic style) in small-scale overview maps, because this one resembles the most on school maps. World topographic map is used for medium-scale country maps because this design facilitates the clearest data interpretation. In the case of large-scale city maps, any of them can be used, because their streets and blocks are properly designed for large-scale mapping.

As the maps mostly show places with address, point type elements dominate the map symbology. "Points of interest" symbol set provides symbols that can relate to element types like place of birth, place of death, school, military service, important holiday, favourite restaurant, or hotel, etc. In ArcGIS Online, it is easy to link additional information to each point that can be text, photo, media file or a web link (Figs. 6 and 8). The options offered by thematic map making are strongly limited, but it is recommended to create flow maps in cases where we have data on moving from a location to another. Static flow maps work with arrows, and we can add additional texts to them (Fig. 10).

El viaje de estudios de György Faludy a las universidades más famosas de Europa (1928-1933)

Fig. 10 Arrows indicating young György Faludy's movement in Europe

Editing a StoryMap

Once authors have collected all the texts and multimedia elements, as well as have created the maps, they can step further to the phase of StoryMap editing. The goal is to create the most dynamic, attractive, and interactive composition of the above listed elements following the timeline of the author's life and highlighting the outstanding pieces of his/her works. However, custom layouts can be designed; the authors used general templates to test the applicability and easiness for those users without any experience on HTML programming. Each StoryMap contains image galleries, videos and special interactive applications (e.g., swipe, slideshow, sidecar) besides the text. However, a StoryMap-specific feature named Map Tour has a prominent role, because this is the application where connection between the story and the locations (in other terms among Literature, History and Geography) is established (Fig. 11). Text, photo and multimedia can be linked to locations in a web map and the feature automatically guides the reader through the map. In the case of Endre Ady, we can follow the poet's steps along the streets of Budapest, and in the case of György Faludy, who lived in four continents, the reader "flies" over a world map. This feature is the most effective tool to connect elements of the past to the present as we can see the locations on modern maps. In certain cases, when we need a less interactive map than a Map Tour, we can easily import a previously edited (thematic) map from the online map libraries of ArcGIS.

Thinking about poetry, novels or authors, we admire their eloquence in expressing thoughts, their skills of pictorial or sensitive describing of the world around us, but we rarely think on the spatial aspects and the geographic relations that have fundamentally impacted on their words. A StoryMap is a great tool to shed light on the geographic initiation and environment of the greatest literary works, and this way it gives a new perspective to the education of this subject. All the StoryMaps created in the project are accessible on http://www.biomaps.eu/ and on https://arcg.is/1qSn80.

Fig. 11 Section of the map tour feature in ArcGIS StoryMap in editing mode

Conclusions

After studying the current situation of the presence of Geoinformatics in European secondary schools and specifically in Hungary, we can state that there are two trends that can be considered contradictory. The first positive trend is that authors are making an effort to ensure that Geoinformatics is increasingly present in textbooks related to Geography, so that both pupils and teachers can use it in the classroom as an effective tool to illustrate and diversify the teaching of Geography. However, at the same time, we cannot fail to mention a negative trend in a large majority of European countries: the decrease of hours dedicated to Geography teaching in secondary schools, which constitutes an obstacle for the introduction and learning of new technologies in the classroom. For this reason, we consider it important to look for simpler and easier GIS-based solutions that can be used in teaching, even if the time available is reduced.

Based on the results and experiences collected to November 2022, authors can conclude that the online biographical map library offers numerous options to help and make effective contributions to teaching in secondary schools, as well as for learning activities that can be developed by the pupils. Current project also proved that the map, as the main tool produced by cartographers to represent any type of georeferenced data in Geoinformatics, is an unavoidable graphic-spatial solution in any didactic solution whose objective is the easy understanding of the represented content. Storytelling maps are innovative and engaging online solutions to present, teach and pay tribute to the literary heritage of a country in a more interactive way, thus contributing not only to its teaching in schools or to its use by students, but also to the popularization of the selected writers and poets out of the secondary schools too, while at same time constitute a modest but valuable example on the implementation of the Sustainable Development Goal on Quality Education among others.

The project has studied the current options of using Geoinformatics as a tool in secondary education, as well as its teaching in schools. It has also been clarified how the introduction and use of Geoinformatics at secondary level can effectively support the future implementation and further development of the SDGs. The interdisciplinary work developed for the creation of the StoryMaps of European writers and poets reflects how collaboration between different branches of science is a key factor in the fulfilment of the 2030 Agenda for Sustainable Development.

Acknowledgements This research has been funded by the Erasmus+ project "Biographical Map Library of European Authors (BIO-MAPS)" (2020-1-ES01-KA201-082590) of the European Union.

References

Arday I (ed) (2021) Földrajz 9–10. munkafüzet. 1. kötet. Oktatási Hivatal, Budapest

Arday I et al (2020) Földrajz 9 tankönyv. 1. Kötet. Oktatási Hivatal, Budapest

Balassa B et al (2021) Földrajzi Atlasz Középiskolásoknak. Educational Office (Oktatási Hivatal), Budapest

De Lázaro-Torres M-L, Buzo I et al (2021) Las biografías de literatos en StoryMaps: una herramienta de aprendizaje en la enseñanza secundaria. Avances en Ciencias de la Educación I(54)

De Lázaro-Torres M-L, Buzo I et al (2022) Biographical map library of European authors. In: Proceedings of EUROGEO 2022 conference on Re-visioning geography for sustainability in the post-Covid era, pp 30–31

Educational Office (Oktatási Hivatal) (2020) Földrajz kerettanterv középiskolák részére. OKTATÁS. https://www.oktatas.hu/pub_bin/dload/kozoktatas/kerettanterv/Foldrajz_K.docx. Accessed 13 Nov 2022

ESRI (2022) Map and interact with your location data. ArcGIS Online. https://www.esri.com/en-us/arcgis/products/arcgis-online/overview. Accessed 18 Nov 2022

Höhnle S, Schubert JC, Uphues R (2011) Barriers to GI(S) use in schools—a comparison of international empirical results. GISPOINT. http://gispoint.de/fileadmin/user_upload/paper_gis_open/537510015.pdf. Accessed 21 Oct 2022

Reyes Nunez J-J, Irás K, Cselik A, Varga M, and Kis D (2022) Presentation of Hungarian writers and poets using ArcGIS Online. Proceedings of EUROGEO 2022 conference on Re-visioning geography for sustainability in the post-Covid era, p 45

Singh S, Singh B, Kleeman G, Van Bergen P (2012) Opportunities to implement GIS in teaching and learning geography: a survey among smart schools in Sabah, Malaysia. Procedia—Soc Behav Sci 69:884–889

United Nations (2023) The 17 goals. https://sdgs.un.org/goals. Accessed 21 Feb 2023

Wheeler P, Gordon-Brown L, Peterson J, Ward M (2010) Geographical information systems in Victorian secondary schools: current constraints and opportunities. Int Res Geogr Environ Educ 19(2):155–170

Education, Geography and New Geospatial Technologies and Tools

Teaching and Learning Geography with a Web GIS Approach

Joseph J. Kerski

Abstract The advent of Web GIS tools, with their ability to collect, map, and analyze field data, to access spatial data as services covering a wide variety of scales, and to build web mapping applications such as dashboards and story maps, allows students of all ages to understand the SDGs and why they matter. This chapter discusses the societal, educational, and technological advancements that led to the advent of the modern Web GIS architecture. It then assesses the benefits of Web GIS in primary, secondary, and university educational institutions around the world. This chapter offers practical strategies in which the SDGs and other topics can be taught with interactive maps and analytics powered by modern GIS. Web GIS tools and spatial data can be effectively used in face-to-face, hybrid, and online instructional settings in geography. Furthermore, Web GIS is enabling the geographic approach to be integrated in disciplines outside of geography—in mathematics, history, environmental sciences, civil engineering, language arts, data science, business, and beyond.

Keywords Web GIS · SDGs · Geotechnologies · Fieldwork

The Advent of Web GIS and Its Implications in Instruction

Geography continues to face challenges: Inadequate grasp of the importance of the discipline in education, lack of societal awareness of geography's ability to solve problems, slow uptake of innovative technological tools, and underused connections that geography offers to other educational content areas. The rigorous use of Web-based Geographic Information Systems (GIS) technology and methods can elevate, strengthen, and connect geography in key ways. To teach with GIS is *applying geography.*

Teaching about GIS and teaching with GIS began in the late 1980s in higher education, spreading to secondary schools during the 1990s. By the early 2000s, GIS had become embedded in most higher education institutions worldwide, largely

J. J. Kerski (✉)
University of Denver, Denver, CO, USA
e-mail: jkerski@esri.com

A. Klonari et al. (eds.), *Re-visioning Geography*, Key Challenges in Geography,
https://doi.org/10.1007/978-3-031-40747-5_7

in GIScience and geography programs (Kerski 2008). By the 2020s, it manifested as a diverse set of courses, degree and certificate programs, and as a key instructional tool in a widening array of disciplines.

GIS not only *endured* numerous evolutionary changes in Information Technology (IT), from mainframe computers, microcomputers, personal computers, to web and mobile technologies (Sinton and Kerski 2020), but it *thrived* thought these changes. Each change brought new capabilities, applications, and audiences to GIS. During the evolution of GIS to the web, GIS increasingly connected with mainstream IT trends: cloud computing, machine learning, and distributed architecture. Web GIS hastened the adoption of spatial thinking in disciplines beyond geography. However, using GIS still embraces a unique set of research questions and methodologies: Its practitioners ask "whys of where" questions as they examine change over space and time at a variety of scales. Hence, "spatial" is still rather "special." These simultaneous trends have a deep impact on how, when, and where GIS is taught and learned, worldwide.

Because the world is complex, GIS, as a way to understand that world is also complex. Teaching GIS has thus depended on faculty skilled in these complex methods and tools, which are traditionally geography faculty. But because of its smaller number of and easier-to-grasp tools, ability to run online, and connections to data services, Web-based GIS expanded geotechnologies and geographic thinking in higher education, technical, community, and tribal colleges, and primary and secondary schools beginning in 2010. GIS simultaneously spread from standard introductory through advanced courses and "capstone project" courses to remote sensing programs, GIS coding, web mapping application development, spatial analysis, GIS systems administration, and others.

Societal expectations about the value of and outcomes that education should deliver have changed. Education resulting in employable graduates is demanded, causing technical colleges, which focus on workforce development, to thrive. GIS became seen as a valuable skill set to empower students. Web GIS led to an increased demand for coding and systems architecture skills. Geotechnologies as applied, problem-solving tools increasingly integrate into other information technologies: 3D visualizations, Jupyter Notebooks, JavaScript, APIs and SDKs, statistics tools (R, others), imagery including drones, and augmented reality tools. GIS is now seen as important to universities offering new programs such as sustainability, data science, and geodesign.

The COVID-19 crisis and ongoing global natural disasters made it clear that interactive maps, dashboards, and infographics are valuable decision-making tools and that geography was an important unifying framework. University health programs added "health informatics" courses. GIS became a key tool for ethical and socially responsible operations of modern businesses and integral to resiliency initiatives in planning, energy, and environmental studies. GIS was added to new courses and plays a key role in spreading the geographic approach and spatial thinking to a wider variety of disciplines in academia and in the workplace as never before.

In secondary education, GIS is sometimes taught as a standalone course, substituted for advanced geography or coupled with a school's computer and technology education (CTE) focus. In primary schools, GIS is mostly used as an instructional and content-rich map source in science, social science, and language arts. The analytical, problem-solving, critical thinking aspects of GIS are considered valuable for students at all ages (Kerski 2003; Jo et al. 2016). Primary and secondary educators typically must teach content according to their national educational content standards. Educators increasingly see that the rich data sets and tools GIS offers enable them to teach core concepts—change, scale, ratios, historical events, biomes—and are tied to these standards. The report from the US National Academies of Sciences (2006) lent support to the use of GIS in schools as did studies showing the effectiveness of GIS in learning (Schulze 2021).

Web GIS also thrives within new educational delivery models. GIS-focused Massive Open Online Courses (MOOCs), from Penn State, Esri, and elsewhere, have attracted over 100,000 students who now consider maps as more than simple reference documents (*where* something *is*); rather, as analytical tools (*why* something is where it is). New methods of achieving an agreed-upon standard or degree, such as (micro)credentials and certificate programs, align with Web GIS. Instructors focus on specific tools rather than having to embrace all of GIS as they had to in the past. For example, field surveys, interactive maps, dashboards, and infographics are combined to study public art, water quality, noise, walkability, and other topics (Kerski 2021b). Lessons can be nimbly adjusted as technologies and institutions evolve.

The rise of Learning Management Systems (LMS) supporting online instruction and increasing LMS sophistication for embedding multimedia such as web maps have widened the scope of GIS educational offerings. LMS and Web GIS are both cloud based and thus form the perfect union for many educators. Tools to visualize, classify, symbolize, and analyze data can be performed on any device, anytime, on standard web browsers (Manson et al. 2014). Maps, tools, and data are accessed, analyzed, and presented online. The cloud infrastructure eliminates the need to transfer files via physical media; mapping applications can be saved and shared, enabling collaborative learning. The integration of Web GIS and LMS tools proved to be critical as educational institutions shifted to online and hybrid during COVID-19 (Lukinbeal 2022). Some institutions use virtualization software (VMWare or Citrix) so students can run GIS software on their own devices (Geraghty and Kerski 2020). Instructors increasingly viewed GIS as a resilient approach—suitable for face-to-face, hybrid, and online instruction.

Why Is Web GIS Revolutionary in Instruction?

During the 2010s, Software-As-A-Service saw tools and data migrating to the cloud. People streamed music, stored data online, and shared documents via services such as Google Drive and Dropbox. However, even more revolutionary than hosting music and documents in the cloud was the advent of cloud-based GIS.

All issues of our twenty-first-century world—climate, hazards, social, equity, habitat, sustainable agriculture, human health, ocean acidification, water quality and quantity, energy, public safety, supply chain management, and others—are spatial in nature. These, including the UN SDGs, are complex, serious global problems that increasingly affect people's lives. These problems transcend physical, political, and disciplinary boundaries, lending credibility to the use of GIS as an interdisciplinary toolset. Solving twenty-first-century problems requires collaboration enabled by Web GIS. It requires a wide diversity of thought leaders and practitioners, because every discipline asks the whys of where (Kerski 2019).

In a Web GIS, data collection, mapping, and spatial analysis are performed in the cloud, on any device from tablet, laptop, to smartphone, nothing to install, requiring only modest bandwidth. Data as services through portals, ArcGIS Hubs, and other data provisioning technologies provide instant access to spatial data, scalable from local to global, on themes from agriculture to zoology (A to Z). Sharing maps and layers fosters collaboratory group work. Field tools allow data to be collected via crowdsourcing, then mapped and analyzed within Web GIS. Students and instructors can move from collection to analysis for invasive species, water quality, litter, historical homes, or other phenomena they wish to study. Multimedia web mapping applications enable the communication of research using these same tools. Story maps, dashboards, and infographics are configured in drag-and-drop mode, fed from real-time data streams, and shared. As they blur the lines between maps and visualizations, these tools expand the array of educators using them. These tools fit perfectly into peer-to-peer and student-to-instructor presentations, student portfolios, assessment instruments, and community communications tools beyond the classroom.

Past complex desktop software with challenging user interfaces and user experiences required much time and effort to master. Trailblazing educators forged through these challenges by creating lessons, textbooks, and professional development opportunities. Web GIS changed instruction—decreasing the time required to become confident in using tools. Web GIS as a platform made it possible for professionals in any field—including education—to include GIS on their toolbelt alongside other tools.

Web GIS enables researchers, instructors, or students to use specific components of the platform pertinent to their needs, unlike the past where all of GIS had to be mastered. Certain fundamentals remain important—symbology, classification, modeling, map projections, and metadata. When, for example, business instructors seek to use GIS to teach vehicle routing, map consumer behavior, or determine drive times to competitors, tools such as Business Analyst Web can be used in marketing, risk assessment, or supply chain courses.

GIS is not simply a way of sharing of spreadsheets and maps. Web GIS allows sharing source data, methods, and results. Data is typically not shared when research is published (National Academies of Sciences 2018), hindering scientific advancement. Web GIS empowers the open science movement. Web GIS brings the "big data" world with feature services, maps, imagery, 3D scenes, and real-time data feeds. While licensing and open access policies still vary around the world, the data-as-services model means that more time can be spent on analysis, rather than downloading and manipulating data.

Web GIS changes how hands-on exercises can and should be structured and delivered. GIS instruction long depended upon step-by-step scripted lab exercises laden with screenshots, because these were the *only* resources available for students to work through tools and problems. Web GIS allows students to learn more quickly, using videos, graphics-enriched documentation, hands-on activities, and the online GIS user community. Modern GIS tools are more intuitive, reducing the need for over-scripted activities. Because modern Web-based GIS updates often, instructors will be trapped in a never-ending curation cycle if these are tied to screenshots and tools. One solution is assigning students a problem and directing them to relevant data in brief, focused lessons.

Web GIS is a leap forward for sharing content. When GIS relied on standalone workstations, the challenges of sharing and processing large spatial data sets relegated much instruction to GIS labs and to higher education. Graphics cards had difficulty rendering maps; imagery was often too large to fit onto physical storage devices. To access the education community's wisdom, one had to "know people" via online networks or conferences. Web GIS encourages the tenets of Community of Practice—a common domain of interest with a commitment to that domain, wherein participants learn with and by each other, where members are practitioners, producing a shared repertoire of resources (DeMers et al. 2020).

Changing expectations of society, institutions, and instructors are changing teaching GIS methodologies and approaches. An oft-heard complaint about GIS instruction is that it focuses on "buttonology"—about teaching tool X using version Y software. The long-time focus on teaching tools was understandable because the tools were challenging to learn, but Web GIS is changing this as well. Calls are heard to "stop teaching GIS" and instead "teach how to learn GIS" (DiBiase 2018).

Yet the "line" between teaching about GIS and teaching with GIS was always blurry. Even when collecting data required much time, instructors were keen to teach "higher goals." Higher goals of teaching core geographic concepts about water, business patterns, or climate mesh well with GIS instruction: GIS always involves the use of real-world data about the Earth. GIS is a tool for solving real problems in real places. Thus, when instructors teach about proximity analysis, they use real-world data—buffering rivers into riparian zones for planning urban greenways or wetland protection. Points are geocoded so that students can view results as store sales volume, traffic accidents, or earthquakes.

Hence, instruction with GIS has always encouraged geoliteracy: The "geoliteracy stool" (Fig. 1) is held by content knowledge, skills, and the geographic perspective. Content knowledge encompasses core disciplinary theory, tenets, and themes. Geography is the most natural fit for Web GIS, given geography's focus on space, place, patterns, and the dynamic Earth. Content knowledge in geography includes scale, cycles (carbon, water, others), the interconnectedness of Earth spheres, climate and weather, population dynamics, energy, water, and natural hazards, human–environment interaction. Skills include geotechnologies (projecting, georeferencing), applying spatial statistics, and communicating using maps. The geographic perspective includes the ability to ask questions, frame problems, synthesize information about change, and articulating recommended actions. Hence, GIS instruction is never *just* about tools or skills.

Fig. 1 Stool of geoliteracy (Kerski 2019). *Source* Joseph Kerski

Web GIS has further changed instructional methodology for two groups: It enables instructors teaching GIS to devote more course content to analysis, rather than preparing data or teaching software. It enables geography and other instructors to use GIS as an instructional tool. Instructors use interactive map layers in the ArcGIS Living Atlas of the World and other libraries to tie potential sea level rise to population density. Students use trace downstream analysis to understand hydrologic systems.

Teaching with Web GIS is conducive to problem-solving learning environments using real data (Milson et al. 2012). Challenges remain: Data is often messy, laden with uncertainty. Teaching with Web GIS requires instructor comfort posing "what if" questions with multiple answers. Teaching with Web GIS means using a professional toolkit in instruction. Instructors must also have "Plan B" when a needed map layer is slow or offline. Instructors must also be comfortable "learning alongside" the students when working with new, unfamiliar tools. Another challenge is the continued rapid evolution of GIS, such as 3D voxels and LocateXT for unstructured data.

Geography's contents—dynamic, addressing real-world issues, scale-dependent, its high-tech nature, and its holistic approach—are all supported by the rigorous use of Web GIS in instruction. How can Web GIS be most effectively taught and learned?

Approaches and Best Practices to Address Key Challenges in Geography Through Web GIS

Which Web GIS tools and applications should be included in the geography curriculum, where should they be included, and how should they be taught? These can be answered in part by investigating the approaches in three courses that use Web GIS—cartography, physical geography, and cultural geography. Each course builds student content knowledge, skills, and geographic thinking in a sequenced manner, adhering to methods advocated by Bednarz et al. (2013).

Web mapping applications (https://livingatlas.arcgis.com/en/apps/) provide excellent introductory tools, rich content, and engaged immersion without sign-in. These include the ArcGIS Wayback imagery, where high-resolution satellite imagery over a 10-year time period can be compared using swipe and playback tools, and the Landsat Explorer app, where 50 years of deforestation, urbanization, agriculture, construction, and other natural and human causes can be compared. The Water Balance app can be used to compare precipitation and evapotranspiration across 20 years from the Sahara to the Amazon. The Living Atlas indicators of Planet Earth (Fig. 2), which provide a monitor or "heartbeat" of many variables in real time—wildfire, air quality, cyclones—(https://experience.arcgis.com/experience/003f05cc447b46dc8818640c38b69b83) can serve as a weekly check on the state of the planet throughout the semester.

Next, the ArcGIS Online (www.arcgis.com) 2D map viewer and 3D scene viewer can be used to access, symbolize, classify, filter, and share geographic data.

An advantage of teaching with Web GIS is that the tools can be scaffolded. Students begin with existing content and then build their *own* content. For example, existing story maps such as the "Age of the Anthropocene" can be used as engaging multimedia content. Then, students can be asked to build their own story map on a

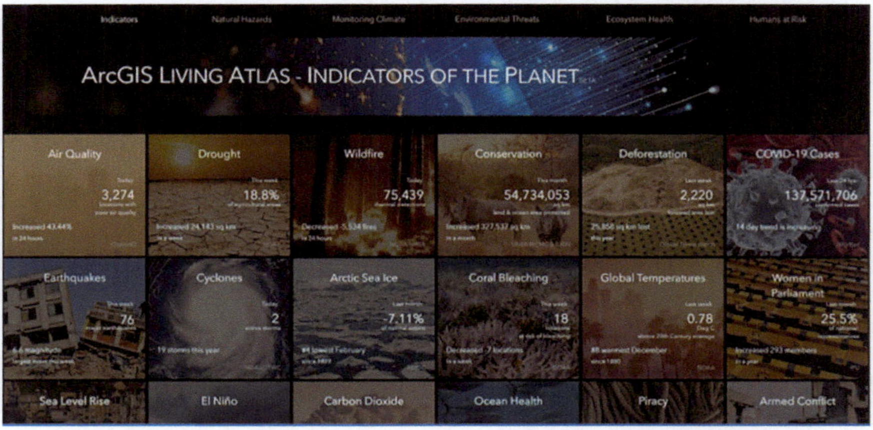

Fig. 2 Examining major indicators of Planet Earth in real time. *Source* Esri

theme according to a prescribed lesson and present their findings about tsunamis of the Pacific. Asking students to build their own story map as a *first* activity exposes too many choices and uncertainties, from "what should I create," to "do I have permission to use image x" to "how do I create the web maps to embed in the story map?" The scaffolded approach enables students to understand how the Web GIS platform fits together before they construct their own maps and apps. After a short time on foundational activities using these intuitive tools, students are ready to create their own maps on a topic of their own choosing.

The same approach can be done with field survey tools such as ArcGIS Survey123. In the first week of a course, students are asked to fill out a survey about their attitudes about geography or to collect species, height, and condition of 3 trees on their campus. Later, students are asked to create their own survey of something *they* care about using Survey123, collect data into their survey, and share their survey with 2 other classmates. Thereafter students map their survey results and build a dashboard that indicates, in real-time, historical homes, weather, noise, or other attribute they collect in their survey. The same approach works with ArcGIS Online: Work first in the platform without signing in, and follow this with spatial analysis where a signing in is required to analyze and save results.

Spatial analysis tools can be introduced later: Routing, overlay, and aggregating points change students' views of maps from *what is where* to *why things are where they are*. The focus is on geographic relationships, patterns, and trends. Unlike in the challenging "desktop only" days of GIS, Web GIS analysis functions are accessed from a browser menu as wizard-driven tools with intuitive prompts, such as, "With what distance would you like to buffer the river to create a riparian zone?"

Each issue examined through Web GIS should be performed not because the technology exists, but because the technology is used to foster meaningful *inquiry*. And because of the media and data fluency that these tools and methods foster, spatial skills instruction in geography can also improve outcomes in science, technology, engineering, and mathematics (STEM) as well (Sorby et al. 2018).

An introductory geo-visualization and cartography course (https://experience.arcgis.com/experience/61b85d6e210b4dae8844ba534de183da/) that the author created shows how to integrate Web GIS in instruction. The 30 videos, 250 pages of readings, 50 activities, 7 quizzes and answer keys, and final implementation project make use of ArcGIS Online as the primary toolset, plus Gapminder for analyzing demographic change over time, Worldmapper for analyzing agriculture and other variables via cartograms, ColorBrewer for choosing map colors, plus additional tools provide engaging hands-on activities. This course focuses on analyzing patterns and relationships, and uses Web GIS to present content: Each week is comprised 4 ArcGIS story maps—readings, hands-on activities, quizzes, and quiz answers. The story maps are stitched together using the ArcGIS Experience Builder.

The course also uses content maps (Fig. 3) to help guide students in each week's work. Each content map contains content, skills, and societal issues. Cutting across are themes covered in the readings and activities.

The big data section illustrates how readings and discussions are tied to interactive hands-on activities. The readings and discussions support the course's objectives of

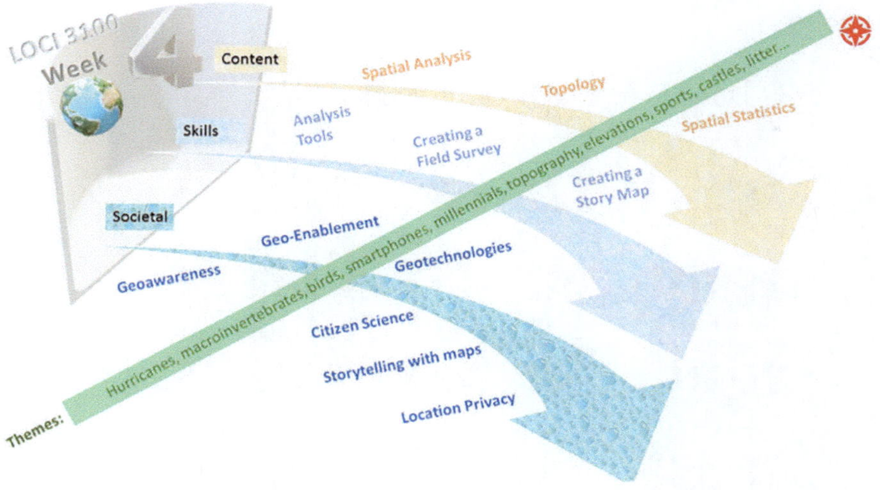

Fig. 3 Sample content map for Week 4 of a cartography and geo-visualization course. *Source* Joseph Kerski

ways that maps and geovisualizations provide a common language and framework for communication and solving problems, analyzing data spatially, including mean center, and identifying how society influences mapping, and how mapping influences society. Students consider how storing information changed over time, and how GIS handles big data. Students work with a large geodatabase of every building (Fig. 4), learning how the layer was extracted from imagery and segmented using Deep Neural Networks via machine learning and artificial intelligence.

After students add buildings from a GeoJSON file to ArcGIS Online, they assess spatial accuracy by comparing the building outlines to satellite imagery and evaluate the completeness of the building attributes. Later, students describe how they will implement what they have learned.

A key course goal is to foster critical thinking about data. Students read sections of a public domain data textbook and blog (https://spatialreserves.wordpress.com), asking: Where can I find geospatial data? How can I know if I can trust that geospatial data? What are key societal issues surrounding spatial data—ethics, location privacy, positional accuracy, and scale? Today, where everyone is a map user and a map creator, fostering critical data thinking is important in *every* course. Instead of providing students with canned "perfect" data sets, work with real data and real portals. Students are challenged with difficult data portals but express end-of-course gratitude that they dealt with real-world uncertainties.

A physical geography course created by the author, presented as a story map, (https://storymaps.arcgis.com/stories/dc91fa76f21649d5b8f34df8eeae2849), provides a second example of how Web GIS tools and data can be used to teach core topics (geomorphology, agriculture, weather, ecoregions) and themes (spatio-temporal change, scale, systems thinking). The course uses 2D and 3D maps for

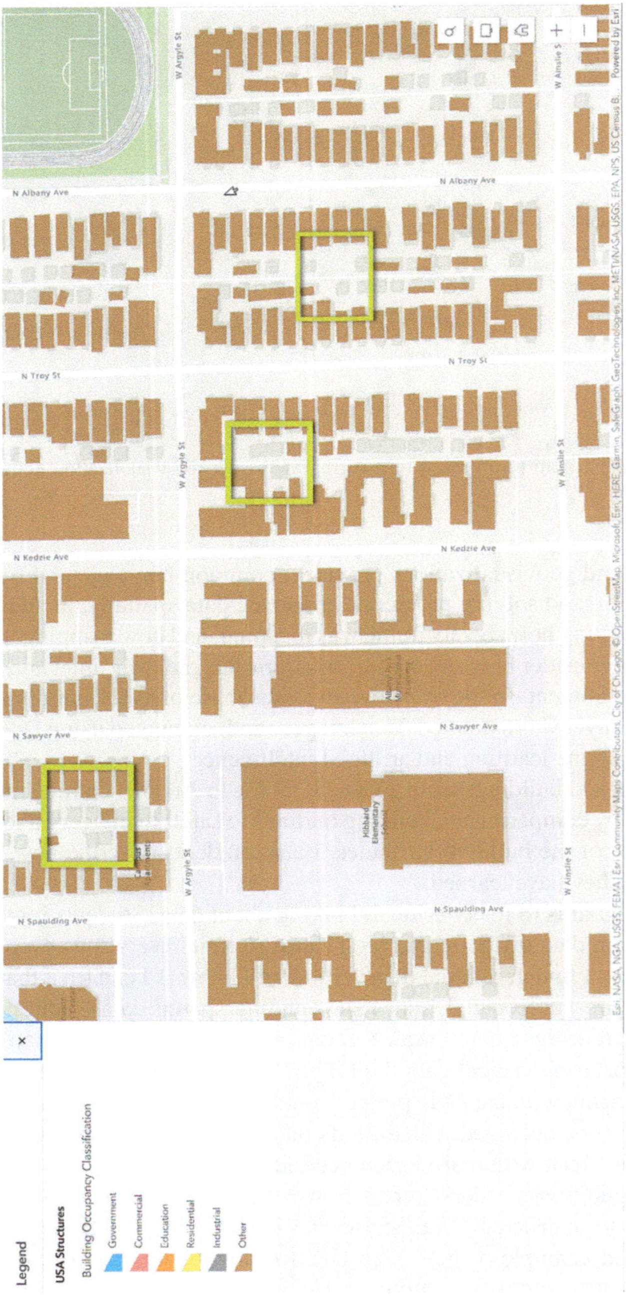

Fig. 4 Using Microsoft's big data layer of 100,000,000 buildings in ArcGIS Online. *Source* Esri

studying geomorphology. Students investigate ecoregion locations by examining proximity to oceans, latitude, and elevation using layers in ArcGIS Online. They use the ecological marine units explorer (Fig. 5) to investigate water quality variables from ocean surface to floor at 10-km resolution and the green infrastructure asset finder for ideal locations for plant and animal species studies using slope, elevation, direction of slope, and land cover. They integrate mathematics and geography to determine direction and distance that water travels from their campus to the ocean. They use real-time feeds (river height, weather, wildfire perimeters); they use spatial analysis tools to determine which regions are susceptible to both landslides and floods. They study data in the field using Survey123 in physical geography, collect data, map the results using ArcGIS Online, and create a dashboard that includes gauges and charts. Patterns in land cover versus soil erosion or temperature versus proximity to coasts can be detected. Variables are filtered, sorted, and summarized within a table and the map features associated with those attributes are examined on the map. These include the deepest earthquakes over the past week or lakes in the students' region with the lowest pH. Through these activities, students learn content, develop mapping and analysis skills, and apply the geographic perspective to understand change over space and time.

A cultural geography course presented as a story map, https://storymaps.arcgis.com/stories/3c606dfa01c04f44ac37cc1279cefae0, provides a third example how Web GIS tools and data can be used to teach core topics (here, population change, land use, energy) and themes (temporal and spatial change, scale, systems, patterns). Students investigate incoming and outgoing migration using a 3D web map, examining push and pull factors. They use 3D globes to learn about population density, patterns, and proximity to coastlines. Using Urban Observatory, students compare age, land use, population density, and other variables across 100 cities around the world. They use ArcGIS policy maps to investigate racial equity and social justice using home ownership, historical redlining, and other variables for selected communities. Students investigate cultural regions through multimedia story maps, including music archives and the digital humanities collection. Students study population through the US Census Bureau's county flows mapper, (Fig. 6) and maps showing in-migration and out-migration.

Using spatial analysis tools, considering the electrical grid, population, and wind speed, students determine the optimal location for a wind farm. They study walkable neighborhoods around the world and in their own community through a crowdsourced field survey. They create photo tours of their campus using the Mapillary application. Through these investigations, students learn content and develop mapping, analysis, and communications skills, applying the geographic perspective to understand spatio-temporal change.

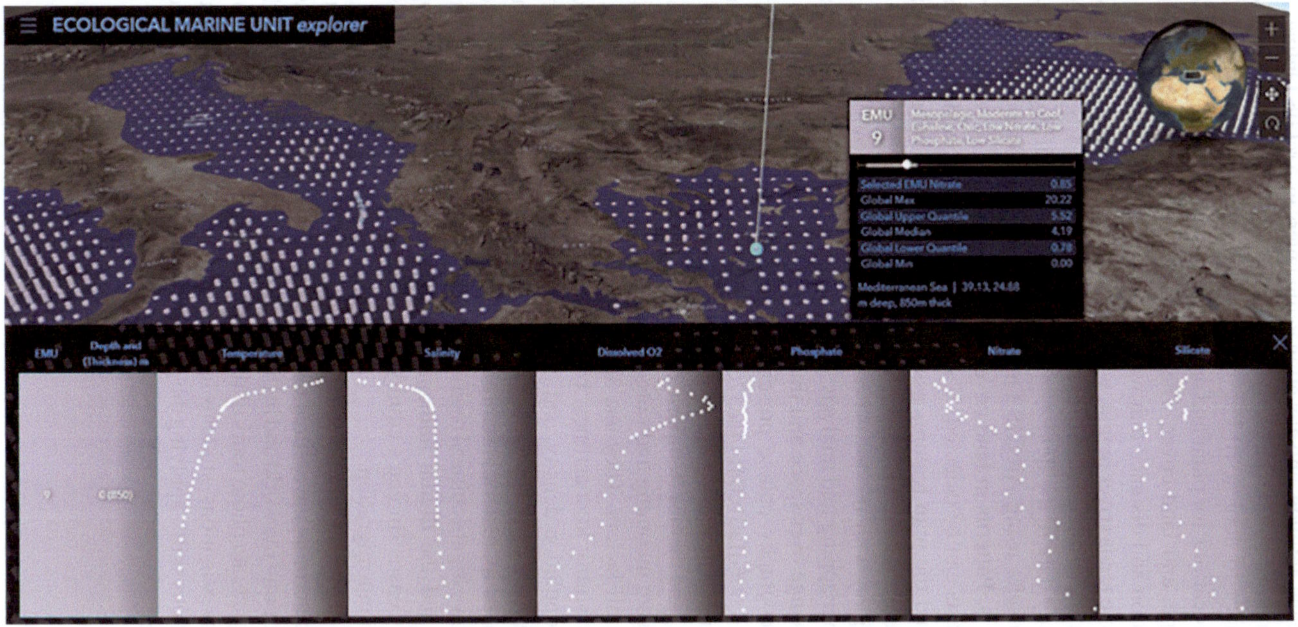

Fig. 5 Using the Ecological Marine Unit Explorer in the Aegean Sea. *Source* Esri

Fig. 6 Examining population inflow to Colorado, 2015–2016 with ArcGIS Online *Source* Esri and US Census Bureau

Addressing Key Challenges in Geography with the Benefits of Web GIS

Working with GIS offers many benefits to teaching and learning geography. "Doing GIS" is not "just making maps," but to help understand patterns, relationships, and trends. Understanding, not mapping, is the goal: Understanding the planet and its people in a deeper, richer, more holistic way, leading to the goal of taking action through the smarter decisions that result from that understanding. As TED founder Richard Saul Wurman said, "Understanding precedes action."

The first benefit that Web GIS offers is spatial thinking. Sometimes called graphicacy, scholars have argued its importance alongside numeracy, articulacy, and literacy (Balchin, 1971). Maps have always been appealing conduits of large amounts of information in small spaces. This space was always constrained in the past by physical media—stone tablets, wood blocks, silver plates, silk, film, and paper. Today's digital maps are all around us, embedded in dashboards, story maps, text, video, and multimedia. They are viewed as 2D and 3D representations, with time viewed through animations and space–time cubes. Spatial thinking is key to understanding the increasingly interconnected, complex world, and to solve problems, and is enabled through GIS. Through GIS, students consume maps created by *others*, but also create their *own* content, connecting with the creative process: Students are active learners using GIS. They perform some of the same work with the same professional tools that a climate scientist uses when analyzing climate data, a biologist when examining ecoregions, or a demographer when investigating population trends. They are not simply reading and responding—they are *doing*.

Spatial thinking through Web GIS fosters holistic thinking—seeing the world as more than just the sum of its parts—a system of systems—the biosphere, lithosphere, atmosphere, hydrosphere, cryosphere, and anthroposphere—the human sphere. Spatial thinking fosters understanding key Earth cycles (the carbon cycle, the hydrologic cycle). How? Students "see through layers" of data not just visually but by creating multivariate maps, flow maps, and 3D scenes, which foster the understanding the connectivity between scales, variables, and themes, and how those elements change over space and over time.

Web GIS offers the benefit of *critical thinking*: About data, methods, and maps. Students cannot help but grapple with questions as they work through GIS-based lessons and projects: What difference would changing themes, resolution, time, or scale make in the final analysis? Or changing the data source to one collected by a different organization? Or changing the arithmetic expression to affect the map's appearance? Or using erase rather than union to affect the final sites? Can I trust this map to help me make decisions? Is this map or layer suitable for my project? What are the inherent errors, from map projections or missing attributes? How can I manage this error?

The third benefit that Web GIS offers is *Problem-based Learning* (PBL). GIS was created to solve problems, and this remains its chief use today. Teaching with GIS helps students frame, visualize, and grapple with problems. It enables students to craft solutions to the UN SDGs. PBL implies active learning: GIS is a natural fit: Students become actively engaged as scientists, planners, and other professional decision makers. Students are not just learning about supply chains or fisheries; they are *doing* supply chain management and habitat analyses. Students learn by actively engaging in real-world and personally meaningful projects. Using GIS, students can choose projects and problems that they see that are complex, serious, and need addressing. These may be at the local level (historical preservation, litter, water quality) or global (health or climate). Given the interactivity of Web GIS, students can analyze topics at the local and at the global scale in one interface, often with scalable data. Then, they understand that global SDG and other issues often affect lives and places at the local scale. GIS fosters skills in initiating and managing projects, and integrating systems that these projects entail.

The fourth benefit of instruction with Web GIS is fostering geographic and scientific *inquiry*. Inquiry involves asking questions, gathering data, assessing data quality, evaluating methods, analyzing results, making decisions and recommendations, and taking action. This sparks additional questions, continuing the process. GIS has always been "a thinker's tool," requiring students to ask questions. A good map teaches students to ask a better, more thoughtful, question: "What if the area south of the city is examined instead of north? What if the scale is changed from neighborhood to postal code? What if I change the classification from quantile to standard deviation? What if I change the variable from dissolved oxygen to conductivity?" One of the appeals of using GIS is that some of the deep questions students ask when using it *cannot* easily be answered. Asking questions leads to tenacity and problem solving. Those who ask meaningful questions are the type of employees who employers want to hire to help their organizations meet their missions.

Fifth, teaching with Web GIS fosters data *fluency* (Jukes et al., 2010). Fluency conveys lifelong learning, such as becoming so in a language. The five fluencies are solution fluency, whole brain thinking (creativity and problem solving), information fluency (accessing digital information sources to retrieve information and critically evaluating the quality of information), collaboration fluency (cooperating with partners to create original digital products), creativity fluency (artistic proficiency through design, art, storytelling), and media fluency. Media fluency is the ability to look analytically at communication media to interpret the message, determine how the chosen media shapes thinking, and evaluate the message's efficacy and the ability to publish original digital products matching the media to the message. Students using GIS engage in all fluencies (Kerski, 2015). Students access and evaluate real-time data feeds, images, maps, photographs, and layers, from many sources. With community data portals such as ArcGIS Hub, and Internet of Things (IoT) real-time feeds, and students' ability to collect their own data, there is no shortage of data to map and examine. Creating story maps requires creative fluency, using photographs, maps, audio, video, and text. Thinking critically fosters media fluency which helps in many aspects of modern life. One way to engage students is to begin with personal competencies on the Geospatial Technology Competency Model (DiBiase et al. 2010) from the GeoTech Center: "Are you organized? Are you ethical? Can you work with data?"

The sixth benefit of teaching with Web GIS is building *community connections.* GIS is used for global problems (climate, education, water, other UN SDGs), but it can also be used to engage with local issues (planning a new bike trail, nurturing community gardens, tackling traffic accidents or graffiti). To understand different stakeholders' needs in these projects, the community is required. Students can present their results using GIS to local authorities.

Fieldwork is the seventh benefit fostered using Web GIS. Methods, data, and getting outside are a routine part of professional research projects and can be readily incorporated into schools and universities. Fieldwork is essential to understanding the changing world and local communities, and is essential for nurturing an "Earth Ethic"—a caring for the planet. Successful fieldwork involves planning, executing, and analyzing results. It hinges on organized project planning, dealing with uncertainty, incorporating high and low-tech methods, and dealing with data—the units that will be used, variables, data tables, images, geodatabases, and maps. It may involve collaborating with community groups to access lands or to garner their support and participation. ArcGIS Survey123, iNaturalist, noise apps, or Artificial Intelligence (AI) enabled apps to identify plant species or bird calls are used. Resulting data is brought into the Web GIS environment for spatial analysis. Fieldwork was an integral part of the Data Citizens Project (Kerski 2021a), where students mapped storm drains and learned about their community's water treatment. Even fieldwork limited to the school or university campus involves *observing* with technology but also with one's own senses. It is key to overcoming "nature deficit disorder" (Louv 2008).

Using Web GIS offers the eighth benefit of helping students to blaze their own career *pathways.* "Are we ever going to use this after we get out of school?" is an oft-heard question. While instructors should not use Web GIS in education just because it is in workplace demand, GIS provides students with key career skills that will never "go out of style." Students who use GIS become valuable employees for nonprofit, academia, government agencies, and industry. They make decisions, work with data, and see things holistically. The "Whys of Where" will be asked in *all* workplaces given ever-increasing pressures on our planet. Sustainability and resilience will be in every organization's plan, and GIS will have a key role. Students' behavior tends to be more professional with the awareness that they are using the same tools as are used in the professional community (Kerski 2021b).

The use of Web GIS also offers the ninth benefit of helping students gain *content knowledge.* When an educator teaches with GIS, they are teaching core content knowledge. GIS was never about "buttonology"—memorizing where tools are on the interface. Understanding how to use GIS is important, but the focus should always be on the data that is at one's fingertips. Every GIS workflow is backed by real data, and so educators teach plate tectonics, ecoregions, climate, hydrology, transportation, energy, historical events, or other aspects of the natural and human-built world. GIS is fundamental to science, social science, and many other fields. Students learn content in fields such as data science, biology, mathematics, history, and others in addition to geography and environmental science. This will accelerate as GIS expands to fields such as health, business, civil engineering, humanities, and sociology. Students learn about cause-and-effect and human–environment interactions upon health, supply chain management, building energy efficiency, and other topics in those disciplines. In "teaching GIS" and "teaching with GIS" approaches, content knowledge is acquired more rapidly than memorizing large volumes of information, because students actively *engage* with the data and methods as a practicing professional.

The use of Web GIS also offers the tenth benefit of helping develop "*students as change agents*." Students empowered with the geographic skills, content knowledge, and geographic perspectives gain the confidence and ability to become change agents in their future. Given many examples in schools (Fitzpatrick 2020) and in a community college (Kerski 2019), students *already are* change agents even before they graduate. GIS also enables young women and other underrepresented populations to step into technology-based careers. Educators from primary to university level and those in informal educational settings (museums, libraries, after-school programs) have affinity for using GIS because it embodies why they all became educators in the first place—making a positive difference in the world through caring, engaged students.

The 10 Most Important Strategies in Teaching Geography with Web GIS

What are 10 key strategies, or approaches, for teaching with Web GIS in an effective manner that engages students?

First, instructors should anchor the use of Web GIS *anchored* to their program goals. Instructors are not teaching GIS so that students will simply become more proficient with GIS. Rather, instructors foster critical thinking, data fluency, spatial thinking, communication skills, community connection, field method proficiency, and more. GIS skills are important, but maps are a means to a greater goal of understanding and taking action.

Second, instructors should make their instruction with Web GIS *holistic*. The geographic perspective through GIS fosters consideration of the lithosphere, hydrosphere, biosphere, atmosphere, and anthroposphere as a complex, interacting "system of systems." Instructors cannot model the world in all its complexity in any single lesson or even a single course or program. Researchers spend their entire careers trying to understand even small aspects of the Earth system. However, instructors can ensure that their courses invite students to consider the "bigger picture." Decisions in one sphere cause ripple effects, positive and negative, in other spheres. With multivariate maps and visualizing themes across 3D scenes and across time periods through swipe and animation tools, instructors can foster holistic viewpoints given weekly themes such as "water" or "hazards." Because of the interconnections between natural systems and human-built systems, GIS helps students understand patterns, relationships, and trends. By varying the themes studied, students emerge with a sense that "everything is spatial" and interconnected.

Third, instructors should make their instruction with Web GIS *focused*. Focus each week or another period of time on a problem in one sector of society. Crime, energy, water, historic or sacred structures, biodiversity, land use, climate, urban forms, hazards, social, racial, and economic inequalities, land use, agriculture, transportation, utilities, and supply chain management provide many examples. Focusing readings, videos, discussions, and activities builds content knowledge around specific knowledge domains. Doing so over time also helps students realize that GIS and the spatial perspective are relevant to every twenty-first-century problem faced by humanity.

This focused advice does not conflict with the holistic advice because instructors can include thought-provoking questions to students in online discussion boards, face-to-face conversations, and quizzes that "go beyond" the lesson. On siting a business using GIS, ask, "If you were really doing this site selection as a consultant for this specific convenience store chain, what other data would you consider?" In this activity, students use traffic volume, demographic characteristics, consumer preferences (with fuel and lottery tickets figuring prominently in convenience stores), and drive times to competitors to determine the ideal location for a convenience store. This question helps students to consider additional variables such as zoning, left turns vs. right turns, employment centers, and commuting patterns.

Instructors should ask similar types of questions during other lessons: "If you were really doing this study as a consultant, what other variables would you pursue?" These questions encourage students to think about other aspects of the physical or cultural environment and that the world is complex with variables that change over space and time. It also reminds students that in their own future workplace, time, budget, and staffing will always cause constraints to factor into problem solving.

Fourth, instructors should make teaching with Web GIS *multiscale*. Most pressing, relevant issues, including the UN SDGs, are global in nature and foster planet-scale investigation. For maximum student engagement, focus some Web GIS activities on one's own community. Students may have never been asked what they care about in their community; many may feel as if they do not have a voice. GIS infusion represents an opportunity for them to start investigating and start caring about a skateboard park, urban greenway or garden, historical sites, dangerous intersections, handicapped access, or other issues in which they care about and want to take action. Many of the lessons in the ArcGIS lesson library (https://learn.arcgis.com) can be modified for one's own area.

Fifth, make instruction with Web GIS *varied* and *interesting*. Interested instructors strongly influence student engagement. By varying the instructional methods used, including online discussion boards, active assessments including presentations with using story maps, hybrid and face-to-face meetings, embedding maps in videos or dashboards, using interactive quiz tools such as Kahoot, mixing video and audio, group vs. individual projects, and courses remain lively and fresh. And even though using GIS is "not just about the tools," ask students occasionally to investigate engaging, fun maps (such a map of "hats around the world") or new GIS visualization (such as the Mars 3D viewer).

Sixth, make instruction with Web GIS *relevant*. Given serious, daily issues, this strategy may be the easiest strategy of all to incorporate. Creating a short lesson from a current event is a powerful way to keep courses interesting, timely, and relevant. Employ "GeoNews" in which a team of students in a newscast format presents a topic currently making headlines, using interactive maps.

Seventh, make GIS instruction *field-based*. Geography is field driven and inherently tied to space and place. Working with maps, satellite imagery, and visualizations can help foster "topophilia" (Tuan, 1990). Students can immerse in real spaces, using their own senses. Therefore, include activities where students collect, map, and analyze something in the field–invasive plant species, light poles, trees (species, height, and condition), or other themes. Analyze something that changes often (noise, pedestrian counts, weather) so that students can compare these observations to different places and over different times of day or seasons. Consider collecting the same themes each semester to build a long-term database of phenomena.

Even without an established field study center or an arboretum on campus, fieldwork on one's own campus is valuable. For virtual courses, ask students to perform fieldwork in *their own* neighborhoods, using ArcGIS Field Maps, Survey123, QuickCapture, and other tools such as PictureThisAI or iNaturalist. If international students are collecting data, comparing housing types, plants, weather, and other variables across countries will be fascinating.

Eighth, make Web GIS instruction *multi-level*. Web GIS offers multiple levels of engagement, such as the Drought Aware app. Students visually study data via these web mapping apps in an introductory setting. In intermediate settings, students save layers covering 1 theme and use them in ArcGIS Online with its spatial analysis tools. In an advanced setting, students use the image classification tools in ArcGIS Online or ArcGIS Pro to measure changes on the landscape in 2D or 3D. This multi-level approach is implementable with most Web GIS tools.

Ninth, make Web GIS instruction so students can *shine*, *explore*, and *grow*. Ask students to reflect upon their learning with questions, even inside quizzes, such as "What is the most valuable thing you learned this week? What was most frustrating? What is 1 thing that you read about that you would like to learn more about?" Encouragement will help students become more reflective learners (see Brookfield (1995) for additional ideas).

Include assignments where students create story maps and other web mapping applications (such as using the ArcGIS Experience Builder). Students share them with their instructors via a URL and with their peers. Students use these apps in online or face-to-face presentations to instructors and classmates, and even with external stakeholders. These story maps also extend beyond specific courses or programs and become a key part of a professional portfolio that students take into the workplace.

Tenth, make Web GIS instruction *visionary*. Include space–time cube mapping, artificial intelligence and machine learning, the blurring of the lines between mapping and visualizations, 3D analytics, the meshing of BIM, CAD, and GIS tools for inside-and-outside buildings views, coding via Jupyter Notebooks, virtual reality, and other cutting-edge GIS trends and capabilities. This affirms that the tools rapidly evolve, and thus, their users must be lifelong learners. Include podcasts and video interviews with visionaries, such as Esri's Virtual Job Shadow videos and the Directions Magazine Geoinspirations series. Keep the students mindful about ways they can make societal contributions.

Keep the higher, more noble goals in mind: Using GIS is ultimately about building a better world. Instructors and their students have a key role to play in that world!

The 10 Most Important GIS Skills that Are Fostered in Teaching Geography with Web GIS

As this chapter articulates, the most important benefits in teaching with Web GIS include fostering spatial and critical thinking. However, a chief attraction for educators in teaching with professional tools, such as GIS, is fostering job skills that are in high demand in the workforce. Despite economic upswings and downturns over the history of GIS, and despite changes in Information Technology (IT) infrastructure, the demand for GIS skills has steadily risen. The deepening and widening of GIS in education and society ensures that the "whys of where" questions will always be asked.

Educators and students can do *anything* in the dynamic platform that is modern Web GIS. Teaching students these 10 skills is similar to the analogy of teaching students "how to fish" versus "giving them a fish." By focusing on these skills, students develop perspectives and problem-solving abilities transferrable to different issues and regions. Examples (Kolvoord 2017) of students doing innovative work exist from primary to university level.

The first GIS skill is working with map layers. GIS relies on looking "through" a series of raster and vector layers to understand the "big picture"—the world. Know how to search for, open, and save map content. Be organized about managing content. Know how to interpret and create metadata, especially today where everyone is a potential mapmaker. Know how to create features, spreadsheets, multimedia, and other content. The second skill is to know how to share mapped content with a group of people, the global community, and also *when not* to share, due to location privacy, health, or other valid reasons.

The third skill is navigating the map interface or GUI of whatever GIS software is used. Since 2000, the GUI has been the fundamental way in which users work with GIS. As GUIs change, these skills need to be kept current with an adaptable mindset. Navigation includes changing map scale and map projection, finding and saving locations, and measuring perimeters, volumes, and areas. Know how to select map features (cities or biomes) and how to access the information about features in data tables or information popups.

The fourth critical skill is the ability to visualize information on a map in a GIS: Know how to set and change the symbology of mapped features or image layers, in 2D and 3D. Know how to create dot density, choropleth, and heat maps; how to create relationship, multivariate, and predominance maps. Know how and when to use classification methods such as Standard Deviation, Quantile, and Natural Breaks. Because the number of features that Web GIS can map is often too large to be understood, know how to filter—reducing the number of features in a layer or bands in an image to aid interpretation.

The fifth critical GIS skill is work with tabular data: Know how to create fields and tables, select rows and columns, sort, summarize, and create charts and popups. The sixth skill is to know how to use several kinds of data collection applications (such as ArcGIS QuickCapture, iNaturalist, others) and how to map data from GPS apps, fitness apps, geotagged photographs, and analog methods including clipboard notes.

The seventh critical skill is to be able to draw and sketch points, lines, polygons, text, and other map objects. The eighth critical skill is to create and use mathematical expressions, to filter, label, and symbolize. Know how to create an expression with ArcGIS Arcade, Python, SQL, and other languages.

The ninth critical skill involves creating maps that can communicate well: As web mapping applications are viewed more frequently than the web maps themselves, know how to thoughtfully and clearly create story maps, infographics, and dashboards. The tenth critical skill is to understand when, why, and how to perform analysis. Know how to create proximity zones, summarize data, perform map overlay, join spatial and tabular data, create routes, use spatial statistical tools, trace downstream, create viewsheds, and other analytics.

These 10 skills can be used to create student assessment instruments. All or portions can be used as a pre-test for a planned course or workshop, and results can be compared to a post-course evaluation. Student-created story maps can be used as content that faculty can evaluate from a content or technology standpoint. Instructors can assess the maps online and/or assess a live or recorded video presentation given by students.

Given the web platform upon which modern GIS rests, developing coding skills in JavaScript and Python merits inclusion in the GIS skills list. Editing mapped features remains important. Other core skills worthy to be considered include these core elements of GIS: Create geographic data, analyze data, and visualize data. The University of Southern California (https://gis.usc.edu/blog/top-gis-skills/) grouped skills into data collection and evaluation, visualization, analysis, and modeling. One university's advice includes understanding workflows (https://dusk.geo.orst.edu/gis/PPTs/essential_skills.pdf) and the GIS Lounge skills list (https://www.gislounge.com/gis-skills/) focuses on data.

Recommendations

Web GIS presents educators with a stunning array of resources and tools with which to create and teach innovative online and face-to-face courses in geography, including the teaching of the SDGs. These approaches can also be used to help instructors in other disciplines to see that geographic content, tools, and methods are valuable for their own courses.

Zhu et al. (2021) argue that existing GIS software targets expert users and does not sufficiently integrate resources for efficient computing. Challenges remain before GIS is sufficiently "easy" and intuitive to enable Rogers (1995) "mainstream adopters" to use. These include determining best curricular fit, dealing with the rapid advance of the technology, and spotty bandwidth in many parts of the world. Uncertainties of what to teach and how to teach (Sinton and Kerski 2020) persist. Because one of the purposes for GIS is to model the real, complex world, GIS will never be as instructionally easy as using worksheets or lecturing. Teaching and learning with any inquiry-driven, problem-based method is challenging, but it is worth investing time and effort. Web GIS presents a new paradigm: The tools are too accessible, the data sets too rich, and the rewards in understanding, engagement, and action are too great to be ignored.

References

Balchin WGV (1971) Graphicacy. Geography 57(3):185–195

Bednarz SW, Heffron S, Huynh NT (2013) A road map for 21st Century geography education: Geography education research. A report from the Geography Education Research Committee of the Road Map for 21st Century Geography Education Project. AAG, Washington, DC

Brookfield S (1995) Becoming a more critically reflective teacher. Jossey-Bass, San Francisco, 320 p

DeMers M, Kerski J, Sroka C (2020) Assessing the effectiveness of a GIS professional development institute: The Teachers Teaching Teachers GIS (T3G) institute. Ann Am Assoc Geogr

DiBiase D (2018) Stop teaching GIS. Esri Education Community. https://community.esri.com/t5/education-blog/stop-teaching-gis/ba-p/893033

DiBiase D, Corbin T, Fox T, Francica J (2010) The new geospatial technology competency model: Bringing workforce needs into focus. URISA J 22(2):55–72

Fitzpatrick C (2020) Fun with GIS 253: teaching underground. Esri Community essay. https://community.esri.com/t5/education-blog/fun-with-gis-253-teaching-underground/ba-p/883758

Geraghty E, Kerski J (2020) The impact of COVID-19 on geography, GIS, and education. J-Reading Journal of Research and Didactics in Geography, Vol 2, Year 9

Jo I, Hong JE, Verma K (2016) Facilitating spatial thinking in world geography using Web-based GIS. J Geogr High Educ 40(3):442–459

Jukes I, McCain T, Crockett L (2010) Understanding the digital generation: teaching and learning in the new digital landscape. Corwin

Kerski J (2003) The implementation and effectiveness of GIS in secondary education. J Geogr 102(3):128–137. National Council for Geographic Education.

Kerski J (2008) The role of GIS in digital earth education. Int J Digital Earth 1(4):326–346

Kerski J (2015) Geo-awareness, geo-enablement, geotechnologies, citizen science, and storytelling: Geography on the world stage. Geogr Compass 9(1):14–26

Kerski J (2019) Supporting community college student research through spatial analysis and story maps. Esri Community. https://community.esri.com/t5/education-blog/supporting-community-college-student-research/ba-p/883517

Kerski J (2021a) GIS and the data citizens project. Esri Community. https://community.esri.com/t5/education-blog/gis-and-the-data-citizens-project/ba-p/1086806

Kerski J (2021b) Online, engaged instruction in geography and GIS using IoT feeds, web mapping services, and field tools within a spatial thinking framework. Geogr Teach 19(3):93–101

Kolvoord R (2017) Breaking down the barriers to geospatial education. ArcNews. Winter. Redlands: Esri. https://www.esri.com/about/newsroom/arcnews/breaking-down-the-barriers-to-geospatial-education/

Louv R (2008) Last child in the woods: saving our children from nature-deficit disorder. Algonquin Books, 416 p

Lukinbeal C (2022) From emergency remote teaching to academic strategic digitalization planning: how a college responded to COVID-19 and what departments need to consider implementing. In: Brunn, Gilbreath (eds) COVID-10 and a world of ad hoc geographies. Springer.

Manson S, Shannon J, Eria S, Kne L, Dyke K, Nelson S, Batra L, Bonsal D, Kernik M, Immich J, Matson L (2014) Resource needs and pedagogical value of web mapping for spatial thinking. J Geogr 113(3):107–117

Milson A, Kerski J, Demirci A (2012) International perspectives on teaching and learning with GIS in secondary schools. Springer

National Academies of Sciences (2018) Returning individual research results to participants: Guidance for a new research paradigm. Consensus Study Report. https://nap.nationalacademies.org/catalog/25094/returning-individual-research-results-to-participants-guidance-for-a-new.

Rogers E (1995) Diffusion of innovations, 4th edn. Free Press, New York, p 518

Schulze U (2021) GIS works—But why, how, and for whom? Findings from a systematic review. Trans GIS 25:716–804

Sinton D, Kerski J (2020) GIS&T education and training. In: Wilson JP (ed) The geographic information science & technology body of knowledge, 1st Quarter 2020 ed. University Consortium for GIS&T, Ithaca, NY. https://gistbok.ucgis.org/bok-topics/gist-education-and-training.

Sorby S, Veurink N, Streiner S (2018) Does spatial skills instruction improve STEM outcomes? The answer is 'yes.' Learn Individ Differ 67:209–222

Tuan Y-F (1990) *Topophilia: a study of environmental perceptions, attitudes, and values.* Columbia University Press, 260 p

Zhu A-X, Zhao F-H, Liang P, Qin C-Z (2021) Next generation of GIS: Must be easy. Ann GIS 27(1):71–86

Education for Sustainability Using Cloud-Based Geographic Information Systems at University

María-Luisa de Lázaro-Torres, Miguel-Ángel Puertas-Aguilar, and Javier Álvarez-Otero

Abstract The eco-social crisis has led to changes in education to address sustainability. Specifically, learning scenarios or case studies have been created using Digital Story Maps and Geographic Information Systems in the cloud. These tools allow sustainability issues to be visualized, which enables students to go beyond the theoretical knowledge of the Sustainable Development Goals' indicators, to transformative action based on critical thinking. This approach was used to create and apply innovative learning processes in teacher training, as part of a geography master's degree for secondary school teachers, as a compulsory topic for the first semester in the academic years 2020/21 and 2021/22. Student evaluations were carried out using blind peer review with the University of Glasgow's tool, Aropä, and a satisfaction questionnaire, which produced a mean of 4.6 over 5. It has also been possible to compare outcomes from the traditional way of preparing a lesson and the use of Web GIS by using an adapted version of Rosenshine's Principles. Thus, Web GIS and Digital Story Maps have been found to be particularly useful tools as they can help in the teaching process, assess learning, and enable participants to share concerns about sustainability issues.

Keywords Geography · Geographic information · Web GIS · Digital story maps · Curriculum sustainability · Sustainable Development Goals

---

M.-L. de Lázaro-Torres (✉)
Geography Department, UNED, Madrid, Spain
e-mail: mllazaro@geo.uned.es

M.-Á. Puertas-Aguilar
International Doctoral School of the UNED (EIDUNED), Madrid, Spain
e-mail: mpuertas48@alumno.uned.es

J. Álvarez-Otero
UNED, Madrid, Spain
e-mail: j.alvarez.otero@madrid.uned.es

A. Klonari et al. (eds.), *Re-visioning Geography*, Key Challenges in Geography,
https://doi.org/10.1007/978-3-031-40747-5_8

Introduction

The vertiginous growth of open geographic information has boosted the use of cloud-based Geographic Information Systems (or Web GIS) in research, scientific dissemination, and in geographic education. It is based on the use of open data, the potential of visualization in Web GIS, its usability, and ease of access. Consequently, the question of whether technology has a place in geography lessons is resolved, and the focus is now on teachers integrating it into the curriculum following their initial university teacher training. As a result, geographical education for sustainability can employ these tools as is done in many other socio-economic, scientific, and cultural fields.

There is a broad academic consensus that Web GIS can facilitate understanding of territorial reality, and its physical and socio-economic attributes and integrate them into university teacher training and secondary education (Álvarez-Otero 2020; Buzo 2021; Buzo-Sánchez et al. 2022; Fargher 2018; Puertas et al. 2022; Zwartjes and De Lázaro, 2019, among others), as well as for example, tourism teaching (Mínguez 2021; Martínez-Hernández et al. 2021), landscape (De Lázaro et al. 2017), archeology (Pons et al. 2019), and fieldwork (Phantuwongraj et al. 2021). While most educators reflect on the most appropriate pedagogies for Web GIS implementation, the inquiry-based method and problem-based learning appear to be the most widely used (Álvarez-Otero and De Lázaro 2018; Digan, 2019; Kerski, 2011; Sebastián and De Miguel 2017). Specifically, in the United Kingdom, Rosenshine's principles (2012) enhanced by Sherrington (2019) are widely employed, as they reinforce all the aforementioned pedagogies. These Principles are based on cognitive research into how the brain acquires and uses information, observation of classroom practice by expert teachers whose students are highly successful learners, and instructional methods based on cognitive support and scaffolding for learning, which is then consolidated by comprehension questions. They are gradually incorporated into increasingly complex tasks, and it is recommended that experiential activities are used after the basic knowledge has been learned. In this way, it becomes evident what the student is learning.

In this context, it is important to highlight the idea of curricular sustainability advocated by Professor Murga and her team, who define it as a process of incorporating sustainability criteria and values in teaching and student learning, so that it permeates all educational spheres (Murga-Menoyo and Bautista-Cerro 2019).

The main aim has been to integrate sustainable development objectives into university education by employing learning situations and/or case studies with interactive mapping and viewers, as well as geo-information relevant to the topic. This provides future geography teachers (trainees) with technological tools, such as ArcGIS Online ESRI Web GIS for learning, teaching, and making the curriculum sustainable. By using learning situations that include activities and content thematically oriented as third-person narratives interwoven with comprehension questions, the learner participates in hypothesis-testing and first-person problem-solving. Modeling accompanies formative assessment and may include simulations and predictions supported by acquired knowledge. These processes integrate teacher and learner perspectives (Chaloupka and Koppi 1998) and are a way to assess motivation, model and test problem-solving, and develop critical thinking (Bearman et al. 2016; Jeffries and Maeder 2005).

Contextual and Theoretical Backgrounds

Educating for Sustainability Using Distance Learning at University

It is worth emphasizing that educating for sustainability is not simply an 'add-on' to existing curricula, but implies a change based on the ability to respond to the current crisis, take advantage of opportunities in teaching practice, emotional aspects, and global thinking (Bautista-Cerro et al. 2019). Consequently, students can change attitudes and improve teaching effectiveness, with an affective mastery of emotions (Puertas et al. 2021). This again demonstrates geography's capacity to design and develop learning content that facilitates understanding of today's world and, in particular, the SDGs (Sustainable Development Goals) (Shannon et al. 2021; Rushton 2021). This is because the process requires students to gain competences, such as: critical analysis; systemic reflection (that is, a responsibility toward future generations to intervene in constructing a sustainable future and promote citizenship and social life); and the ability to make informed and collaborative decisions (Murga-Menoyo 2015) in the framework of interrelationships between territory, landscape, and society, both in the real and virtual world. In this way, SDG education can ensure that all students acquire the necessary theoretical and practical knowledge to promote sustainable development and adopt coherent lifestyles that focus on the whole person (understood to be a 'community of life member who is eco-dependent on the biosphere'); and simultaneously take moral responsibility for life on the planet (Murga-Menoyo 2020). Each of these elements implies societal, economic, and environmental transformation. Therefore, it is vital to include SDG-related content in curricula and to use action-oriented transformative pedagogy (Murga-Menoyo 2021; Rieckmann et al. 2017) (Table 1).

In sum, public policies to integrate Education for Sustainable Development (ESD) in all formal, non-formal, and informal learning contexts are necessary. The Spanish LOMLOE (2020) new secondary school curriculum seems to represent a route to change in non-university education systems, and therefore in teacher training. But it is also essential to integrate it into quality standards to achieve the proposed learning outcomes. There is a political will in universities to work in this direction, by integrating ESD into competencies, professional standards, certification, and teacher-education institutions' accreditation, which will encourage classroom implementation.

Table 1 Educating for sustainability using Web GIS

Teaching approach (Key aspect)	About (Doing things better)	Apply (Doing better things)	Investigate and transform (Doing things differently)
Sustainability	Educating about sustainability (theory)	Educating using concepts related to sustainability (theory)	Educating for sustainability in transformative actions (theory and practice)
Web GIS	Teaching and learning about Web GIS	Teaching and learning using Web GIS	Researching and visualizing in Web GIS to understand and transform
Responsibility (lies with)	The teacher for orientation	The teacher to enable self-autonomy	The teacher and student for autonomous learning and responsible actions

Source Based on Puertas et al. (2021), Abbott (2001), Favier (2013), and Sterling (2004)

Rosenshine's Principles Are Suitable for Integrating Web GIS and SDGs in Education for Sustainability

Rosenshine's (2012) ten Principles of Instruction represent a powerful tool to integrate SDGs and geo-information into teaching using Web GIS for better visualization, avoiding student overload, introducing content gradually, and checking that the student is learning competencies. Sherrington (2019) reformulated them as follows:

A Direct Instruction

There is a daily review to identify and consolidate ideas. Previous learning (key concepts) is retrieved and repeated, as this process reinforces what has been learned and leads to more spontaneous recall. This forms the basis for current learning and can be done by raising questions or issues.

New materials are introduced in small steps, from the basic to the most complex. Web GIS responds very well to this need, for example, by activating and hiding layers to reduce 'web map' overload. The teacher facilitates and eases initiation of the schema construction.

It is essential to formulate good questions to improve student's critical thinking in response. Revision is undertaken with questions, such as: What data are needed? The teacher should accompany students throughout their practice and give regular feedback to avoid misunderstandings. It is essential to connect new learning material with previous knowledge and practice.

B Modeling

Provide models (modeling). Learners can focus on the steps to solve a problem and need to see 'how to do to know how'. Cognitive support is provided to solve more difficult problems (based on previous tasks) and enable independence, but in this case with a role reversal: learners ask the teacher questions.

C Explore

Guide student practice. The best teachers spend a lot of time supervising their students' practice/learning with new material. In this way, they ensure that confidence is maintained and errors and misconceptions, which are essential elements of practice, are minimized. The use of feedback and stimulating questions follow the phases: I do, we do, you do. The 'I do' obliges students to follow a series of steps presented by the teacher, who explains how to accomplish them. Next, in the 'we do' phase, the teacher helps their students by providing support or scaffolding, such as directions or partially completed procedures. In the last stage, 'you do', the pupils implement the procedure or demonstrate their understanding. It is essential to fix this information in the long-term memory through tasks and rehearsals.

D Check

Learner's understanding is constantly checked via the material provided to minimize errors. Systematic monitoring aims to ensure that students' learning is accurate, and misconceptions do not become ingrained. It is especially important to check the acquired schemas or mental maps. The results obtained are contrasted with those of other students.

'Good results' will be approximately an 80% success rate in exercises, questioning…rather than it being too easy (e.g., a 95–100% success rate), so that the challenge is sufficient in terms of difficulty.

The teacher provides temporary scaffolding or support (e.g., guidelines) for difficult tasks, which decreases as students become more proficient and gain confidence in their results. Instructional support, tools, and checklists are given to help learners organize materials by establishing criteria and/or quality standards. Task-solving models are provided so that students can compare their work.

E Solving a Problem with Individual Exploration

Independent practice. Build-in practice time in and out of the classroom to automate the material learned. Students should practice regularly until their learning is consolidated so that they achieve 'automaticity' and 'expertise'. A problem-solving

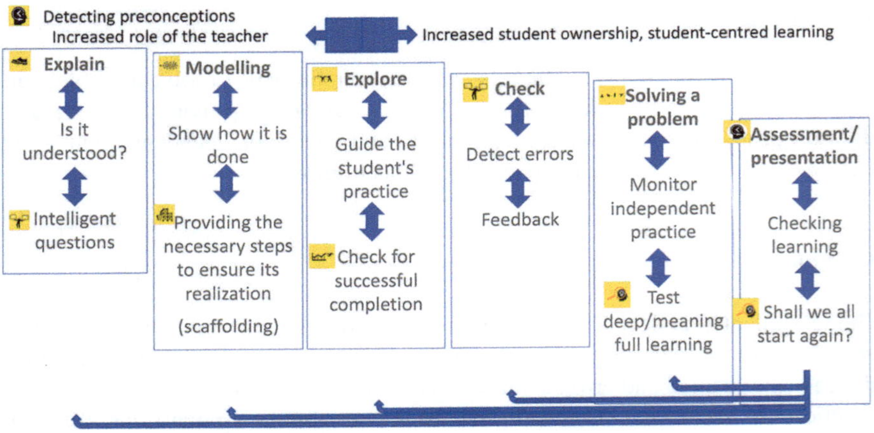

Fig. 1 Adapting Rosenshine's (2012) Principles to a teaching model (Puertas et al. 2022)

activity is proposed to establish the steps for students to work through a problem, guided by the teacher.

F Assessment/Presentation

Weekly and monthly review. Students need to practice intensively to 'automate the material', in order to 'interrupt forgetting' and link previously acquired knowledge to current learning. Results need to be evaluated and shared. This is the phase when learners can feel confident enough to initiate their own exploration and start the cycle again or review some aspects of it.

The Principles have been summarized in Fig. 1, which shows that this process is not linear and it is possible to move from one to another, and even to repeat the same one depending on the requirements of the context, which is expressed by the blue arrows at the bottom of the Fig. 1.

Methodology

Action research method in teaching the subject of Complements for Disciplinary Training in Geography using direct observation, which is one of the benefits of this method, together with the option of aiding teachers to reflect on their lessons practices. Evaluation of the learning process involving Web GIS, SDGs, and Rosenshine's Principles has been carried out using: (a) co-evaluation, using a blind peer review tool (Aropä), developed by the University of Glasgow; (b) the university teacher-trainer qualification (the official final mark); (c) self-evaluation; and (d) a satisfaction questionnaire using the five-point Likert scale. This has then been compared with outcomes from traditional teacher lesson plan preparation (year 2020–2021, $n = 56$) and the trainees' use of Web GIS by following the adapted principles of Rosenshine (year 2021–2022, $n = 58$). All statistics have been formatted and calculated using Excel.

Trainees' Tasks

Within this framework, trainees were asked to carry out a task consisting of two different parts. In the first, it was necessary to design and complete a learning task in accordance with a template provided on the virtual campus, which gave specific guidance following the Principles' steps, aided by a video-explanation illustrating what the work consisted of. In the second part, students uploaded their work to the Aropä platform, within a deadline, in order to run blind peer assessment of the 'learning situation'.

It was clarified that a 'learning situation' is not simply a textbook containing the conceptual content to be taught, but a justification of the activity design with the objectives, concepts, procedures, and attitudes to be worked on in an integrated way, and the assessment that favors the learning and acquisition of competencies. The current Spanish curriculum for secondary schools, LOMLOE, which came into force on 19/01/2021, and its subsequent developments, advises to organize curriculum for acquiring competences. These need methodological approaches of meaningful learning and active methodology to be achieved with the techniques that the future teacher university student considers most appropriate in each case. The trainee will design in detail a learning situation containing at least one prototype from geographical science, to: observe, reflect, locate, define, relate, elaborate and/or comment on graphics, to identify causes, comment on texts and images and learn more. They should provide, on the same document, the didactic resources employed (maps, graphs, texts, images, web pages…) and clearly integrate elements and evidence to educate for sustainability and foster the SDGs (i.e., activities to promote a change in attitudes in relation to prevailing assumptions). Cross-cutting elements (complementarity with other subjects in objectives and competences) and measures to address diversity should also be succinctly formulated. The time dedicated to each of the designed activities could, for example, be a maximum of 50 min, which is common in Spain. The geographical theme and focus are freely chosen by trainees, within their autonomous community's current curriculum (each region in Spain has a curriculum adapted from the National, and some are very different). An appropriate title should define case the study or learning scenario. The chosen topic (containing objectives, content, and challenges) will have SDGs integrated (goals, targets, and indicators), so that education for the SDGs will consist of competences to be acquired in a cross-cutting manner and highlight how sustainability competences acquired will be assessed.

In the second part of the work, trainees use the Aropä platform, which has proven to be very effective as an assessment tool (Coronado-Marín et al. 2018, 2019, 2020, 2022; De Lázaro et al. 2022), to upload their final project or case study. This is done anonymously in their own online place previously prepared by the university teacher. Trainee's next peer blind assess three learning situations or case studies carried out by other students or classmates. Aropä assigns a keyword to each study, since the uploading is accomplished using different tags, which will be, as their name indicates, dedicated to different aspects of geography: GeoPhysics, Population, Cities, Economics, and Others. These tags make it possible that any assignment by the platform will be different to the tag of the uploaded own work. Trainees assess other students' learning scenarios or case studies using the evaluation criteria outlined below and provide comments. This assessment is visible to other trainees on the same platform and, at this time, the satisfaction questionnaire can be completed. An evaluation of how they are assessing will go toward their final grade. In this way, trainees can work on several different topics: that chosen by themselves, which they will work on in greater depth, and those of their classmates.

The details of how to program all these aspects in the tool can be found in the video 'Aropä, online blind peer assessment. A free solution created by the University of Glasgow' (De Lázaro 2021) or in 'Using Aropä for Students' Peer Assessment' (University of Alberta 2020).

Assessment Criteria Used

To ensure a certain homogeneity in the blind peer assessment, the university teacher previously provides a series of indicators to all students, to see if they are following, or not, important work quality rules. A graduated scale following a detailed rubric would be better, but high workloads lead us to give a quick 'yes' or 'no' reply to each indicator:

- Appropriateness of the chosen title. This would represent an umbrella for the content and link to the current official curriculum.
- This provides key information on the development and implementation of the learning situation and uses at least three different assessment systems (e.g., self-assessment, co-assessment, cross-assessment, external assessment, teacher-supervised assessment, etc.).
- Contains activities with the necessary detail (e.g., up-to-date data, variety and quality of resources used).
- A coherent explanation of the discourse, clarity of ideas and argument (i.e., that it is well understood). Focus on student learning (contains elements of meaningful learning).
- Correct application of concepts learnt in the subject. The key ideas are clear and correctly defined (scientific rigor in the concepts handled, in source quotations, adaptation to the current curriculum, etc.).

- It encourages critical thinking. Raises new questions and reflections that open doors to subsequent debates and thinking beyond solutions about pros and cons of possible alternatives.
- It takes advantage of geotechnologies, especially GIS on the cloud, for learning purposes and innovation.
- Content, spelling, and grammar are correct.
- In relation to SDGs content: Has the student made any reference on how to educate for SDGs and provided concrete steps on how to achieve any of the sustainability competences? Have the following competences been addressed: a) critical analysis; b) systemic reflection; c) responsibility toward future generations; d) collaborative decision-making skills, given the learning situation. Have activities been proposed to assess progress in SDG learning? Do they invite transformative actions? (See the United Nations document (2022) for a better understanding of the SDGs and their measurement).
- Do they meet the work standards demanded in the template and use Rosenshine's Principles? Is it appropriately adapted to the proposed outline, standards, and citation of references following the latest American Psychological Association (APA) standards)?
- Has reading the project/essay been enjoyable?

Additionally, each work must include a self-assessment with the score obtained. If this is missing, it will be highlighted in the correction. Further comments should be made that summarize the strengths and weaknesses of the assessed work and an overall evaluation that includes how education for sustainability (SDG achievement) has been integrated. The assessment should be based on clear and objective data.

In this way, the learning situation or case study will be evaluated from several sources: the external assessment of the university teaching team, co-assessment from the blind peer reviewers using the Aropä platform, and self-assessment by the students themselves.

University teacher offered a number of mandatory guidelines for revision/correction of their peers' work:

- Collaborate in improvement of the text (and not in its disqualification).
- Justify the grade in a rational way. The aim is not to judge. The use of periorative adjectives is forbidden and will be penalized.
- Detect errors in content, spelling, expression, and grammar.
- Consider that there are no absolute truths, and that everything has a temporal and spatial context.
- Look at the innovative approach of the experience/tool, and whether it is possible to educate for the SDGs from the learning situation being evaluated.
- Identify the strengths and weaknesses of the learning situation and how sustainability education (SDG achievement) has been integrated into it.
- Trainees should be aware that the deadline for the first part differs from the second, and that if they do not submit on time, they will be outside of the current call, and cannot get grades. In addition, a short questionnaire on the student's work

is asked for at the end of the activity. However, in practice, this questionnaire is usually returned by around a third of the class.

With this work, trainees will demonstrate that they have learnt to:

- Teach the contents of the geography curriculum by integrating elements of educational innovation.
- Educate for the SDGs and collaborate in the sustainability of the curriculum.
- Take advantage of different resources for teaching geography and evaluate other trainees' anonymous reflections made via the Aropä tool.
- Increase secondary school students' learning.
- Assess geographic content with a variety of instruments that measure learning outcomes.

Results and Discussion

Results have been compared in Excel by contrasting official marks with evaluations made by the peers themselves and teacher evaluations using two ways of working: the first academic year without Rosenshine Principles, and the second with them (Table 2). A satisfaction questionnaire was also administered at the end of the course (Table 3) using a five-point Likert scale.

The grade given by the teacher in the final report is weighted by the average of the quality of the peer assessments. In general, trainees rate each other slightly higher than teachers, but there is a very small difference, with a maximum that some authors set at plus or minus two points. These data are congruent with those obtained by the creators of the platform (Hamer et al. 2015). However, the results of peer or teacher assessment are much higher than the self-assessment by individual students. Most trainees find it useful to develop a learning situation at the start of the Masters' degree and believe that it brings efficiency in learning geography and rubric assessment. However, following correction of the learning situations (developed by the trainees), we can affirm that while an increase in geographical competences has been noted in relation to the previous course, integration of the SDGs in most cases has been nominal, without specific SDG competencies, such as critical thinking. We can therefore say, in relation to Table 1, that the learning situations have remained in the early stages of SDG integration, without promoting transformative actions.

Direct observation in the teacher training course and frequent discussion of each result obtained by the university research group and teachers' network have validated this way of teaching, as the results have bettered the previous year, when a conventional learning plan was carried out.

Table 2 Academic results in the subject of Complements for Disciplinary Training in Geography

Academic Year	Students presented (enrolled)	Average mark (submitted)	Standard deviation of grade	Self-assessment	Aropä Peer rating	Standard deviation grades	Peer discrepancy out of 10 (Average)	Work done in Aropä integrating SDGs
2020–2021	56 (64)	8.02 (51)	1.47	-	7.98	1.61	1.25	Traditional lesson plan
2021–2022	58 (75)	8.74 (58)	0.90	7.48	8.84	0.88	1.33	Learning situation

Table 3 Responses to the questionnaire on satisfaction with the working method in Aropä (Likert scale from 1 to 5). Subject of Complements for Disciplinary Training in Geography

Academic Year	Indicators	Viewing other students' teaching units has been useful for my own learning	Feedback provided on my own work has been useful	The review has given me a better understanding of how I am progressing in this course	It has helped me improve my ability to reflect on my own learning and skills	The reviews conducted have helped me improve my analysis and critical skills	It has improved my learning about rubric assessment	I agree with the grade received	Overall, the peer review experience has been helpful
2020–2021 (17)	Arithmetic mean	4.35	4.06	4	4.06	4.12	3.71	3.88	4.12
	Mode	4	4	4	4	4	4	4	4
	Standard Deviation	0.49	0.75	0.79	0.83	0.60	0.77	0.78	0.60
2021–2022 (22)	Arithmetic mean	4.5	4.64	3.84	4.76	4.88	4.07	4.76	4.25
	Mode	5	5	5	5	5	4	5	5
	Standard Deviation	0.73	0.5	1.12	0.44	0.33	0.60	0.44	0.87

Conclusions

Education for sustainability using cloud-based Geographic Information Systems (Web GIS) in learning situations and/or case studies has proved useful in geography teacher training. Trainees have understood importance of verifying that secondary school students are learning what they are being taught and integrating Sustainable Development Goals (SDGs) in their teaching. Thus, this way of using Web GIS and viewers for curricular sustainability has been implemented at the UNED in the subject of Complements for Disciplinary Training in Geography of the Master's degree in Teacher Training for Compulsory Secondary Education and Baccalaureate, Vocational Training and Language Teaching since the academic year 2021–2022. Trainees developed their own learning situation or case studies focused on geography and SDGs using these tools and later evaluate other trainees learning situation, to conclude with an assessment of the results obtained and their responses to the questionnaires. Some of these learning situations have been included in a book published to help trainees for next academic year (De Lázaro and Puertas 2022).

This means that formative assessment has been encouraged as a fundamental part of students' future teaching. Therefore, the presentation of lesson objectives, linking to prior knowledge, feedback, modeling, scaffolding, and connecting the content to subsequent lessons, created a teaching sequence that enabled trainees to build a schema and create better lessons for their future secondary school students.

Additionally, the results show an improvement across several important competencies in education. Namely, these include information and geo-information management, as students must select reliable sources, detect basic content to define threshold concepts (those that will allow students to move to the next level), and be able to create a sustainable curriculum that leads them to foster civic and citizenship competences, as well as critical thinking.

Further, this approach shows efficient learning with the use of primary sources using Web GIS, interactive digital maps, and other ways of visualizing geo-information. The trainee teachers who have used these materials and developed new ones, have experienced the potential of cloud mapping and web maps, or web GIS, to learn about the territory and its sustainability issues in a reasoned and powerful way, and have subsequently highly valued these activities.

Thus, education for sustainability using interactive cartography and web GIS, make it possible to observe trends and interrelationships through the map layers that represent territory. These visualizations also make it possible to both broaden knowledge and promote transformative action supported by critical thinking, systemic reflection, responsibility toward future generations, and collaborative decision-making skills beyond the theoretical knowledge of the Sustainable Development Goals (Puertas et al. 2021).

Acknowledgements This research has been part of the project 'Web GIS literacy boosts sustainability competencies' (2022I/PPRO/004), and the project GI-Pedagogy, 'Innovative Pedagogies for Teaching with Geoinformation' (2019-1-UK01-KA203-061576).

References

Abbot TD (2001) An evaluation of geographic information system software and its utility in promoting the use of integrated process skills in secondary students [Tesis doctoral, University of Connecticut]. https://opencommons.uconn.edu/dissertations/AAI3034002

Álvarez-Otero J, de Lázaro Torres ML (2018) Education in sustainable development goals using the spatial data infrastructures and the TPACK model. Edu Sci 8(4):171. https://doi.org/10.3390/educsci8040171

Álvarez-Otero J (2020) El uso educativo de las Infraestructuras de Datos Espaciales (IDE) para mejorar la responsabilidad social de los ciudadanos del siglo XXI sobre el territorio. [Tesis doctoral. Universidad Complutense de Madrid]. https://eprints.ucm.es/id/eprint/64592/1/T42185.pdf

Bautista-Cerro Ruiz MJ, Murga-Menoyo MA, Novo MA (2019) La Educación Ambiental en el S. XXI. (página en construcción, disculpen las molestias). Revista de Educación Ambiental y Sostenibilidad 1(1):1103. https://doi.org/10.25267/Rev_educ_ambient_sostenibilidad.2019.v1.i1.1103

Bearman N, Jones N, André I, Cachinho HA, DeMers M (2016) The future role of GIS education in creating critical spatial thinkers. J Geogr High Educ 40(3):394–408. https://doi.org/10.1080/03098265.2016.1144729

Buzo I (2021) Aprendizaje inteligente y pensamiento espacial en Geografía [Tesis doctoral, Universidad Complutense de Madrid]. https://eprints.ucm.es/id/eprint/70029/1/T43019.pdf

Buzo-Sánchez IJ, Mínguez C, De Lázaro-Torres ML (2022) Expert perspectives on GIS use in Spanish geographic education. Int J Digit Earth 15(1):1205–1219. https://doi.org/10.1080/17538947.2022.2096131

Chaloupka M, Koppi T (1998) A vignette model for distributed teaching and learning. Res Learn Technol 6(1):41–49. https://doi.org/10.1080/0968776980060107

Coronado-Marín A, Murga-Menoyo MA (2018) Una propuesta metodológica en contextos digitales: el uso sinérgico de la 'evaluación por pares' y la 'autoevaluación' para la formación de competencias en sostenibilidad. In: 6th International Congress of Educational Sciences and Development. 21–23 de junio. Setúbal (Portugal)

Coronado-Marín A, Murga-Menoyo MA (2019) La evaluación como estrategia para la formación de competencias en sostenibilidad. Una práctica docente innovadora en contextos digitales. En X Jornadas de Investigación en Innovación Docente en la UNED. Innovación educativa en la era digital. 28 y 29 de noviembre, UNED. Madrid

Coronado-Marín A, Bautista-Cerro MJ, Murga-Menoyo MA (2020) Students and university teachers facing the curricular change for sustainability. Reporting in sustainability literacy and teaching methodologies at UNED. In: Leal Filho W, Borges de Brito P, Frankenberger F (eds) International business, trade and institutional sustainability. World Sustainability Series. Springer, Cham. https://doi.org/10.1007/978-3-030-26759-9_60

Coronado-Marín A, Murga-Menoyo MA (2022) Self-assessment in the training of competences for sustainability: virtual learning environments in Higher Education. In: 11th World Environmental Education Congress (WEEC), 14–18 March Prague Czech Republic

De Lázaro ML, Coronado A, Murga MA (2022a) Enseñanza personalizada y activa en contextos virtuales: posibilidades de la plataforma Aropä. XII Jornadas de Investigación en Innovación de la UNED. https://canal.uned.es/video/6283612e6f3c002a0742a7c2?track_id=628374656f3c002d0506cb52

De Lázaro Torres ML, De Miguel González R, Morales Yago FJ (2017) WebGIS and geospatial technologies for landscape education on personalized learning contexts. ISPRS Int J Geo Inf 6(11):350. https://doi.org/10.3390/ijgi6110350

De Lázaro Torres ML (2021) Aropä, evaluar por pares ciegos en línea. Una solución gratuita creada por la Universidad de Glasgow [Archivo de vídeo]. YouTube https://youtu.be/ZCZzpWPsH0M

De Lázaro ML, Puertas MA (Coord) (2022b) La geoinformación en la educación para la sostenibilidad integrando los principios de Rosenshine. Situaciones de aprendizaje. (Geoinformation in education for sustainability integrating Rosenshine principles. Learning situations). UNED, Madrid. https://www.une.es/la-geoinformacion-en-la-educacion-para-la-sostenib

Digan S (2019) Integrating GIS in experiential fieldwork. Geogr Educ (Online) 32:26–29. https://search.informit.org/doi/abs/10.3316/ielapa.882183728570404

Fargher M (2018) WebGIS for geography education: towards a geocapabilities approach. ISPRS Int J Geo Inf 7(3):111. https://doi.org/10.3390/ijgi7030111

Favier T (2013) Geo-informatietechnologie in het voortgezet aardrijkskundeonderwijs: Een brochure voor docenten [Tesis doctoral, Vrije Universiteit Amsterdam]. http://www.edugis.nl/lesmodules/Portaal/assets/Geo-informatietechnologie%20in%20het%20voortgezet%20onderwijs.pdf

Hamer J, Purchase H, Luxton-Reilly A, Denny P (2015) A comparison of peer and tutor feedback. Assess Eval High Educ 40(1):151–164. https://doi.org/10.1080/02602938.2014.893418

Jeffries C, Maeder DW (2005) Using vignettes to build and assess teacher understanding of instructional strategies. Prof Educ 27:17–28. https://files.eric.ed.gov/fulltext/EJ728478.pdf

Kerski JJ (2011) Sleepwalking into the future—the case for spatial analysis throughout education. In: Jekel T, Koller A, Donert K, Vogler R (eds) Learning with GI 2011. Herbert Wichmann Verlag, VDE VERLAG GMBH, Berlin/Offenbach, pp 2–11

LOMLOE Ley Orgánica 3/2020, de 29 de diciembre, por la que se modifica la Ley Orgánica 2/2006, de 3 de mayo, de Educación (BOE 30/12/2020)

Martínez-Hernández C, Yubero C, Ferreiro-Calzada E, Mendoza-de Miguel S (2021) Didactic use of GIS and Street View for Tourism Degree students: understanding commercial gentrification in large urban destinations. Investigaciones Geográficas 75:61–85. https://doi.org/10.14198/INGEO2020.MYFM

Mínguez García C (2021) Teaching tourism: urban routes design using GIS Story Map. Investigaciones Geográficas 75:25–42. https://doi.org/10.14198/INGEO2020.M

Murga-Menoyo MA (2015) Competencias para el desarrollo sostenible: las capacidades, actitudes y valores meta de la educación en el marco de la Agenda global post-2015. Foro de Educación 13(19). http://www.redalyc.org/html/4475/447544537005/

Murga-Menoyo MA, Bautista-Cerro MJ (2019) Guía PRADO. Sostenibilizar el currículo de la Educación Secundaria. UNED. https://doi.org/10.5944/catedra.eads.501120

Murga-Menoyo MA (2020) El camino hacia los ODS: conformar una ciudadanía planetaria mediante la educación. Comillas J Int Relat 191:1–11. https://doi.org/10.14422/cir.i19.y2020.001

Murga-Menoyo MA (2021) La educación en el Antropoceno. Posibilismo versus utopía. Teoría De La Educación. Revista Interuniversitaria 33(2):107–128. https://doi.org/10.14201/teri.25375

Phantuwongraj S, Chenrai P, Assawincharoenkij T (2021) Pilot study using ArcGIS Online to enhance students' learning experience in fieldwork. Geosciences 11(9):357. https://doi.org/10.3390/geosciences11090357

Pons JJ, Armendariz J, Andreu J (2019) El story-map como herramienta didáctica y divulgativa en el ámbito de la Arqueología. In: Ortega C, López-González MA, Amor P (eds) X Jornadas de Investigación en Innovación Docente: innovación educativa en la era digital. UNED, Madrid, pp 311–316. http://arcg.is/0frOXW

Puertas-Aguilar MA, Álvarez-Otero J, de Lázaro-Torres ML (2021) The challenge of teacher training in the 2030 agenda framework using geotechnologies. Edu Sci 11(8):381. https://doi.org/10.3390/educsci11080381

Puertas-Aguilar MA, Conway B, de Lázaro-Torres ML, De Miguel González R, Donert K, Lindner-Fally M, Parkinson A, Prodan D, Wilson S, Zwartjes L (2022) A teaching model to raise awareness of sustainability using geoinformation. Espacio, Tiempo y Forma. Serie VI. Geografía 15:23–42. https://doi.org/10.5944/etfvi.15.2022.33687

Rieckmann M, Mindt L, Gardiner S (2017) Educación para los Objetivos de Desarrollo Sostenible. Objetivos de aprendizaje. UNESCO

Rosenshine B (2012) Principles of instruction research-based strategies that all teachers should know. Am Educ 36(1):12–39. https://www.aft.org/sites/default/files/periodicals/Rosenshine.pdf

Rushton EAC (2021) Building teacher identity in environmental and sustainability education: the perspectives of preservice secondary school geography teachers. Sustainability 13(9):5321. MDPI AG. http://dx.doi.org/10.3390/su13095321

Sebastián López M, De Miguel González R (2017) Geographical education 2020: Iberpix and Collector for ArcGIS as teaching resources for spatial learning. Didáctica Geográfica 18:313–316. https://didacticageografica.age-geografia.es/index.php/didacticageografica/article/view/402

Sherrington T (2019) Rosenshine's principles in action. John Catt Educational, Limited

Shannon J, Hankins KB, Shelton T, Bosse AJ, Scott D, Block D, Fischer H, Eaves LE, Jung J-K, Robinson J, Solís P, Pearsall H, Rees A, Nicolas A (2021) Community geography: toward a disciplinary framework. Prog Hum Geogr 45(5):1147–1168. https://doi.org/10.1177/0309132520961468

Sterling S (2004) Higher education, sustainability and the role of systemic learning. In: Higher education and the challenge of sustainability: problematics, promise and practice. Springer, pp 49–70. https://doi.org/10.1007/0-306-48515-X_5

UN (2022) Global indicator framework for the Sustainable Development Goals and targets of the 2030 Agenda for Sustainable Development https://unstats.un.org/sdgs/indicators/Global%20Indicator%20Framework%20after%202022%20refinement_Eng.pdf

University of Alberta (2020) Using Aropä for Students' Peer Assessment. Canada. https://www.youtube.com/watch?v=z6lH4Dib9gQ

Zwartjes L, De Lázaro ML (2019) Geospatial thinking learning lines in secondary education: the GI learner project. In: Geospatial technologies in geography education. Springer, Cham, pp 41–61. https://doi.org/10.1007/978-3-030-17783-6_3

Comparative Dimensions of Teaching in Austria and in Israel: Holocaust Education Using Primary Sources

Sonja Danner, Tal Yaar-Waisel, and Israel Ben-Dor

Abstract Using primary sources, such as geographical or historical documents, "revive" teaching, encourage students' involvement in class, and thus lead to a deeper understanding of the subject. The *Shoah* of Austrian Jewry was taught both in Israeli and in Austrian teacher training institutions, and, in addition, to a not directly "involved" group of teacher training students in Ireland by using primary sources like a passport, pictures, maps, and other personal documents. All students (160 BA and MA students) were exposed to the same lesson and story by the same lecturer, on zoom. The lesson was based on documenting the personal story of Mrs. Karoline Bloch, who managed to escape from Vienna to the Land of Israel in 1938. This story is a micro-history that reflects the macro-history: The extinction of Austrian Jewry in WW2. In this Collaborative Action Research (CAR) conducted during the COVID pandemic, the main goal was to detect and explain similarities and differences between the responses of Israeli and Austrian students to the lesson. It was found that national narratives greatly impact students' responses and attitudes. At the same time, similarities were found in the necessity to discuss actual problems of human rights, immigration, and refugees.

Keywords Comparative study · *Shoah*/Holocaust · Narratives · Immigration · Collaborative Action Research · Teacher training

---

S. Danner
Private Church University College of Teacher Education KPH Wien/Krems, Wien, Austria
e-mail: sonja.danner@kphvie.ac.at

T. Yaar-Waisel (✉)
Oranim College of Education, Kiryat Tiv'on, Israel
e-mail: yaar.tal@gmail.com

I. Ben-Dor
Private researcher, Tal-El, Misgav, Israel
e-mail: bendoric@gmail.com

A. Klonari et al. (eds.), *Re-visioning Geography*, Key Challenges in Geography,
https://doi.org/10.1007/978-3-031-40747-5_9

"After every war
someone has to clean up.
Things won't
straighten themselves up, after all..."

Wislawa Shimborska, The end and the beginning[1]

The Story of Mrs. Karoline Bloch

A single document or picture sometimes becomes a "historic treasury" through which we can explore the past, in a way that leads to valuable learning. Sometimes such "diamonds" are within our reach. In 2012, Tal Yaar-Waisel was sorting out photographs and documents in her parents' home and came across the passport of her great-grandmother, Mrs. Karoline Bloch, issued in Vienna in 1938. The "*Deutsches Reich*" passport, stamped with the Nazi seal, allowed Karoline Bloch to leave Vienna just two days before *Kristallnacht*, in November 1938, saving her life and enabling her to join her daughter's family in Mandatory Palestine. Within nine months, after her husband's loss in February 1938, followed by the German annexation of Austria in March, the "*Anschluss,*" Mrs. Bloch lost everything she had, and her survival depended on her passport. Miraculously, because her children had left Austria before those tragic events, she could choose between two destinations: Palestine or Brazil, at a time when only a few people had any option at all. This personal story reflects the persecution and expulsion of Austria's Jews. Austria was united with Germany by force, and the passport is, therefore, one of Nazi Germany, with the words "The German Reich" emblazoned at the top, under the Reich Eagle of Nazi Germany, and a swastika. In red ink, the letter "J" is on the left-hand side of the passport, showing that the passport's owner was a Jew (Jude). Besides, the date 12.10.1938 appears vertically (see Fig. 1).

The passport was issued on September 29, 1938; from October 1938, Jews had to carry special passports which identified them as Jews. Mrs. Bloch received the passport just before the directive came into force; the letter "J" and the date (12.10.38) were added later.

Karoline Bloch was born in 1877. When she was 19 years old, in 1896, she married Moritz-Moshe, and the couple had three sons and a daughter: Heinrich, Rudolph, Alfred, and Gertrude. Two of the boys served in the Austrian army during World War I. After they had come back to Vienna, they couldn't find a job and all emigrated from Austria: Heinrich to the United States, Rudolph to São-Paulo, Brazil, and Alfred to England. In 1934, Gertrude and her husband, Dr. Rudolph Steinhertz, reached the conclusion that there was no future for Jews in Europe, and they immigrated to the Land of Israel, along with their daughter, the infant Zuzanna. One year later, they

---

[1] Wislawa Shimborska, *The End and the beginning*, Translated by Joanna Trzeciak Poetry foundation https://www.poetryfoundation.org/poems/52955/the-end-and-the-beginning.

Fig. 1 Evidence of the "Anschluss" in Karoline's passport (This fig. belongs to authors)

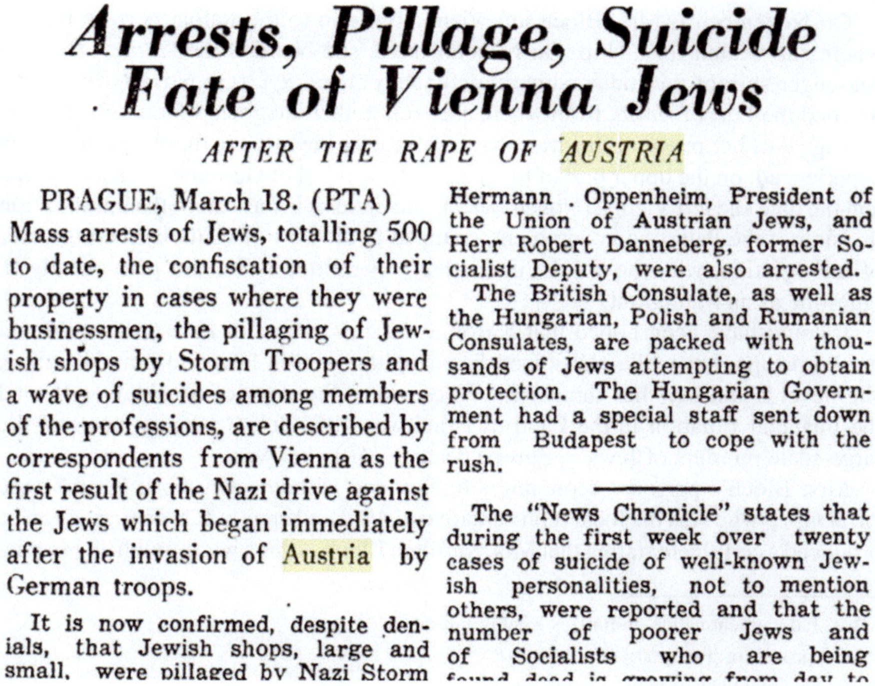

Arrests, Pillage, Suicide Fate of Vienna Jews

AFTER THE RAPE OF AUSTRIA

PRAGUE, March 18. (PTA) Mass arrests of Jews, totalling 500 to date, the confiscation of their property in cases where they were businessmen, the pillaging of Jewish shops by Storm Troopers and a wave of suicides among members of the professions, are described by correspondents from Vienna as the first result of the Nazi drive against the Jews which began immediately after the invasion of Austria by German troops.

It is now confirmed, despite denials, that Jewish shops, large and small, were pillaged by Nazi Storm

Hermann Oppenheim, President of the Union of Austrian Jews, and Herr Robert Danneberg, former Socialist Deputy, were also arrested.

The British Consulate, as well as the Hungarian, Polish and Rumanian Consulates, are packed with thousands of Jews attempting to obtain protection. The Hungarian Government had a special staff sent down from Budapest to cope with the rush.

The "News Chronicle" states that during the first week over twenty cases of suicide of well-known Jewish personalities, not to mention others, were reported and that the number of poorer Jews and of Socialists who are being

Fig. 2 The *Anschluss* in the media, *The Palestine Post*, March 22, 1938

settled in *Nesher*, a working-class neighborhood in the south of Haifa. Rudolph, a chemist by profession, established a laboratory at the "*Nesher*" cement factory.

Moritz and Karoline Bloch stayed in Vienna. On February 13, 1938, when he was about 70, Moritz passed away, and Karoline, who was then aged 61, remained alone.

One month later, just after the "*Anschluss*," Mrs. Bloch's children asked her to make haste and leave Vienna. Thanks to them, she had two good and very rare options for that time: Her son, who had immigrated to Brazil, obtained a visa for her to go to Brazil; her daughter, who had emigrated to the British Mandate in Palestine, obtained a certificate enabling her to immigrate to the Land of Israel. One of the pages in Karoline's passport states in Portuguese that her son is legally in Brazil. On October 27, she was thus able to obtain a visa for Brazil, signed by the deputy consul of Brazil in Vienna. On the next page, there is a visa for Palestine, issued by the British passport office in Vienna on October 15. This visa allowed a single journey and was valid only until the November 30, 1938. The visa's recipient was entitled to remain permanently in the land with the status of an immigrant. The visa which enabled Mrs. Bloch to immigrate to the Land of Israel was issued about five months before the third "White Book"[2]; the visa for Brazil was even rarer (Dror 2002). On September 27, 1938, the Immigration Division in Haifa informed Rudolph Steinhertz, Mrs. Bloch's son-in-law, that she had received a "type-D" certificate.[3] Mrs. Bloch was also asked to provide a medical certificate stating that she did not suffer from any serious mental or physical illness.[4]

On November 7, Mrs. Bloch left Vienna by train to the Italian port of Trieste, where she waited for a ship. On November 18, she boarded the "*Marco Polo,*" a passenger ship of the Italian shipping company *Adriatica*.[5] On November 24, she reached the Port of *Haifa*, from where she went to her daughter's family in *Nesher*.

Page 9 of her passport illustrates with simplicity the miracle that Karoline Bloch experienced: on the upper part of the page is the symbol of Germany with a swastika and the date she left Vienna (November 7), and under it is the seal of the immigration division of the Palestine government, stamped at the Port of Haifa by the inspector of immigration, with the words, in English: "permitted to remain permanently in Palestine as an immigrant."

It could have been hoped that Karoline Bloch's arrival in Palestine would end her hardships, but while she joined her daughter, son-in-law, and granddaughter, she never again saw her three sons. These were the days of World War II, and the financial situation in the Land of Israel was difficult.[6] Gradually, news of the large-scale murders of Jews in Europe became partly known.

Mrs. Bloch's personal economic situation was difficult. On April 2, 1940, the "*Mukhtar*" who was the local representative of *Nesher-Yagur*, a sub-district of Haifa, produced a certificate stating that Mrs. Karoline Bloch, "was a widow with no income,

---

[2] British document which minimizes legal immigration.

[3] An approval for immigration to Palestine.

[4] This document has been used in the lesson.

[5] *Haaretz*, of November 25, 1938.

[6] Bombings by Italian planes and fear of German occupation marked the years 1940 – 1942.

who therefore needed regular financial assistance from her son, Rudolph Bloch, in São-Paulo, Brazil." In 1940, Karoline attempted to obtain a visa to the United States where her son Heinrich lived with his family. On May 3, 1940, the American consul in Jerusalem informed Karoline that her request had been rejected and "would not be discussed again in the coming years."

Karoline Bloch died in 1943 and was buried in Haifa. Her daughter and granddaughter often noted that "it was good that she died without knowing the true scope of the disaster suffered by the Jewry of Europe."

Persecution of Jews in Austria and the Destruction of the Community

On March 12, 1938, German military units annexed Austria to Germany in the "*Anschluss.*" At that time, there were around 185,000 Jews in Austria, of whom about 170,000 were in Vienna (Rosenkrantz 1990, p. 24). Pogroms against the Jews of Vienna had begun even before the entry of German soldiers, with Germany seeking to implement in the city all the methods of persecution practiced against Jews in Germany, from 1933 onwards. Within a short time, their anti-Jewish policy became even more far-reaching and brutal than in the "*Old Reich*" (Bata 2002). Teachers and leaders of the Jewish community in Vienna were sent to the *Dachau* concentration camp. Jewish property was confiscated within the framework of accelerated "Aryanization." On August 20, 1938, leading Nazi Adolph Eichmann moved to Vienna, acting rapidly, with the enthusiastic help of friends in the Austrian Nazi party, to dispossess and deport the Jews. His staff humiliated, beat, and robbed Jews who came to arrange the necessary forms for emigration (Barley 2007, pp. 322–323). During September 1938, when the blackout was enforced in preparation for aerial bombings, Jews were pushed under the wheels of tram cars; synagogues in Vienna were attacked on Yom Kippur, and Torah scrolls, and religious books were burned in the streets. Before the outbreak of the war, more than two-thirds of Austria's Jewish population (around 128,500) had immigrated to 89 countries. The main destinations were Britain (30,850), North America (28,700), China (18,124), Central and South America (11,580), and Palestine (9,195) (Gilbert 1982). More than one-third (more than 65,000) died in the ghettos and camps in Eastern Europe. At the end of the war, about 1,000 Jews survived in Vienna, one-third of them in hiding. The rest were employed by the Gestapo to sort out the tremendous quantities of property confiscated from Jews (Rosenkrantz 1990, pp. 27–28).

Holocaust Teaching in Contemporary Contexts

The most relevant issue in Holocaust-*Shoah* teaching is getting closer to the era of "post-survivors." The survivors had tremendous importance in the creation and maintenance of Holocaust education, and we must find a solution to the future of both Holocaust memory and Holocaust education after the survivors pass away (Pearce et al. 2020).

The problem of the passing away of the survivors becomes essential for the spread and maintenance of Holocaust consciousness (Jean 2011). Holocaust education has become a major tool for the transmission of memory and numerous national institutions and transnational initiatives are working for this aim in recent decades. The negative implications of the passing away of the survivors are augmented when we consider the recent increase in Antisemitism. The new rise of anti-Semitism is mostly seen within circles of Islamic extremists or those who incite anti-Israeli feelings.

Another threat to the memory of the Holocaust is of "fake news," when truth and facts have often become under attack (Kansteiner 2017). This social reality is encouraging Holocaust denial (Pearce et al. 2020, pp. 5; 8–12; 23–24). Using social media and digital possibilities can serve Holocaust education on the one hand, but at the same time, they enable fake news of *Shoah* denial. Patterns of Holocaust remembrance as museums, films, or books existed before the digital revolution, but today, because of the abuse of social media for Holocaust denial, researchers and institutions detest the possibility of using this new media for enhancing and distributing Holocaust memory. As younger generations are used to social media, we should adapt to the digital age and use all means for this aim for the future of Holocaust memory (Kansteiner 2017).

Teaching History by Primary Sources

One of the key questions of this study was to find out how the use of primary sources can improve the teaching and learning of history. A passport, as a document to be used in specific circumstances, is loaded with many meanings because of the symbols and the texts produced by states, officials, and institutions (Jean 2011). To extract the full data and implications, the teacher should "intrigue" the students, arouse their interest and sympathy, and make them see the historical document as an enigma that is worth the efforts needed to decipher (Gulddal and Charlton 2017; Keshavarz 2018).

Through research based on a primary document, unlike the study of history through textbooks, the student acts as a historian, draws conclusions, and comes to understand history (Kobrin 1996). Working with primary sources enables students to construct their own understanding of the past. The process involves the creation of links between the source and their prior knowledge, thus turning the primary source relevant to them. After the research, students provide feedback and describe the significance of their learning (Harrison 2005).

By using *Inquiry-based teaching*, the students become critical readers and experience the first steps in historical research. In the internet era, the availability of primary sources from archives around the globe creates the possibility for everyone to study history through primary sources. Inquiry-based teaching, using primary sources, encourages students to build their own images and knowledge, and supports and develops students' "higher order thinking" skills that are critical to the survival of democratic states. This might be described as a circular process, "that involves asking meaningful questions, finding information to answer the questions, drawing conclusions based on evidence, and reflecting on possible solutions" (Woyshner 2010, p. 37). The reflection opens the way to new questions. Teachers should teach students to interpret primary sources within the context and within a repeating cycle that includes connecting their lives to the topic, wondering about it and its importance, and reflecting on what was learned and its significance (There).

The model for Inquiry-based teaching adopted in this research is Woyshner's "Stripling Model of Inquiry," named after its creator, Barbara K. Stripling, and adopted by the Library of Congress for its Teaching with Primary Sources (PTS) initiative. The model includes six stages: Connect, wonder, investigate, construct, express, and reflect.[7]

Methods

In a previous study (Ben-Dor and Yaar-Waisel 2020), it was shown that the use of primary sources, such as personal documents, may "revive" teaching, result in students' deeper involvement, and lead to a better understanding of the subject. A study unit, developed for teaching the destruction of Austrian Jewry in the *Shoah, using* authentic documents, and focused on the life story of Karoline Bloch, proved to be very fruitful. In January 2021, the positive results encouraged the Israeli researchers to expand the study and join researchers from Austria. The main aim of this research is to find and explain similarities and differences between the responses of Israeli and Austrian students learning the *Shoah* of Austrian Jewry by the above study unit. Irish students served as an experimental control group.

---

[7] The Stripling Model of Inquiry (Stripling Model of Inquiry, Emerging America, http://www.emergingamerica.org/teaching-resources/stripling-model-inquiry (Woyshner 2010: 39).

Action Research (AR)

Action Research (AR) is a model of professional development that promotes collaborative inquiry, reflection, and dialogue. Whereas Kurt Lewin (1890–1947) first coined the term, John Dewey established the use of AR in education.

There are two main types of action research: *Practical action research* and *Participatory action research*. Practical action research usually involves a small-scale research project, directed at a specific problem or issue and undertaken by individual teachers or teams (Lesha 2014, p. 380). Participatory action research is used when the research project is on a large scale.

The Action Research process enables educators to learn about their own teaching practices and monitor how new practices improve students' learning. The core of it in education is seeing educational problems best researched in the classroom, factual or virtual; it regards teachers as most suitable and capable of setting research goals, defining findings, and applying them immediately. One of the most important principles of action research is sharing ideas openly and accepting the results with mutual support. The research described in this article was conducted as Collaborative Action Research (CAR) focuses on studying a problem or issue in some classrooms, by several teachers from many schools (Lesha 2014, p. 382).

Collaborative action research has many benefits: Improving students' achievements, enhancing teachers' openness to new ideas, and improving willingness to communicate with colleagues (Lesha 2014, p. 383; Eikland 2019).

Teachers feel empowered as professionals and willing to implement the changes they recommend (Lesha 2014, p. 384; Vaughan et al. 2019, pp. 133–136). Students are revealed as a new source of data, unveiling information that teachers are normally unaware of (Vaughan et al. 2019, pp. 135–138).The partnering of teachers from universities and colleges of different countries, as was done in the collaborative action research discussed in this article, is unique. Virtual tools in general and especially in times of crisis enabled teachers to cope with the limitations of time and distance (Yaar-Waisel and O'Reilly 2022). Dr. Tal Yaar-Waisel taught the lessons on zoom to classes in Austria and in Ireland. "*Padlet*" and "*Google forms*" were used to gather material from the students. The meetings between the researchers were frequent, all conducted on zoom.

Research Questions

- What are the similarities and differences in reactions of students from different societies to the same study unit and what are the causes of these similarities and differences?
- Do primary sources get different responses from students with different backgrounds and why?

Table 1 Collaborative Action Research, database summary table

Country	Date	Number of Students	Institution and Degree
Israel	Oct. 2020–Jan 2022 (3 lessons)	72	Oranim College of Education BA, MA
Austria	May 2021 (3 lessons)	66	The University of Vienna and the KPH* College, BA, MA
Ireland	March 2021–Oct. 2021 (2 lessons)	22	Dublin City University (DCU) BA
Total	Oct. 2020–March 2022 (8 lessons)	160	

*Kirchliche Pädagogische Hochschule: Ecclesiastical College of Teacher Education

- How and in what way should we discuss moral dilemmas when teaching history and geography?
- What current issues are raised by students when they learn about the destiny of Austrian Jewry in the *Shoah* and how should we discuss them in class?

The lesson was delivered 8 times on zoom to 160 students, between October 2020 and January 2022: 22 from Ireland (two lessons), 66 from Austria (three lessons), and 72 from Israel (three lessons). All are BA and MA students: 55% (88) are MA students and 45% are BA students. All the students prepare to become teachers and educators (see Table 1).

Students study through micro-history, the personal story of Mrs. Bloch, the macro-history, and the destruction of Austrian Jewry in the Holocaust. After reflecting on their own inquiry work, the students formulate new questions that are gathered on "*Padlet*" boards.

In addition, after the lesson, each student fills in a "*Google-form*" questionnaire, which includes 11 questions. The questions invite the students to share their feelings and thoughts with the researchers. All the material is gathered anonymously; identifying details are general basic details such as the country, the course, academic degree, and general idea of the syllabus. All "*Padlet*" boards and "*Google forms*" were translated into English (from German and Hebrew) and have been analyzed.

Findings

At first, the intensity of the emotions that the classes evoked. When asked about their emotions during the study of Karoline Bloch's story, Austrian students used a wide range of adjectives with expressions such as "*interested*" or "*impressed,*" which tend to describe the rational approach, and then go on to stronger emotions: "*excited, touched, empathized, shocked, horrified….*" The terms do not stand alone, however, but are examined more closely by relating them to one's own family history. This reveals a reference to the present, which is also applied to other topics such as refugees. There is emotional identification with the suffering of the Jews, especially stimulated by the biographical work on Karoline Bloch:

In contrast, both Irish and Israeli students limit themselves to "*sad*" and "*moved*" to express their feelings, although they also felt touched and appreciated studying a real biography.

In general, it seems that the Austrian students were much more touched by the story than the Israelis. The Israelis remained mainly on the basic level of appreciating the story. Many Austrian students were shocked that such a scenario of persecution, escape, and displacement could have taken place in Vienna not so many years ago. Some imagined what they would have done if they had lived in that awful period.

> The question: "Do primary sources get different responses from students with different backgrounds and why?" aimed at the national level on which students approached Karoline Bloch's story. Here, Austrian students show a range of thoughts that can be summarized: The biography is suitable for learning from history, dealing with the victims—whereby "victims" does not only mean Jews who were killed in the concentration camps—and showing that this shocking story happened in the middle of Vienna. This builds a bridge to the present and to questions about options for action: "*Perhaps the most existential question: what if I had been in this situation myself ...? What if I found myself in a similar situation today?*"

Finally, the narrated story is connected to one's own family history.

> *I recently found such papers myself. However, these were from the Nazi side. There were many different postcards, most of which were signed by Julius Raab* (the Austrian Kanzler 1953-1961). *We also found many other documents from this time at home. So, I almost had a direct comparison. After hearing the story from another side, I tried to picture the story of my relatives.*

Israeli and Irish students, on the other hand, focus on the story and fate of Karoline Bloch and hardly relate it to the present. Some asked what should have been done to prevent the catastrophe. They expressed an obligation to act so that the catastrophe does not happen again. For some, the story provided a reason for exposing their own family story, at times the story of the perpetrators.

To encourage students to engage with the story, they were asked to address questions that arose from the narrative to Karoline Bloch.

The impression is that for the Austrian students, the story is much more than "a story." For them, the past and present are bound together in the current scenery of Austria and the city of Vienna. They are looking for answers to questions referring to anti-Semitism and racism in Austria today, asking "*How can one draw conclusions from what happened in the past regarding the dangers ahead?*" They have a criticism about the ways the Austrian state and society handle the memory of the *Shoah* and WW2 and the students would have liked to consult Mrs. Bloch about those issues if it were possible. In a way, Dr. Yaar-Waisel, as a descendant of Karoline Bloch, can relate to their dilemmas from current reality. They also want to know how they should teach the theme in schools:

> More precisely, how she is teaching the theme and exchanging experiences with how we teach in Austria and what she thinks about it. What does she think about anti-Semitism and fighting against it?

From Israeli students' point of view, the story of Karoline Bloch is one of many they were exposed to. Some of the Israeli students are Arabs who have a historical interest in the emigration story of Mrs. Bloch; they deal with the question: Why did she become a Zionist? They also discuss major current dilemmas in Israeli society: Can Israel be at the same time Jewish and democratic? Students in Ireland ask mainly questions about the suffering and destiny of Karoline and her family.

Moral and Global Dilemmas

Addressing the moral dilemmas the lesson evokes, Israeli students have broad reference to family dilemmas: Family versus life, family versus Zionism. The Irish students return to Nazi Germany and that period and focus on the value dilemmas raised by Nazism, and human moral dilemmas: "*Right and wrong, the abuse of power.*" The connection to current moral dilemmas in the Land of Israel is also mentioned. It is repeated several times by Arab Israeli students but appears only once in Austria, without using the word "Palestinians":

> The difficult history of Israel and the dispute over the West Bank are difficult to grasp. On the one hand, the Jews of the world are undoubtedly entitled to a homeland; on the other hand, the creation of this homeland has sparked a major, almost insoluble dispute in the region at first, we were impressed by the intensity of the emotions that the classes evoked.

Austrian students have a wider range of moral dilemmas, from personal ones "*During the session, I constantly felt the need to apologize. Is it appropriate if I do this? Is it my duty to feel this way?*" to dilemmas on the national level, many about "unsolved issues." It seems these questions derive not only from Mrs. Bloch's story, but also many are about Austria during WW2 and the post-war time.

What was the learning effect in relation to current global dilemmas?

The most important issues for all groups of students, nearly one-third, are discussions about the possibility to learn from history or from the past and the connection with human rights. Maybe they focus mainly on that because human rights were explicitly mentioned in the question. Some students used the expression "*human rights*" explicitly in their answers; others mention one human right indirectly, as an example. Many students from Austria and students from Israel deal with the controversial question if human beings learn from history or from the past. Israeli students feel that human rights are not protected—instead, they are violated. The Irish students focus more on the right to movement and the right to travel while the Austrian students are aware that they enjoy some rights while in other states people are deprived of them. Human rights are not observed everywhere; the main claims are for equal rights for everybody, because "*all people are equal because they are all human beings regardless of the color of their skin, religion, gender, etc.*" While half of the students believe that it is possible to learn from the events of the past, the other half represent the opinion that things stay the same and "*people don't learn from past mistakes.*"

An Israeli student calls it "*amazing that there are things that have not changed,*" while another has hope that "*one can draw conclusions from other people's experiences and learn from the results of their decision.*" More Austrian than Israeli students also see the relevance of the story to global dilemmas in ongoing discrimination which is still taking place. In Ireland, the last two points are not mentioned at all.

In all groups of students, there is considerable emphasis on the relationship between Karoline Bloch's story and migration/refugees. In Israel, students stress the dangers and the economic situation involved in the dilemma of immigration, and the motivation of people to leave their home countries is enquired. One student goes deeper and says, "*you can learn about the plight of immigrants and refugees and various minorities in the world.*" When Irish students look at migration, they focus on the point that immigrants are forced to leave their homelands due to persecution. People who don't want to stay in their home countries anymore due to the feeling of uncertainty must have the possibility to move: "*[…] right after Mrs. Bloch left, travel for Jewish citizens was blocked, which is a violation of freedom of movement.*"

Austrian students also see that there is always a reason why people leave their homelands and claim that there is a responsibility to take in refugees, especially in times of a pandemic. One student learns from the story how Karoline Bloch must have felt without belongings and without money. He/she draws the conclusion that: "*Refugees today feel similarly when they come to Austria or wherever and we should treat them accordingly, instead of antagonizing and fearing them.*"

In Austria and in Ireland, students find parallel mechanisms established by Nazi society and additional societies today; mainly Austrian and Israeli students reflect upon it in a more personal way, wondering how people should live together and what the difficulties in interpersonal relationships are. As personal insights, the Israeli students write down that "*hatred of the other leads to severe harm to the person and his environment.*" In Ireland, the banalization and the normalization of evil are extracted as the core of the story. Austrian students instead are advocating the possibility of peaceful coexistence. Many keywords such as *respect, helpfulness, esteem,* and *appreciation* are mentioned.

Only in Austria do students feel the need to talk more about the subject in general and within the family. One student thinks that the question is rather difficult and that she/he has "*to think about [it] a little more.*"

Learning Effects and New Insights

The affirmation of the fact that the students learned something important was very high, especially in Austria, as only one Austrian student denied having learned anything significant. One-third of the students give a more detailed description of what they learned in general. Especially for Austrian students, the focus is on biographical research in general and the necessity to learn from their own family history. They found out that family documents are very important to trace their own family histories. The lesson showed them that it is important to explore one's own family (hi)story by talking to grandfathers and grandmothers, exploring family documents, and visiting sights in Vienna:

> The unit showed me how important it is to deal intensively with your own family history. As soon as possible, I would like to talk more intensively with my grandparents about their experiences from the war and record them in some way.

Israeli students see the connection between past and present similarly. They also want to explore their own family history and want to make sure that such branding and exclusion won't happen, as the Israeli slogan says: *"Leolam lo od"* (in Hebrew, never again). The teacher plays an important role in understanding the relevance of the past:

> I learned that a teacher could share something personal, and that does not undermine his authority; rather, it strengthens his bond with the students.

In this context it is interesting that only students from Israel focus explicitly on the importance of Zionism in the story of the Jews:

> It is important not to forget the journeys that the Jews made, the atrocities they went through, and the strong connection to a foreign and undeveloped country that they knew only from stories [Zionism, love of the Land of Israel].

Israeli students reflected through the story on the meaning of life and the importance of building connections with other people. They think it is important to tell your own stories and to listen to the stories of others. Through these, we can learn empathy and we *"become human beings."*

Irish students, from their perspective, learned what oppression and persecution can cause and concluded that discrimination should never be tolerated. Another lesson learned from them is aware of the damage fascism causes. In Austria, students summarize the learning effect also as follows:

> It can be learned from it that slander is not beneficial, that historical learning is important to understand how people feel and that history should never be reduced to memorizing dates, facts and figures, that human rights are valued [and] respected….

One may not "label" or condemn somebody because of his/her origin, religion, etc. *"We should try to avoid such dilemmas by seeking dialogue with other people so that it will never ever happen again."*

This lesson made students see the *Shoah* from another perspective—not only from historical texts and facts but also through the perspective of a fugitive's biography. Being taught by a descendant of a survivor helps a lot. Students now realize that anti-Semitism spread beyond Europe and see how important it is to talk about it. In conclusion, they understand that society must prevent the exclusion and persecution of people, and individuals in society shouldn't let it happen.

Irish students reveal broad consensus with Austrian and Israeli students:

> The Holocaust happened in stages. Branding Jewish peoples' passports was a small act in comparison to the concentration camps, but it is important to see these steps in history and recognize them in the world today to avoid the same horrible acts occurring.

Among all the themes mentioned in the lesson, Israeli and Austrian students focused on emigration and learning from a document—the passport marked "J." They emphasized the distinction between past and present:

> Documents present and teach a whole world without words - from the passport with the marking of Jew, through the visa issued both in Brazil and in the Land of Israel to the application letters written from the Land of Israel and presenting the difficulties that were here.

Irish students, on the other hand, expressed interest in human fields like oppression, injustice, politics, and family.

Most remarkable is the connection students make between the story of Karoline and their own family stories. They don't speak about their personal connection but about the connection of their descendants. Fourteen Israeli students—some Arab—and six Irish students find personal connections but with different focal points. Jewish students link the story to the migration of their families from different places from Europe and the Arab countries and the description of departure and restart. The students from Ireland see the story in connection to the immigration trauma of the Irish people. Austrian students cannot find a connection to their own family stories, *"because none of my [their] ancestors were Jewish,"* but they speak about their personal (emotional) involvement in being ashamed: "*As an Austrian, my past, like my family history, is linked to National Socialism and as a person who lives in Austria, one should also deal with one's past.*"

Israelis and Irish students draw the conclusion that we should sympathize with human suffering, no matter when or who suffered. Austrian students, instead, express a clear distinction between the "Holocaust of European Jews in World War II" and any other suffering, even if it was in World War II, and included unwanted persecution and immigration.

For the students, teaching by using a life story was a novelty, because they didn't see this way of teaching before. Most of the answers express the value of family stories and students' liking of the biographical approach. *"I realized how important family history is."* The life story appeals to emotions, while history in general turns to logic. They mention some facts of the biography they heard as new to them. The fact that the family members never met again, because of their flights to different countries, is mentioned by Irish and Austrian students. Listening to an individual and "*concrete family story*" of an escape is also very meaningful for Austrian students. Two of them are also excited by the fact that it was a different story, *"another fate of a Jewish family at the time of National Socialism*" that ended with salvation and not with deportation to a concentration camp and extermination. The passport itself and its specifics are described as an innovation, allowing insight into a Jew's life at the time. Further, students get to know a Jewish Viennese woman's life in the twentieth century, the procedure of leaving Austria at that time, escape routes, etc.

But besides this, the acquisition of historical knowledge is mentioned by Israeli and Austrian students, knowledge that came along with the story. Israeli students focus on the history of Great Britain, especially "*the struggle between Italy and Britain.*" One Austrian student learned about colonial history: "*I was not aware that*

Palestine was still a British colony at that time." In Ireland, one student is taken by the fact that some people were able to predict that Hitler had planned to persecute Jewish people in Europe.

The way the lesson was taught was especially new for Austrian students. They like the teaching of a professor from a foreign university,[8] which they find moving and educational, especially because the teacher is involved, as it is in this case. Some of the students reflect also on a personal level. The story also opens an exciting *"link between the knowledge [he/she] acquired in school and the knowledge [he/she] can connect here with a real story.*"

Students' personal reflections go further and take into consideration also their individual roles as teachers and their search for ways of teaching the subject of the Holocaust. Students in Israel, training for teaching, learned that sharing personal experiences and feelings by teachers can strengthen the connection with their students and provide better learning. The Austrian students became more interested, and the impact was to read more about the subject. They wanted to know more about how to teach, what could be told to children, and what should wait until an older age. They wanted to be prepared well for that task and thought that *Shoah's* education should start as soon as possible, in elementary schools. But students also learned something for their lives and for the future; they appreciate their own family stories and want to explore them. The Irish students see the *"importance of memory and history in current life"* and feel the urge to be vigilant for any kind of populism which often has a binary logic that makes one side "pure" and the other "evil." In Austria, a student reflected on how precious it is to spend one's life in freedom together with family and friends every day and be able to choose the way of life. One should enjoy each moment.

It is evident that all the students involved in this experiment learned a lot from the lesson about Karoline Bloch, taught with the help of personal documents. Influenced by narratives of their own countries, Israeli students focus mainly on the story of Zionism and Great Britain in connection to Israel, while students from Ireland realize the problem of freedom of movement for Jews. Austrian students are touched by the fact that the story of Mrs. Bloch took place in their hometown and that they are mainly descendants of the perpetrators and must cope with the heavy burden of the past. The narrative "never again" is very present for them as well.

Narratives and Their Influences on Students' Responses

The findings obtained in this study were examined in parallel with the relevant national narratives. Our basic assumption was that students' responses are based on systems of beliefs, images, and values they absorbed from their national narratives. Education systems in different countries make efforts to develop and pass on national narratives that contain values and justifications for the existence of the national state.

[8] Normally only professors at their own university are teaching.

The national narrative interprets history, selecting facts and events with the aim of preserving the nation. These narratives are reflected in many aspects of local culture, starting from kindergarten to all stages of life, and become a crucial part of social consensus (Anderson 1991).

The Irish Narrative

Irish students from Dublin City University were not a directly "involved" group when they were exposed to the same story and lesson. This is more significant, as this group has no direct connection with the fate of Vienna Jewry before World War II.

> Additionally, the story opened my eyes to a different side to what I originally thought about Jewish migratory efforts during the 1938-1941 period.
>
> I learned a lot more about World War II and the way in which the different religions were treated. I also learned the importance of family and love.

It has been presupposed that when Irish students heard the story, they would relate to it based on the Irish narrative (Brady 2012) in which civil rights are very significant, and immigration is a survival option. Issues like religious uniqueness in one's surroundings, return to culture, language, and names, the renewal of historical narrative, and valuing of independence—are not to be taken for granted.[9]

Irish students dock into their country's history when confronted with Karoline Bloch's family history. Many parallels are drawn here. First, the dead from their own families are named—in the struggle for independence. *"There are similarities to many Irish family histories, as many Irish people have family members who died during the struggle for independence."* However, that is not the only connection they make. The ethno-religious conflict that had been simmering for centuries and the Penal Laws enacted in 1700, which discriminated against the Catholic population and excluded them from public office and elections, are also associated with Karoline Bloch's story. *"I felt upset at the story, the fact that Mrs. Bloch had to get her passport stamped with a J from the Nazis because of her religion, was outrageous."* The issue of migration also plays a major role in Ireland, as the country had to cope with a large wave of emigration—caused by impoverishment and famine—before becoming a multicultural immigration country itself. *"I'm sure many people of many nationalities relate to essential/forced migration due to religious or ethnic reasons."* The conflict caused by the division of the island (Northern Ireland conflict) and currently Brexit does not allow the population to rest and raise questions about border demarcation or border crossing. Freedom of travel has become a highly valued commodity.

---

[9] We thank Prof. O'Reilly for his contribution on the subject. Please see acknowledgement at the beginning of the article.

> People should have the right to travel where they want to, no matter what religion they are. And if they do not want to be in a country or do not feel safe somewhere, they should have the right to move.

These quotes strongly show the connection to the Irish narrative and history, the human vision, and the connection to human suffering; they attest to students' ability to connect the past with the present. The distance from the story itself did not alienate the Irish students from emotional involvement as part of "humanity."

The Israeli Narrative

The Israeli narrative (Shindler 2012) as it probably influenced the students' comments is very complicated since Israeli students represent different national affiliations within Israeli society: Non-Orthodox Jews, religious Jews, and Arabs—mainly Muslims but also Christians and Druze, all studying together in one teacher training program. The "Israeli Narrative" might represent mainly non-Orthodox Jews, who are the majority, with the most important influence on the national narrative:

Jews should be in their homeland—in the Land of Israel. Jews cannot feel safe anywhere else in the world where they would feel as if "in exile." Sooner or later, they will suffer badly in exile. The only way to prevent the Holocaust from happening again is to maintain a strong, independent Jewish state in the land of Israel.

> I learned that we must not forget the journeys that the Jews made, the atrocities they went through, and the strong connection to a foreign and undeveloped country that they knew only from stories.

Jews in Europe suffered from anti-Semitism and persecution and did not always understand the outcomes. Maybe it's not much different from what's happening today. *"Jews lived in a completely routine way and did not necessarily feel the future to happen."*

All peoples of the world "stood aside" and did nothing to prevent the Holocaust of the Jews; not only the ones who were active in the extermination camps are to blame.

It is important not to stand aside and let events happening in the world pass by.

> Everyone who immigrated to the Land of Israel before World War II made a smart move because he acted as if he could predict the danger. At the same time, this option was not open to many because the other countries did not want refugees and there was a small number of immigrant permits known as 'certificates' [given by] the British Mandate in the Land of Israel.

Even today, as the world progresses and modernizes, it is very difficult to stop countries that commit crimes against their people. The "survival" mode is when you do what you can to be saved and escape wherever possible. Israeli students did not respond like others who asked: *"Why did the brothers choose to separate?"* as it is understandable one is doing what he can to survive.

> It is important to learn that the Jews had few opportunities and that those who managed to emigrate and flee Austria were in fact brave and adventurous. It is important to learn to be alert to the environment and follow one's gut feelings.

The *Shoah* of the Jews in Europe during WW2 cannot be compared to any suffering of other people (in evilness, in scope, in dimensions, in cruelty, in the organization and in systematizing, in the indoctrination of the masses, in the participation of the peoples and in the wide geographical space). The parallel word in Hebrew to the word *Holocaust* is *Shoah.*

Involuntary immigration of people causes human suffering. Similarities can be found between different family stories. From the earliest days of Zionism, there was a rift in many Jewish families, when younger people emigrated and left the older generation behind. Sometimes it went along with leaving behind the religious way of life. Other families didn't see emigration as their choice.

Those who did not perish in the Holocaust were very lucky, so even Jewish students were surprised that this story does not have a "*happy end*" …

One must learn about the Holocaust to learn lessons from it. The historical memory of the Jewish people is of great importance. Only studying the past will guarantee that a *Shoah* will not happen again in the future—even if there are those who doubt that you can learn from history. "*It is impossible to learn too much from the past.*"

Israeli narratives appear in students' words clearly: *You need to be strong; the Jewish state and Jewish suffering are important.*

The Austrian Narrative

Austrian students live in the area where the *Shoah* took place. As early as 1993, Tony Judt found that in practically all countries in which National Socialism had influence, the narrative is shaped by two major narratives: the victim myth and one's own resistance (Judt 1993, pp. 87–120). In Austria, this image was shaken when, in 1986, the discussion about Kurt Waldheim's candidacy for the post of Federal President raised questions about his involvement in war crimes. Suddenly, guilt and responsibility are the focus, which is dealt with in emotional debates. Linked to this is a new culture of remembrance that—influenced by political discourses—produces new places of remembrance. On the recommendation of the ministry, schoolchildren are brought in masse to *Mauthausen* to be taught about the *Shoah* there. Who remembers how, when, and where? are questions that are visibly negotiated through monuments and other visible sights in public space. In large parts of Austria, the victims of the National Socialist regime are commemorated next to those of the fallen soldiers of the German *Wehrmacht*. As early as the late 1940s, there was less talk of perpetrators in political and public discourse. The topic is pushed more and more into the background, as more people with National Socialist sentiments hold public and political offices again.

> I asked how she found Austria's way of dealing with its past. I asked this question because her family was affected by National Socialism and Austria has often denied its past.

Interest in coming to terms with complicity is falling in favor of a narrative that focuses on the identity-forming features of Austria's new democracy. The national narrative emphasizes resistance to German occupation and the efforts to build a democratic state which is committed to neutrality, human values, and international cooperation (Federal Press Service Austria 1990, pp. 23–36). This also included the fact that the resistance of Austrians was claimed by the two major parties (Christian Social and Social Democratic) and the Communist Party in different ways.

"By 1949 at the latest, the split between a foreign policy and a domestic politically functionalized narrative became clear: In relation to the Allies, the women and men of the resistance were the ones who embodied actual Austria; compared to the voters, on the other hand, those who returned home and those who were killed in the bombing were the ones who represented Austria." (Pelinka 2007, p. 15) From 1955, however, the resistance issue became a niche topic for a small group. According to the Austrian political scientist Anton Pelinka, the two groups that shaped the resistance significantly are the communists and the monarchists. Neither played a significant role in the post-war period and so the narrative of the resistance is still shaped by the major parties today (Pelinka 2007, p. 18).

As many perpetrators didn't want to talk about the NS Regime and their active participation in WW2 within their families till now this period remains hushed up by a big part of the Austrian population.

> In my environment, the Holocaust seems like something that is not entirely real. To hear about it from such a directly affected person is eye-opening for me.
>
> There should be more people who deal so intensely with the past of their ancestors.

Students identify talking about the *Shoah* and WW2 as important in order to become aware of the importance of tolerance and openness. "*I am reminded once again how important it is to talk about this time. It becomes clear to me once again how important it is to be tolerant and open.*"

Today, on the one hand, the memory of the *Shoah* is characterized by the official slogan: "Never again!" This is intended to draw attention to the fact that preventive measures must be taken so that genocides are a thing of the past. "*Today we have to deal differently with people who are excluded from society.*"

On the other hand, Austria was downgraded to an electoral democracy and plummeted 14 places in the Reporters Without Borders ranking and is now ranked 31st on the Press Freedom Index.

> One always wonders how something like that could have happened. That so many people took part. But on the other hand, something like that can happen very quickly if people don't stand up and precisely point out injustice and racism.

The "never again" degenerates into mere lip service on the part of many politicians, in that they specifically stand against taking in more refugees and give the wrong impression that Austria has taken in a larger number of people, while the facts are different. It can be felt in the answers of the students. "*I learned that we do not have to remain silent.*"

There is tension between ban and discrimination "on the basis of color, race or ethnic origin" and, at the same time, the desire for a stricter immigration policy. We could also speak of a tension between the assessment of "multiculturalism" and unlimited immigration. "*[It] Reminds of today's refugee problems (no state wants to accept refugees)*" and today we are in "*Post multiculturalism era.*"

The debate about religious affiliation and religious freedom is also embedded in the refugee discussion. Islam, to which many of the refugees adhere, is often portrayed as a threat. Students reflect on that and question the place of religion in life. They legitimize the existence of different religions, within the secular state. *"Having a different religion doesn't mean being a bad person."*

> Why do people have to be treated differently because of their religion?

The answers of Austrian students show that in our study young people seem to be more influenced by today's official narrative they absorbed in school than by older narratives.

Summary

As this study's findings demonstrated, familiarity with the same story and learning the same lesson led to different responses regarding feelings, thoughts, and reactions.

Students' different narratives and responses to the same lesson led the research team to look for similarities and differences between them, thus answering the question: Is teaching based on primary sources perceived differently by students from other countries? This research indicates differences and similarities between Austrian and Israeli perceptions of the lesson taught in both countries.

Irish and Israeli students who were emotionally touched or sad focused mainly on the biography of Mrs. Bloch. Austrian students used stronger expressions for their feelings, such as shocked, horrified, etc.

Israeli students see this story as "one more" history lesson; they don't see this as a special event or unique family story, while Austrian students find it special.

As the story is related to Vienna, the place where the students live, Austrian students connect the story of Mrs. Bloch closely with their present. They feel that it has something to do with their lives and with the dilemmas they experience today, such as growing anti-Semitism in Austria and Europe. For the students from Vienna, the fact that the events took place in thier city, was the most important detail but for the Israelis, it was not important at all—could have been in any other European city. Therefore, in Austria students think that for descendants of Austrian Jews, it must be very emotional to visit or return to this city and they were interested to hear about it from the storyteller.

Austrians are also interested in the apartment left behind in Vienna. Israelis hardly pay any attention to that. All Austrian Jewish families have similar stories about property that was left behind.

Moreover, Austrian students see this story as a Jewish story and therefore cannot see a close connection to their own family history. What is interesting here is that apparently Judaism and Christianity are related and compared to each other, and the conclusion is drawn that there is no connection between them. The idea that it could possibly be a perpetrator-victim story, i.e., the involvement of one's own family in the atrocities against the Jews, is completely left out.

For the Israelis, it was a story about immigration.

All the students acknowledge the importance of talking about the past and exploring their family stories. While in Israel students are used to dealing with their family's stories already in their schooldays in Austria it is still a taboo for many young people to ask their ancestors questions about their family history in WW2 and (grand)parents didn't talk about their experiences. On the side of the perpetrators as well as on the side of the victims, the silence was kept about the time of the Second World War. Some did not tell anything because they did not want to burden their descendants with the trauma of what they had experienced, and others kept silent because they wanted to leave their complicity behind and would have preferred to see themselves as victims of the war.

But hearing and talking about the stories is more than only getting new information. It can also contribute to a change of perspective.

Many family members who could be interviewed about the WW2 period are no longer alive today. Nevertheless, the students recognize the importance of dealing with their own family history. They conclude that they need to talk to previous generations in their family about the past as long as it is still possible.

Israeli students also mentioned Zionism which was an important movement for the establishment of Jewish state. Israeli and Irish students found similar themes between this story and the story of their ancestors. This is not the case with the Austrian students, who find themselves in a conflict between their values today and the history of Nazi Austria in WW2. All students learned much about the suffering of refugees and relate it to current international problems of war and displacement. Some historical facts were new for many of them.

In addition to personal interest, students also have a professional interest and are interested in how to address the *Shoah* topic well in the classroom. Some of them express insights into their role as teachers.

It is new for them that a teacher tells his/her personal story, and they are amazed by the teacher's readiness to tell and share a private family story. Some of the students were shy and afraid to offend a family member with a question, in interviews.

A teacher's choice to share his or her personal story is not taken for granted. Students greatly appreciate this personal approach as a contribution to teaching and it strengthens the relationship between teacher and students. *"A teacher can share something personal, and that only strengthens his/her connection with the students."*

Telling a personal story makes it come alive in the classroom and makes it seem even more believable. *"Stories can have a huge impact on students; it is learning history through concrete people and fates. Learning through empathy."*

In addition, the students are amazed at the effect the use of an artifact—in this case, a passport—can have. For them, this is an extraordinary method, as is the personal approach of the teacher, both of which can have a great impact and students have discovered both as a new way of teaching.

> It is unbelievable that an "everyday object" like a passport makes tracking VISIBLE in a vivid way. I am impressed that artifacts have the power to evoke moods and create history in the mind and they can help to understand family histories. What I find most important is that personal engagement and storytelling make it possible for us to understand parts of history and our own lives and to learn from them. It is wonderful that these symbols/stories are being used in the classroom and for generations to come so that they can reduce stigmatization and make history learning individually meaningful.
>
> In school, only facts and superficial knowledge were presented. I got an inside perspective that I would never have gotten otherwise.

Human values, humanism, and human rights are common denominators of all students. All were educated in free democracies and appreciated it highly.

Epilogue

The researchers were amazed by the level of interest and cooperation shown by the students from the three countries. Humanistic approaches were explicit in the urge to discuss actual problems of human rights, immigration, and refugees. It was proved that *Shoah* is not only a "chapter in history" but is also gaining meaning for the younger generation.

As teachers and researchers coping with the challenge of Holocaust education, the writers believe that the current digital age must be embraced. The change becomes even more acute especially because witnesses to the *Shoah* disappear (Kantsteiner 2017, pp. 331–336). New ways of teaching the memory and the moral impact of the *Shoah* must be adopted by future generations. Our CAR is a modest step in this direction.

Further research could focus on the development of teaching materials by teaching students. In addition, the teaching of the story can be extended to other places in the world: North and South America, and other European countries. Holocaust education should also focus on bringing students together to learn from the history of their countries. It is important that exclusion, tracking, racism, etc. in relation to Jews did not happen only in Germany and Austria during the Nazi era. It is a worldwide phenomenon and students should be aware of that and know how widespread anti-Semitism is and how they can fight it together.

The writers believe this research will encourage teaching the *Shoah* in high schools, colleges, and universities, and more explicitly in teacher training education.

Acknowledgements We would like to thank **Mr. Avi Blumenfeld** who kindly started this journey with us, made this research commence, and let us teach these lessons in his classes at KPH College, Vienna.

We are grateful to **Prof. Gerry O'Reilly** from Dublin City University, who let us teach this lesson to his students and allowed us to enrich the study by encountering an Irish point of view.

References

Anderson B (1991) Imagined communities. Verso, London/New York

Barley M (2007) The third Reich: a new history. Or Yehuda, Zmora Beitan/Yavneh (Hebrew)

Bata M (2002) The Re-Establishment of a Jewish Community as the Archetype of the Judenrat. In: Zilberklong D (ed) Yad Vashem, A collection of research studies, pp 359–370 (Hebrew)

Ben Dor I, Yaar-Waisel T (2020) Passport for life: teaching using primary sources. GI Forum 8(2):29–44

Brady C (2012) Ireland—In the shadow of the fond abuser. In: Furtado P (ed) Histories of nations—how their identities were forged. Thames & Hudson, pp 0–81

Dror G (2002) The Catholic Elite in Brazil and the Attitude towards the Jews, 1933–1939. Yad Va-Shem Studies 30:187–219

Eikeland O (2019) Action research-applied research, intervention research, collaborative research, practitioner research or praxis research? Int J Action Res 8(I):9–44

Fast K, Jansson A, Lindell J, Bengtsson LR, Tesfahuney M (2018) Geomedia studies. Routledge

Federal Press Service (1990) Austria facts and figures. Wien

Gilbert M (1982) The Macmillan atlas of the holocaust. Scribner

Gulddal J, Charlton P (2017) Passports. On the political and cultural impact of modern movement control. Symploke 25(1–2):9–23

Harrison L (2005) Political research. The Open University (Hebrew)

Hecht DJ, Lappin-Eppel E, Raggam-Blesh M (2018) Topography der Shoah zweite auflage. Gedachtnidorte des zerstorten judischen Wien, Mandelbau (German)

Jean Y (2011) Construction of a catastrophe: Jews and travel documents in Europe between the two world wars. In: Berg N (ed) Konstellation Uber Geschichte, Erfahrung und Erkenntnis, Vandenhoeck & Ruprecht, pp 231–249

Judt T (1993) Die Vergangenheit ist ein anderes Land. Politische Mythen im Nachkriegseuropa. Transit 6:87–120 (German)

Kansteiner W (2017) Transnational Holocaust memory, digital culture and the end of reception studies. In: Sindbæk Andersen T, Törnquist-Plewa B (eds) European studies, vol. 34. The twentieth century in European memory. Brill, pp. 5–343

Keshavarz M (2018) The design politics of the passport—Materiality, immobility, and dissent. Bloomsbury, Visual arts.

Kobrin D (1996) Beyond the textbook: Teaching history using documents and primary sources. Heinemann.

Lesha J (2014) Action research. Educ Eur Sci J 10(13):379–386

Library of Congress. Teaching with primary sources. http://www.loc.gov.

Mareste J (2020) Commemorative culture and approaches to remembrance in Israel; Museums and memorials of the Shoah. Friedrich Ebert Stifung

Naveh E, Yogev E (2002) Histories towards a dialogue with the Israeli past. Bavel (Hebrew)

Ofer D (2013) We Israelis remember, but how? The memory of the Holocaust and the Israeli experience. Israel Stud 18(2):70–85

Pearce A, Foster S, Pettigrew A (eds) (2020) Holocaust education, contemporary challenges, and controversies. UCL Press

Pelinka A (2007) Der österreichische Widerstad im Widerspruch der verschiedenen Narrative 1. In: Dokumentationsarchiv des österreichischen Widerstandes (Hrsg.). Wien, p 15 (German)
Rosenkrantz H (1990) Austria. The Encyclopedia of the Holocaust, A, Yad Va-Shem / Ha-Poalim library, pp 24–28 (Hebrew)
Shindler C (2012) Israel—The Zionist experiment. In: Furtado P (ed) Histories of nations—How their identities were forged. Thames & Hudson, pp 294–303
Vaughan M et al (2019) Positioning teachers as researchers: Lessons in empowerment, change and growth. Florida J Educ Res 57(2):133–139
Woyshner C (2010) Inquiry teaching with primary source documents: An iterative approach. Soc Stud Res Pract 5(3):36–45. http://www.socstrp.org/issues/PDF/5/3/2/pdf
Yaar-Waisel T, O'Reilly G (2022) Small but Smart: international shared virtual class link-up during the pandemic 2020–2021—third-level students in Ireland and Israel. Eur J Geogr 13(5):58–74. https://doi.org/10.48088/ejg.t.yaa.13.5.58.74

Geographic Education and Spatial Citizenship: Collaborative Mapping for Learning the Local Environment in a Global Context

María Sebastián López, Ondrej Kratochvíl, and Rafael De Miguel González

Abstract This work recapitulates the foundations and didactic principles of geography teaching by means of resources based on Geographic Information Technologies and geoinformation, but specifically in the examples of collaborative cartography at different scales, as these present the necessary actions to respond to the need to improve digital competencies and increase the adoption of digital education methodologies. Through its implementation in the Geography Teacher Training Master's Program for Secondary Education, but also in the schools involved in the internships, the digital literacy of teachers and students has been developed in a comprehensive manner through the use of collaborative mapping. Digital literacy is implemented by the use of Geographic Information Technologies (GIT), in such a way that five competence skills are developed: (i) instrumental; (ii) cognitive-intellectual; (iii) socio-communicational; (iv) axiological; and (v) emotional. The ultimate goal of this pedagogical approach is the acquisition of (digital) spatial citizenship capable of facing current challenges, and promoting local changes, with the purpose of making global impacts.

Keywords TPACK · Geographical competences · Digital competences · Geoinformation · Democratic participation · Spatial citizenship

M. Sebastián López (✉) · O. Kratochvíl · R. De Miguel González
Faculty of Education (Geography Education), University of Zaragoza, Zaragoza, Spain
e-mail: msebas@unizar.es

O. Kratochvíl
e-mail: onkra@unizar.es

R. De Miguel González
e-mail: rafaelmg@unizar.es

A. Klonari et al. (eds.), *Re-visioning Geography*, Key Challenges in Geography,
https://doi.org/10.1007/978-3-031-40747-5_10

Introduction

During the last two years, the measures taken to combat COVID-19 helped us to verify that there is a significant gap in digital literacy issues in the countries of the European Union. Education and training systems around the world were affected and stressed like never before by this pandemic, which has radically changed the way we learn, teach, communicate and collaborate in our education and training communities and between them. This fact had repercussions on students, their families, teachers, trainers, school management and society in general. It is promoting a transformation in education, activating the use of online information and forcing us to reflect on teaching tasks from a completely new perspective. And also we discovered new opportunities in this process: tools and methodologies that can improve learning and motivation or new directions of advance in our field of research.

Therefore, at an educational level, we can talk about the role of the pandemic as an accelerator of trends, with changes in the contents and in the way of doing science and teaching. During these two years, we have attended to an effort in collaboration and a truly unprecedented volume of scientific-educational information exchange. Not only the most immediate obstacle such as the mobility restrictions imposed by the confinement was possible to overcome by online collaboration tools, but also the limitations in resources and capacities inherent to small groups of educators. In this context, geography has acquired special relevance since the pandemic has caused an increase in the demand for geoinformation—statistical and cartographic—by society, registered both in access to specific web portals and in monitoring and interactions of socio-geographic profiles in social networks. All of this is accompanied by an encouragement of spatial citizenship, at the level of both the country and the European Union.

In such circumstances, geography and in particular geographic education are essential in the current paradigm of global change and sustainable development, since they are the disciplines that provide us with knowledge about society and the territory at all educational levels. In addition, in this area, the technologies (particularly geospatial technologies, Geographic Information System (GIS), digital atlases, remote sensing, geolocation-based mobile apps, etc.) have had great impact on innovative learning and educational benefits: didactic resources, pedagogies, inquiry, problem-based learning, learning and service, etc.

In this way, we can affirm that geography is an essential subject for the consolidation of spatial citizenship and digital literacy, as indicated by several recent studies (Georgeson and Maslin 2018; Liverman 2018; Nightingale 2018; Sultana 2018; Fu 2020). At the same time, the transversal approach to understand and combat this crisis requires good learning of school geography for the sake of digital geographic education that advocates change, as shown by both previous international works (De Miguel González and Sebastián-López 2022; Sebastián-López and De Miguel-González 2020) as examples of good practices in European digital geographic education (Sprenger and Nienaber 2018; De Lázaro y Torres et al. 2020; Jeronen 2020), but

Table 1 Integrated framework of geographical competences for international curriculum and assessment (De Miguel 2021a)

geographical knowledge	⟷	knowing/reasoning	⟷	geographical thinking/knowledge	⟷	critical thinking
geographical skills	⟷	applying	⟷	spatial thinking	⟷	analytical thinking
geographical practices	⟷	practicing	⟷	spatial citizenship	⟷	lateral thinking

especially for the setting of an integrated framework of geographical competences for international curriculum and assessment (De Miguel 2021a) (Table 1).

However, it should be noted that, even though this process of democratization and use of geospatial technologies and open data have increased exponentially in the last two years during the pandemic, we have also been able to verify that there is a significant gap in matters of digital literacy in the countries of the European Union. Spain has also reopened the debate on educational renewal and the promotion of critical and democratic thinking through the efficient use of Information and Communication Technologies (ICT) and open data. This fact, at the same time, caused an increase in the demand for geoinformation—both statistical and cartographic—by society, registered in the number of accesses to specific web portals, as well as in the monitoring and interactions of socio-geographical profiles in social networks. Nevertheless, are teachers capable of critical and reflexive use of this information? Are geographic educators prepared for this change?

Methodology

In order to answer these questions, a survey was designed whose objective was to measure the degree of digital skills of geography teachers and their inclusion in their teaching praxis. This survey was sent to different middle and high schools—chartered and public schools—in Aragon (Spain). The sample size was $n = 136$ teachers (with 96.3% specializing in geography and history and 3.7% in language and social studies). This sample is not homogeneous in terms of gender or age. In the case of the perception of their digital skills as citizens and teachers (Fig. 1), the obtained results were very high, since 88.2% of those surveyed considered that they had plenty of digital skills and that they could promote these skills in his students (80.9% of them answered in the affirmative way with arguments).

The measured aspects were (Mishra and Koehler 2006; Kampylis et al. 2015; Gómez 2016; Drummond and Sweeney 2017; Mishra 2019; Hidalgo Cajo and Gisbert Cervera 2020): (i) degree of knowledge of the European and national reference

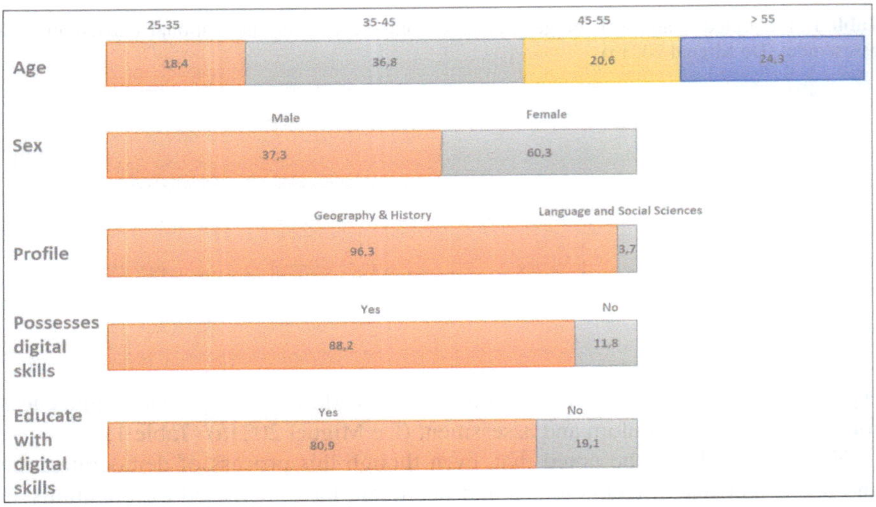

Fig. 1 General data of the respondents and their perception of the degree of their digital skills as a citizen and adequacy of their teaching practice to promote them. *Source* Own elaboration

frameworks to carry out the teaching project; (ii) use of reference standards to assess digital and geography-specific skills (Fig. 2); (iii) use of the methodology that serves to acquire the geographic skills through the efficient use of GIS and geoinformation (Table 2).

Later, a series of didactic proposals were designed and implemented in different educational contexts. These proposals were mostly based on collaborative mapping, since it is an active learning approach that: (i) fosters the multi-causal explanation of spatial distributions and phenomena through an inquiry-based learning (Marrón Gaite 2011); (ii) increases the understanding of the spatial information and promotes the formulation of a working method and the subsequent presentation of geographic information (Souto González 1998); (iii) helps to put into practice the interdisciplinarity of knowledge, not only to geographic education but also to education for citizenship or connectivism (De Lázaro y Torres et al. 2017).

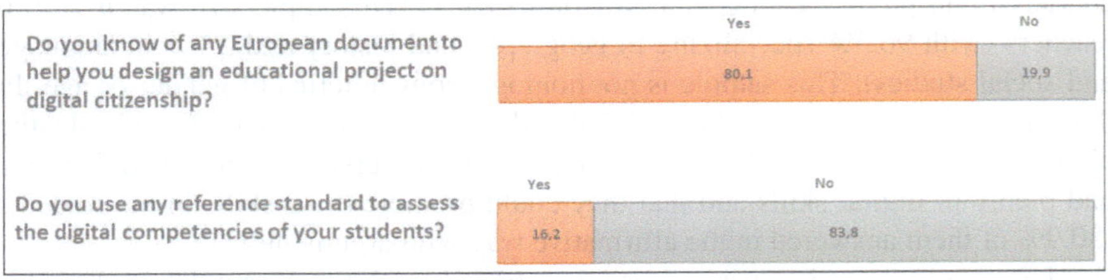

Fig. 2 Teachers' perception of knowledge of the European reference frameworks to address their teaching project and assess their students according to standards. *Source* Own elaboration

Table 2 Example of the most common responses on how teachers consider that the acquisition of specific and digital skills works in their subject

Use of the computer and the mobile phone to carry out tasks, search for information, prepare documents…
Working with different programs, working with classroom, email…
Working with spreadsheets, using email and Classroom
Interactive activities and programs like Edpuzzle, Genially, podcast, etc.
Realization of activities where the use of digital tools is mandatory
Presentations, Internet searches, resolution of online questionnaires
Knowledge of different specific programs of the module
Search for information online, work on the subject or presentation
Webquests, tasks and presentations. Using: Genially, Canva, Google docs, Drive…
Proposing different tasks where they must use the digital resources they consider appropriate
Tasks, presentations, classroom management…
Treatment of information, creativity, understanding and synthesis through the use of instruments such as Canva, Issue, Google documents… Organization of work and promotion of autonomy using Classroom. Cooperative and collaborative using the named tools and others such as Genially, Google presentations, Padlet…
With ICT content

Source Own elaboration

Last, a questionnaire was answered twice (pre- and posttest), in order to validate the collaborative mapping for the acquisition of digital skills in educators, this is, before and after the implementation of the didactic experiences that are using the TPACK teaching-learning model at two different competency scales: (i) one that is based on digital literacy; communication and collaboration; and another whose emphasis lies in (ii) digital content creation and problem solving.

In this way, an experimental design was used in both the quantitative and qualitative approaches using questionnaires created from instruments already developed for the TPACK model (Schmidt et al. 2009; Valtonen et al. 2019; Cabero-Almenara and Llorente-Cejudo 2020; Cabero-Almenara and Palacios-Rodríguez 2020; Ortega-Sánchez and Gómez Trigueros 2020). The questionnaires contained a first part of general identification and a second part that included closed, open and Likert-type questions (Cejas et al. 2016; Lopera et al. 2020), which corresponded to the self-perception of the types of knowledge of the TPACK model. PCK (pedagogical content knowledge) in the questions that are oriented to the application of various aspects of geographical thinking and spatial competence and TPK (technological content knowledge) in the questions that are focused on the knowledge of the geographical information technologies (GIT) and its application in the different educational contexts.

Case Studies of Didactic Proposals for the Promotion of Digital Citizenship

It is evident that geography during the pandemic has had a certain advantage over other disciplines, since, as we have already seen, the use of geotechnologies in the classroom has proliferated exponentially in recent years (De Miguel 2019; Álvarez and De Lázaro y Torres 2018). However, the challenge of COVID-19 has forced teachers to reinvent themselves and to adapt not only to the need to acquire digital skills, but also to develop basic transversal skills in students (Carretero et al. 2018; Redecker and Punie 2017; Vuorikari et al. 2022), which are essential to the success of the educational process. Autonomy, critical vision and the capacity for self-learning stand out among these competences. In an educational context where daily contact in the classroom has been lost, these skills, together with work habits, self-regulation capacity, and self-discipline when tackling tasks, come to play a fundamental role.

In this context, it is really interesting to address an education that fosters spatial digital citizenship. Addressing the concept of citizenship implies establishing educational practices that develop critical and democratic thinking and promote the ability to interpret and evaluate information in each place (De Miguel 2021b). The didactic use of collaborative cartographies through web-GIS platforms, such as those presented in this article, involves training autonomous, efficient, responsible, critical and reflective citizens when selecting, processing and using information and its sources, as well as the different technological tools. It also entails having a critical and reflective attitude in the evaluation of the information available, contrasting when necessary. Finally, it promotes the participation of young people in the democratic system and integrates them as actors of change to improve their environment (living space) or to make their city's street map more equitable and egalitarian.

The following are two case studies that, as we have been able to verify during their practice and through the results of the pretest and posttest survey, promote the acquisition of citizenship skills by the students, developing didactic strategies that allow them to be active citizens capable of fully participating in their communities and making intelligent decisions online and in life. In such a way that it can be affirmed that students can better and more reliably understand the political, economic, social and cultural complexity of the world in which they live, and thus be actors of the change postulated by 2030 Agenda.

Collaborative Mapping of Your Living Space During a Pandemic (Confinement Period and Immediately After) "My World in 15 Minutes"

This initiative of the University of Zaragoza and the GEOT Chair of Territory, Society and Geographic Visualization aims to be a collective reflection experience that allows us to tell the story of our city during pandemic period. This didactic experience is part of other international works that debate about the necessary transformations of the urban world, focusing on the relationship between space and useful time, the urbanism chrono. Our hypothesis is: How did my living space change during the first months of COVID-19? What measures did we adopt as citizens to be related to our space in a safe way? Do these new habits, taxes and other elements offer a serene city, with more proximity, less stress or less hours of transportation? In the same time, does it satisfy its essential urban social functions? It is about advancing as citizens to propose among all of us a deep transformation of the urban space (still highly monofunctional), toward a polycentric, sustainable city—city that allows increasing quality, easier access to the essential urban social functions such as inhabiting, work, stock up, take care of yourself, learn and rest.

This educational experience also tried to portray emotions. Emotions present the lived space, but also the perceived space (Lefebvre 1974), and they are a transversal object for geography. Society emotionally interacts very differently with places, while territorial practices respond to personal and collective memory. Specifically, emotional mapping allows to understand how one perceives the space where the person is, how does one feel it, and how does one become familiar with it.

Objectives of this initiative were: (i) to experiment with collaborative mapping methodology; (ii) to map how we perceive the different spaces and routes that connect them, how we feel them, how we live them, and how they affect our behaviors and daily movements; (iii) to debate on how our "living space" has changed after the COVID-19 pandemic; (iv) to propose improvements in our day-to-day places to reduce our trips while still enjoying our city.

The urban space where we live our daily lives has been modified in the last years because of the COVID-19 pandemic. This modification takes the form of a reduction in the number of places we go to, in the reduction of the activities that are carried out and in the more intense use of the space near our centers of reference (residence, work, place of studies). Our life unfolded for several months in a small city, which we reached in 15 min. This, in the case of Zaragoza—about 700,000 inhabitants—is good news, because it is a very compact city. People have discovered new places to walk, places that are pleasant to them, that are close to their home and that previously went unnoticed.

The tool used in this proposal is Collaborative Maps of the municipal website of the Zaragoza City Council. Maps are tools to understand the complexity of a territory; they are like photographs that portray geographic or, in this case, sociodemographic realities. Collaborative maps through digital platforms allow us to create these cartographies through collective action. Therefore, they are no longer static photographs of the city taken by a single author, but they are platforms opened for collective participation, being a result of citizens and their interrelation with the environment.

Within the Collaborative Maps section, there was a proposal of the development of three maps (Fig. 3):

(i) Living space: This map proposes geolocation of different spaces that we visit in our daily life: housing, working, shopping, walking, leisure, sports, facilities.

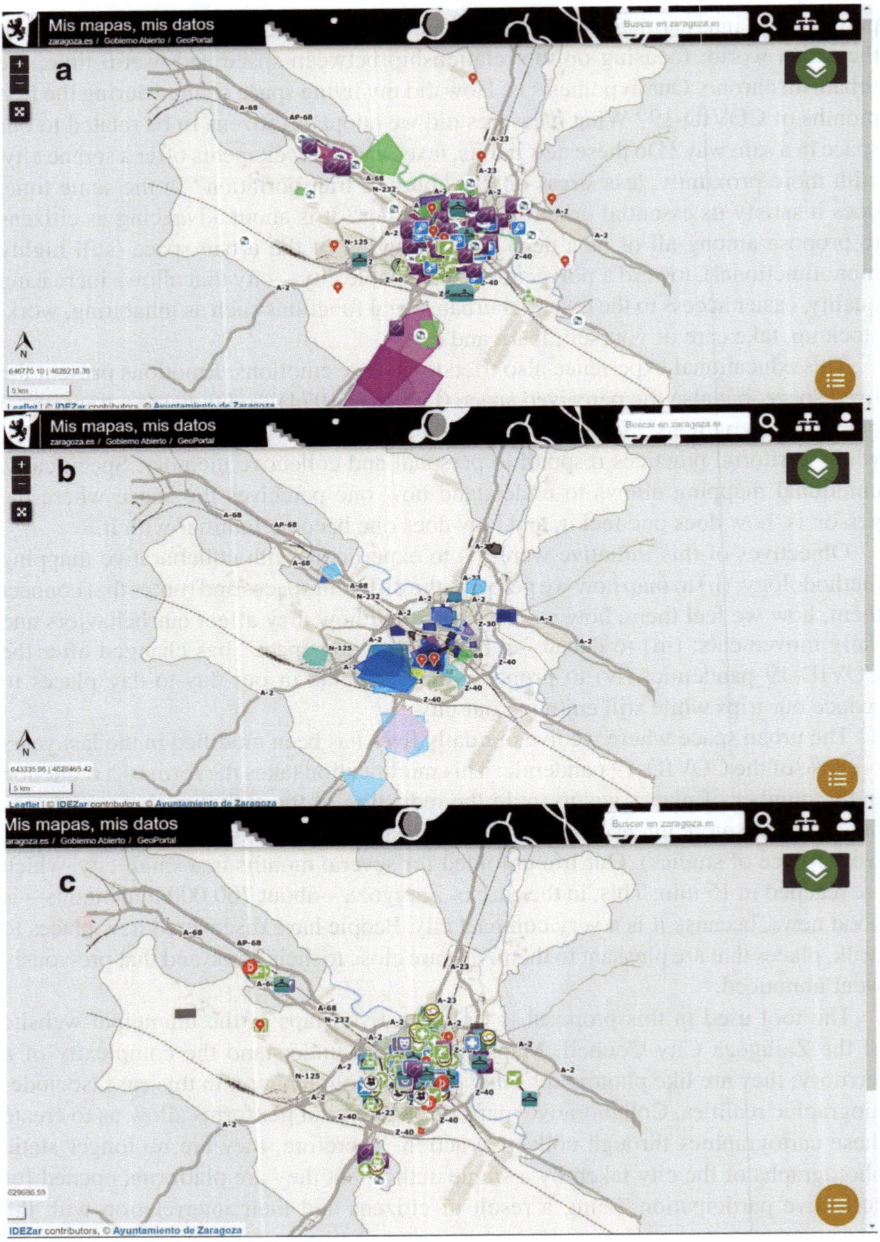

Fig. 3 Maps resulting from the collaborative mapping "My world in 15 minutes": **a** Cartography of the living space. **b** I change house. **c** What do I want near? *Source* Own and collaborative elaboration

(ii) I change my house: This map aims to reflect where would you live if you could choose a place anywhere in the city?
(iii) What do I want nearby? This last map invites to debate about your nearest space, geolocating the proposed location of the equipment that any person would need to improve their living space.

This proposal was initially implemented by students of a course in geographical education, teacher training in Primary School Undergraduate program, Faculty of Education. More than 150 people have participated in this initiative (including school children), so they mapped their story and how they felt during the pandemic period of uncertainty. They also rethought the needs that they have as citizens and how to cover these needs. In short, they present in the map their story about what decisions they made to maintain proximity within the distance.

It seems clear that the society has to progress toward more habitable and sustainable cities, which offers more complete and nicer living spaces. Making citizens reflect on this, how they have reinterpreted and rediscovered their immediate environment, allows them to achieve a more sustainable city and, in turn, empowers them as citizens to ask for facilities and infrastructure that help achieve a more sustainable city for all.

Collaborative Mapping of Women's Streets in Huesca

A joint initiative of the University of Zaragoza, together with the *Geochicas OSM* (GeoGirls Open Street Map) group (Sebastián et al. 2021), has been launched named collaborative mapping of the streets with the name of a woman in Huesca. The experience was carried out in the Ramón y Cajal High School in Huesca during February and March of 2022. The participating groups belonged to the 3rd course of lower secondary education (14-year-old) that allowed us to broaden the spectrum of informants and the conclusions.

One of the groups was made up of students with learning difficulties or situations of difficult adaptation. We will call it A and it was made up of a total of 22 students, of whom three were repeaters, one student lived in a center for under 18 run by social service, but he stopped attending the center at the end of the course. One student suffered a process of bullying in the primary school, one student who had recently joined the Spanish educational system and had serious language difficulties. Also a recently arrived Spanish-speaking student to the spanish education system, one repeating student who had to attend to the class online due to medical prescription that forbids her to attend face-to-face classes and two students with recurring absences whose families were informed about it. In the same group, some negative situations about the experience were detected once it had started, and three students decided to not participate actively in the activity because of ideological reasons. The other group, which we will call B, has different characteristics, although their number was similar, 21 students. Composition of this class was more homogeneous.

There were 14 classes during which the experience was developed and distributed in subject of Social Sciences and Education for Citizenship which was taught by the teacher Lorenzo Mur. The summary of the methodological design is synthesized in following points:

(i) Introductory sessions to raise awareness of the problem in the classroom through an open debate where were presented questions of collective response and close approach. Some examples of these questions: What streets do you know in Huesca? Which ones do you think that are the most important? What places do you avoid to walk through and why? What professions do you think the mentioned names used to perform? Are you missing something?… While the debate was taking place, the most important conclusions were written down on the blackboard and a dialogue was established.
(ii) A second phase proposes the extension of the conclusions to other Spanish, European and American cities and presents the Geochicas project through its viewer. In this way, the students are brought closer to the reality of the nomenclature of our cities to see its masculinization and importance of different professions or activities: politicians, military, saints and virgins…
(iii) Collaborative mapping of the street map of Huesca—during this phase the students are divided in six groups, every one corresponding to one part of the administrative division of the Huesca map. For its correct development, each group of students (students' voluntary composition) is assign to a part of the city where they have to extract the names of the streets and classify them generically into three groups: masculine, feminine and non-gender names. Together (using a shared file) they create an Excel sheet with the extracted data.
(iv) Edition and search for information on Wikipedia about the women's presence in the Huesca street scene, noting the most relevant characteristics. In those cases, that had no entry in Wikipedia a biography would be proposed to fill this gap.

    Once phases (c) and (d) were carried out, the results were analyzed and they verified an evident reality: the majority of Huesca's streets is based on male historical figures dedicated to politics or the military profession. The few examples of women refer mainly to virgins or nuns.
(v) Exhibition of the results of the previous processes and space for a reflection and awareness within a concept of an inclusive and sustainable urbanism where everyone feels represented. From students' initiatives, a consensus document emerged between both class groups, A and B, which reflected their concerns.
(vi) Public exposure of the research process for its dissemination in the city and the claim for actions in search of a more equitable and sustainable street map. Finally, the educational experience had an important dissemination and impact in some interviews in local press, radios and TV channels. Geography students were able to express their citizen values in front of a wider audience, but also in front of the Huesca City Council.

Despite the work done in the search for information, the most important debate and conclusions were drawn once the map of Huesca could be viewed on the *Calles*

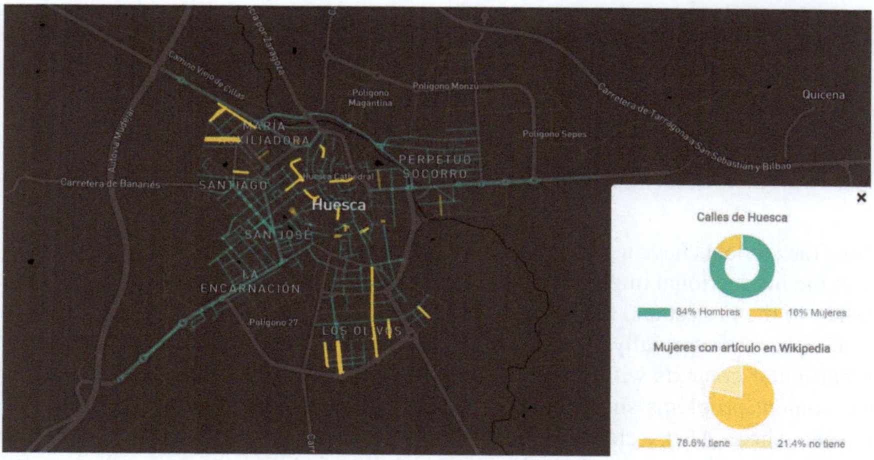

Fig. 4 Street map of Huesca on the viewer of the streets of women. *Source* Own and collaborative elaboration

de las Mujeres viewer (https://geochicasosm.github.io/lascallesdelasmujeres/). The map of the Streets of women of Huesca (Fig. 4) shows evidences: the majority of the Huesca street map is based on male historical figures dedicated to politics or the military profession. The few examples of women refer mainly to virgins or religious life and represent 16% of streets named after relevant historical figures for the city.

This experience has shown that students understand and detect much better the terms of gender through cartography and spatial categories, in particular the street map of their city. In fact, through analysis and reflection of the map, similar conclusions were drawn in both groups: (i) awareness of the existence of a gender difference in the naming of the streets; (ii) historical origin of this disparity intensified in recent times and continues with the same discrimination (the last three new streets returned to follow the criteria of the nineteenth century); (iii) the need to redirect this situation, although there were opinions varying between the urgency of reversing the situation and resignation to this reality, which is waiting for a reflection of the authorities that would stop the perpetuation of this discrimination; (iv) need to propose solutions.

Results and Discussion

Results of the methodological validation, in general, are positive, especially with regard to knowledge about the use of technological tools and resources. This includes the general understanding of how to apply them in a productive way at work and daily life, as well as the ability to recognize changes in the local built environment, so technological information facilitates the achievement of the objective of spatial citizenship learning. As shown in Table 3, the results of the pretest and posttest

Table 3 Comparative summary of the pretest and posttest questions

	Average	Median	Interquartile range	Variance
Pretest knowledge	2,552	2,000	1,000	0.841
Posttest knowledge	2,768	3,000	1,000	0.878

Source Own elaboration based on the answers obtained in the teacher validation questionnaire

show the students have a general positive perception, and also a slight improvement, after the instructional implementation, but presenting a relatively important variance (0.841 in the pretest and 0.878 in the posttest).

This occurs especially when these teaching and learning processes are presented in particular contexts with proposals applied to specific pedagogical, technological and content problems simultaneously integrated, as it is the case of the proposals presented here. Also, a clear and positive trend has been observed in the development of the types of knowledge linked to ICT and digital skills, which greatly improve with the implementation of collaborative mapping, once specified and applied to the particular problem of teaching geography.

However, along with these positive aspects, this study observes some difficulties and limitations related to the integration of digital skills in the teaching-learning processes. In the first place, and especially, there are divergences observed between teachers' self-perception of digital skills and the limitations in their open answers, since an important part of the sample confuses digital skills with ICT tools and provides generic answers and not specific answers for the teaching of geography.

All these elements show that the implementation of specific proposals based on the collaborative mapping model to develop digital citizenship opens an innovative and beneficial field of analysis; however, there still are many challenges. In addition, the evidence shows the need of a knowledge integration process to develop completely the sentiment and attitude of digital citizenship, considering the different types of knowledge (geographical, digital and citizen) but, above all, their integrations in a particular context that converts the students to actors of change. Although research must continue to explore the limitations and problems of the implementation of certain types of knowledge, the integrative proposal continues to be the way forward for teaching-learning processes in which ICT are every time more and more key resource both for teachers and for students.

Conclusions

Geography is an ancient scientific discipline, with an established body of knowledge, but it faces the challenges of a changing world, a technological revolution and the teaching of rapid political, economic, social and cultural transformations that impact space, countries and regions of the world, as we have seen with the COVID-19 pandemic.

Geographic education, and in particular geographic education for a digital citizenship, is committed to innovation and practices that respond to the challenges of the twenty-first century, including the fulfillment of the 2030 Agenda. In this sense, teaching geography involves the design of didactic proposals that allow us to address cognitive domains related to knowing, applying and reasoning and to the acquisition of spatial thinking, geographic thinking and spatial citizenship, in order to develop analytical, critical and lateral thinking.

Collaborative mapping offers teachers and students a very interesting context to address all these competencies, as it allows the entire cognitive process for the analysis, visualization and interpretation of realities, while emphasizing students as active agents in both information gathering and subsequent decisions, empowering them to be agents of change on a local scale (mostly), but with global repercussions. This chapter shows that teaching geography must be re-visioned, taking advantage of the opportunities given by the spread of geotechnologies and the increasing methods for teaching collaboratively and online during pandemic, and promoting spatial citizenship education through collaborative mapping in which secondary school students are able to represent local contexts—as Zaragoza and Huesca—from the perspective of global challenges, like sustainable cities and communities or gender balance.

References

Álvarez-Otero J, De Lázaro y Torres M (2018) Education in sustainable development goals using the spatial data infrastructures and the TPACK model. Educ Sci8(4):171. https://doi.org/10.3390/educsci8040171

Cabero-Almenara J, Llorente-Cejudo C (2020) Covid-19; transofmración radical de la digitalización en las instituciones universitarias. Campus Virtuales 9:25–34. http://uajournals.com/ojs/index.php/campusvirtuales/article/view/713

Cabero-Almenara J, Palacios-Rodríguez A (2020) Marco Europeo de Competencia Digital Docente «DigCompEdu». Traducción y adaptación del cuestionario «DigCompEdu Check-In». EDMETIC 9(1):213–234. https://doi.org/10.21071/edmetic.v9i1.12462

Carretero S, Vuorikari R, Punie Y (2018) DigComp 2.1: the digital competence framework for citizens with eight proficiency levels and examples of use. Publications Office, Comisión Europea, Centro Común de Investigación. https://doi.org/10.2760/38842.Accessed 30 Jan 2023

Cejas León R, Navío Gámez A, Barroso Psuna J (2016) Las competencias del profesorado universitario desde el modelo TPACK (Conocimiento Tecnológico y Pedagógico del contenido). Pixel Bit Rev. Medios Educ 49:105–119. https://doi.org/10.12795/pixelbit.2016.i49.07

De Lázaro y Torres ML, Borderias P, Morales FJ (2020) Citizen and educational initiatives to support sustainable development goal 6: clean water and sanitation for all. Sustainability 12(5):2073. https://doi.org/10.3390/su12052073

De Lázaro y Torres ML, De Miguel González R, Buzo Sánchez I (2017) El Proyecto School on the Cloud: Lecciones Aprendidas = School on the Cloud Project: Lessons Learned. Espacio Tiempo y Forma. Serie VI, Geografía (10):103. https://doi.org/10.5944/etfvi.10.2017.18748

De Miguel González R, Sebastián-López M (2022) Education on sustainable development goals: geographical perspectives for gender equality in sustainable cities and communities. Sustainability 14(7):4042. MDPI AG. https://doi.org/10.3390/su14074042

De Miguel González R (2021a) Geographical and geospatial competences from school education to higher education: the contribution of EUROGEO in the international projects In J Res

Didact Geogr (J-READING) 2(10):57–65. http://www.j-reading.org/index.php/geography/article/view/299

De Miguel González R (2021b) From international to global understanding: toward a century of international geography education. Int Res Geogr Environ-Mental Educ 30(3):202–217. https://doi.org/10.1080/10382046.2020.1809792

De Miguel R (2019) Europe in a global context: EUROGEO and the role of geography and European geographers. Eur J Geogr 10(4):160–176. https://eurogeojournal.eu/index.php/egj/article/view/209

Drummond A, Sweeney T (2017) Can an objective measure of technological pedagogical content knowledge (TPACK) supplement existing TPACK measures? Br J Educ Technol 48(4):928–939. https://doi.org/10.1111/bjet.12473

Fu B (2020) Promoting geography for sustainability. Geogr Sustain 1(1):1–7. https://doi.org/10.1016/j.geosus.2020.02.003

Georgeson L, Maslin M (2018) Putting the United Nations Sustainable Development Goals into practice: a review of implementation, monitoring, and finance. Geo: Geogr Environ 5(1):e00049. https://doi.org/10.1002/geo2.49

Gómez Trigueros I (2016) La inclusión de las tecnologías en la formación incial del profesorado: una intervención de aula a través del modelo TPACK. Tendencias Pedagógicas 28:133–152. https://doi.org/10.15366/tp2016.28.010

Hidalgo Cajo BG, Gisbert Cervera M (2020) Análisis de las competencias digitales del profesorado universitario desde el modelo TPACK (conocimiento tecnológico y pedagógico del contenido). INNOVA Res J 5:79–96. http://obsinvestigacion.unach.edu.ec/sccitys/mod_ob/admin/news.jsp?codid=2818

Jeronen E (2020) Geography education promoting sustainability. MDPI-Multidisciplinary Digital Publishing Institute. https://www.mdpi.com/journal/education/special_issues/Geography_Education_Promoting_Sustainability

Kampylis P, Punie Y, Divine J (2015) Promoting effective digital-age learning. A European framework for digitally competent organisations. Comisión Europea, Centro Común de Investigación. https://publications.jrc.ec.europa.eu/repository/handle/JRC98209 Accessed 30 Jan 2023

Lefebvre H (1974) La Production de l'espace. Anthropos, Paris

Liverman DM (2018) Geographic perspectives on development goals: constructive engagements and critical perspectives on the MDGs and the SDGs. Dialogues Hum Geogr 8(2):168–185. https://doi.org/10.1177/2043820618780787

Lopera Pérez M, Arias V, Jiménez M, Ospina Pineda D, Valderrama Muñoz A (2020) Aportes de la revisión de literatura al diseño de una ruta de apropiación TIC, vinculada con el modelo tecnológico-pedagógico-disciplinar. Rev Virtual Univ Catól Norte 62:276–307. https://doi.org/10.35575/rvucn.n62a11

Marrón Gaite MJ (2011) Educación geográfica y formación del profesorado. Desafíos y perspectivas en el Nuevo Espacio Europeo de Educación Superior (EEES). Boletín de la Asociación de Geógrafos Españoles 57:313–341. https://bage.age-geografia.es/ojs/index.php/bage/article/view/1386

Mishra P, Koehler MJ (2006) Technological pedagogical content knowledge: a framework for teacher knowledge. Teach Coll Rec 108(6):1017–1054. https://doi.org/10.1111/j.1467-9620.2006.00684.x

Mishra P (2019) Considering contextual knowledge: the TPACK diagram gets an upgrade. J Digit Learn Teach Educ 35:76–78. https://doi.org/10.1080/21532974.2019.1588611

Nightingale AJ (2018) Geography's contribution to the sustainable development goals: ambivalence and performance. Dialogues Hum Geogr 8(2):196–200. https://doi.org/10.1177/2043820618780790

Ortega-Sánchez D, Gómez Trigueros IM (2020) MOOCs and NOOCs in the training of future geography and history teachers: a comparative cross-sectional study based on the TPACK model. IEEE Access 8:4035–4042. https://doi.org/10.1109/access.2019.2963314

Redecker C, Punie Y (2017) European framework for the digital competence of educators: DigCompEdu. Publications Office of the European Union, Luxembourg. https://publications.jrc.ec.europa.eu/repository/handle/JRC107466 Accessed 30 Jan 2023

Schmidt DA, Baran E, Thompson AD, Mishra P, Koehler MJ, Shin TS (2009) Technological pedagogical content knowledge (TPACK). The development and validation of an assessment instrument for preservice teachers. J Res Technol Educ 42:123–149. https://doi.org/10.1080/15391523.2009.10782544

Sebastián-López M, De Miguel-González R (2020) Mobile learning for sustainable development and environmental teacher education. Sustainability 12(22):9757. https://doi.org/10.3390/su12229757

Sebastián M, Zuñiga M, De Miguel R (2021) Empoderamiento espacial y de genero a través del mapeado colaborativo en la ciudad de Zaragoza. In José León García Rodríguez (Ed. lit.) Geografía. cambio global y sostenibilidad. Comunicaciones del XXVII Congreso de la Asociación Española de Geografía. Tomo II. Territorio. desigualdad y enfermedades., pp 335–350

Souto González XM (1998) Didáctica de la Geografía. Problemas sociales y conocimiento del medio. Ediciones del Serbal, Barcelona

Sprenger S, Nienaber B (2018) (Education for) sustainable development in geography education: review and outlook from a perspective of Germany. J Geogr High Educ 42(2):157–173. https://doi.org/10.1080/03098265.2017.1379057

Sultana F (2018) An (other) geographical critique of development and SDGs. Dialogues Hum Geogr 8(2):186–190. https://doi.org/10.1177/2043820618780788

Valtonen T, Sointu E, Kukkonen J, Mäkitalo K, Häkkinen P, Järvela S (2019) Examining pre-service teachers' technological pedagogical content knowledge as evolving knowledge domains: a longitudinal approach. J Comput Assist Learn 35:491–502. https://doi.org/10.1111/jcal.12353

Vuorikari R, Kluzer S, Punie Y (2022) DigComp 2.2: the digital competence framework for citizens: with new examples of knowledge, skills and attitudes. Publications Office of the European Union. https://doi.org/10.2760/115376

COVID-19 and Post-COVID -19

The Art of Geographical Analysis of COVID-19-Related Data

Dimitris Kavroudakis, Sofia Zafeirelli, Panagiotis Agourogiannis, and Marios Batsaris

Abstract Almost all sectors of our life have been influenced by COVID-19 pandemic. Informed decision-making regarding pandemic is essential and can be based on credible spatial data. Available COVID-19 spatial data reflect administration areas and in most of the cases used for comparison between countries and regions. These comparisons are tricky and ask for careful consideration of a number of country aspects, especially when the variable in question is dynamic and changes happen very often. Comparison between countries should consider relative numbers (e.g., incidents per capita) and should include information regarding spending for health services. Also, medical provision and climate-related aspects of each country are also important when comparing between countries. Finally, age structure of population is also crucial and needs to be examined. This work illustrates the difficulties when comparing country-data related to COVID-19 pandemic and presents the CRISTINA project. We argue that country-level COVID-19 data ask for standardization in terms of population and geography as well as that correlation of data with country-related characteristics does necessarily not imply direct causation. Finally, this work presents a number of relevant logical fallacies that should be considered when analyzing spatial data of COVID-19 pandemic.

Keywords Spatial analysis · Spatial data · COVID-19 · Logical fallacies · Big data

D. Kavroudakis (✉) · S. Zafeirelli · P. Agourogiannis · M. Batsaris
Department of Geography, University of the Aegean, 81100 Mytilini, Lesvos, Greece
e-mail: dimitrisk@aegean.gr

S. Zafeirelli
e-mail: s.zafeirelli@aegean.gr

P. Agourogiannis
e-mail: geod17010@geo.aegean.gr

M. Batsaris
e-mail: m.batsaris@aegean.gr

A. Klonari et al. (eds.), *Re-visioning Geography*, Key Challenges in Geography,
https://doi.org/10.1007/978-3-031-40747-5_11

Introduction

COVID-19 pandemic has influenced almost all sectors of social life. Informed decision-making regarding pandemic is essential and can be based on credible data which are geo-referenced in most cases. COVID-19 data reflect administration areas and in most of the cases are used for comparison between countries. These comparisons are tricky and ask for careful consideration of a number of country aspects, especially when the variable in question is dynamic and changes happen very often.

Comparison between countries should consider relative numbers (e.g., incidents per capita) and should include information regarding spending for health services. Also, medical provision and climate-related aspects of each country are also important when comparing between countries. Finally, age structure of population is also crucial and needs to be examined.

This work illustrates the difficulties when comparing country-data related to COVID-19 pandemic. We argue that country-level COVID-19 data ask for standardization in terms of population and geography as well as that correlation of data with country-related characteristics does necessarily not imply direct causation.

Globally, from early 2020, with some billions of people under house restrains, in almost all countries and territories of the world, there have been millions of confirmed cases of COVID-19, including thousands of deaths, reported to the World Health Organization (WHO). The COVID-19 virus, which was described as an "enemy of humanity" by the World Health Organization, has infected about 2.3 millions of people. The number does not reflect reality in the absence of extensive diagnostic tests. The effects of the pandemic are incalculable, and thousands of billions of euros or dollars will be allocated by the governments of the world to deal with the deep economic recession that is coming.

The UN is calling the pandemic the worst crisis humanity has faced since 1945. It combines a deadly disease and an economic recession unprecedented in the recent past. While in China the province of Hubei and its capital, the city of Wuhan, are coming out of quarantine, Italy, the country with the most deaths, has crossed the barrier of 15,000 deaths. And people are starting to wonder about the next day, after the easing of restrictive measures. Is there a risk of a possible future wave of this epidemic? Were governments slow to react? How close to reality is the death toll announced by China (just over 3,000)? What will be the dimensions of the economic crisis?

Global tourism industry has been significantly impacted by various types of crises, particularly the pandemic and terrorist crises. These crises have caused travelers to be wary of visiting new places due to the threat of quarantine, fear of using airports, airplanes, restaurants, museums and archaeological sites, anxiety of not knowing what to do in the event of illness in a foreign country, need for cross-border medical insurance, and difficulty of changing tickets and reservations at both hotels and airlines. Media plays an important role in shaping public perception of such situations and can have a large effect on the global tourism market. For example, fear is often cultivated and escalated by news outlets regarding the severity of diseases, health effects and death rates, which has resulted in many people canceling their trips and avoiding travel for long periods of time even after the pandemic has passed. Examples of this include the SARS coronavirus (Holmes 2003), avian influenza A (H5N1) virus (Taubenberger and Morens 2006) and pandemic influenza A virus (H1N1) (Zimmer and Burke 2009, p. 1) all of which caused widespread concern and alerted public health services to the risk of rapidly spreading respiratory viruses with pandemic potential.

Digital Epidemiology and Big Data

The increase in the number of electronic mass media, the massive use of the Internet and electronic social media by most of the world's population, as well as the widespread use of smartphones in recent years have led to the creation of new data sources. Also, new algorithmic techniques have enabled the creation of new tools for data processing such as artificial intelligence (AI) (Hamet and Tremblay 2017), machine learning (ML) (Rajkomar et al. 2019) and natural language processing (NLP) (Locke et al. 2021). All of the above contributed to the emergence of a new branch of epidemiology called digital epidemiology. Digital epidemiology embraces the goals of clinical epidemiology but takes a different approach to their implementation. Instead of relying only on data from the health sector, it makes use of these new data sources. These new sources of data, such as electronic social media, are also called sources of "big" data (Big Data) and are characterized by very large volumes of data which have a complex structure and show great heterogeneity. The big challenge facing digital epidemiology is finding the right tools to process and analyze these data, avoid wrong conclusions and misrepresent the right information as a wrong result can have unintended consequences (Salathe et al. 2012; Salathé 2018; Park et al. 2018).

Nowadays, the term "*big data*" has a double meaning that sometimes refers to the data itself and the sources from which they come, while on the other hand sometimes it refers to the processing methods of these data. We use the term "big data" referring to the actual data itself that is produced by the daily life of people who use the internet, the devices or the sensors for various reasons and activities. These reasons include buying/selling, transferring, using electronic services, as well as social media and location-based services. Today, there are eight characteristic words used to describe big data known as "The 8 V's." The V's can be divided into two groups of which the first contains three which are the general characteristics of the nature of "big" data, while the second contains the characteristics that "big" data acquire once entering an information system. More specific, the basic features are:

Volume: Refers to the very large volume of Big Data referring to the difficulty of collecting and processing large amounts of data. Velocity: It refers to the speed with which the Big Data are produced from various electronic sources. Variety: Refers to the different types of Big Data (such as image, sound, and electronic receipts) and their different structure.

The characteristics they acquire after entering a digital information system include the following topics. Value: It refers to the value that the exploitation of Big Data can offer in various sectors. Veracity (Validity): Refers to the validity of the results that are produced by the utilization of "big" data and the reliability of these data. Variability: Refers to the different forms they can be transformed into, the different models they can be processed with, and the different associations they can be made after entering a system. Virality: Refers to how quickly they can spread through a network to different users. Viscosity: Refers to how much resistance-delay can be observed in the flow-transmission of a certain volume of big data.

"Big" data can also be characterized by the ever-increasing speed of its collection and use in most scientific research today. It is very important to emphasize that the profit of exploiting the "big" data depends entirely on the time interval between its creation and the moment when its use will have given the desired results. The shorter this interval, the more the value of the information derived from it increases. This is true for almost all uses of "big" data and even more so for the field of digital epidemiology, since the time from the outbreak of an epidemic to the moment when the first measures are taken to deal with it is crucial in limiting it. To effectively reduce this time period, the systems that undertake the collection and correlation of these data must be properly designed to achieve the highest possible speeds. Although the health field has long distanced itself from the use of "big" data, the new possibilities that arise with its proper exploitation are enormous. In digital epidemiology, not all sources of "big" data are equally useful, so in the following we will describe the most important of them in terms of their usefulness in monitoring and predicting disease outbreaks, and in terms of the difficulties involved in utilization of data from them (Ali 2019; Bansal et al. 2016; Park et al. 2018; Saecker and Markl 2012).

COVID-19 Data and Countries Comparison

Since the COVID-19 pandemic start, there are numerous sources of official data sharing across the web. There are a number of official sources with statistics, research data and other information about coronavirus (SARS-CoV-2), the disease it causes (COVID-19), the global pandemic and its economic. There are some well-established sources of official COVID-19-related data. One of the most notable source of these data is the World Health Organization, which has posted a special website with the purpose of providing the necessary information on everything related to SARS-CoV-2 (WHO 2022). It also includes international epidemiological statistics, precautionary advice, good practices, debunking relevant widespread fake news, etc. Additionally, the European Center for Disease Prevention and Control (ECDC) has also set up its own special website for information on the novel coronavirus. It also provides international statistics of confirmed cases and deaths (ECDPC 2022). Finally, the last official source of COVID-19 information source regarding Greece, is the Greek National Public Health Organization (EODY) has the official information on the evolution of the pandemic in Greece, useful information, instructions for citizens and businesses of health interest (EODY 2022).

Apart from the above sources of official data, there are a number of related sources about International Epidemiological Statistics. The American Johns Hopkins University has created and daily updates an open data repository with international analytics on the SARS-CoV-2 pandemic. This repository, which is the primary source of statistics for all pandemic-related analyses conducted worldwide, is freely accessible for access (CSSEGIS and Data 2022). In addition, Johns Hopkins University has an interactive map of the pandemic, with data from the same database (Johns Hopkins University 2022). Finally, another source of very informative data related to COVID-19 pandemic is the website "Our World in Data" (Mathieu et al. 2020) which includes a plethora of statistics and informative charts.

There is a need for countries comparison, especially when we need to identify overspread or under-spread of COVID-19 cases. In a very globalized economy, and with full traveling potentials, we need to understand the various country-related COVID-19 statistics. The research project "CoRona vIrus SpaTIal aNAlysis" (CRISTINA) (Kavroudakis 2022) examines the time-series progress of COVID-19 events in 177 countries of the world. We collect daily data for events as well as government interventions and associate them with geographical time-series datasets in order to evaluate the rate of change. We also estimate future events based on assumptions of previous events in a time window of 5 days. Future projections are based on assumptions and should only be used for educational reasons. The webpage of this project is the following: www.dimitrisk.gr/covid19.html. The interventions are grouped by type such as isolation measures, transportation measures and economic activity measures.

Data sources for this project include daily COVID-19 data for 177 countries (confirmed cases, deaths, recovered). Also, government measures for 177 countries (lockdown, business measures, transportation measures, education lockdown, etc.). Population data (age groups, sex, population density), health-related data (hospitals, facilities, intensive care units, doctors, health funds) and finally economic data (GDP, economic sectors, etc.). The main research questions of Christina Project are the following:

- What are the effects of stay-home lockdowns on flattening the curve of confirmed COVID-19 cases by country for 177 countries?
- How did similar lockdown measures affect the curve of confirmed COVID-19 cases by country for 177 countries?
- What is the relationship between: % of GDP for health and % of confirmed population above 65 years old by country for 177 countries?
- What is the contribution of various measures by country: home lockdown, business closure, transportation restrictions 1 and 2?
- What is the contribution of average temperature on COVID-19 cases per week, month?

The results of this project can be found on the website. Some of the most notable results include the following points. The following results are very suitable on the basic understanding on comparison between countries and the avoidance of possible logical fallacies. The following Fig. 1 depicts the relationship between: the ration of deaths over confirmed cases by time. We use Day since first event by country as the main time unit. It is clearly visible that Italy, UK and Spain lines are rising after 58th day, much more than any other country. Germany, on the other hand, shows very low ratios on the same period. On this basis, Fig. 2 depicts confirmed cases of COVID-19, since the day of 1st event. Germany's numbers are among the top countries for this variable. It is now more than obvious that we can use various COVID-19 data sources to focus on different top countries. Germany's numbers are very small in the first figure while they are very high at the second figure. Variable selection can be a form of "cherry-picking" approach while COVID-19 data analysis, for countries comparison. Selection of specific variables which fit our assumptions can be quite confusing if not deceiving. To avoid such misconceptions, we argue on favor of multiple variable evaluation on country comparisons. This is the use of all available variables and then use some sort of voting system before ranking country's progress.

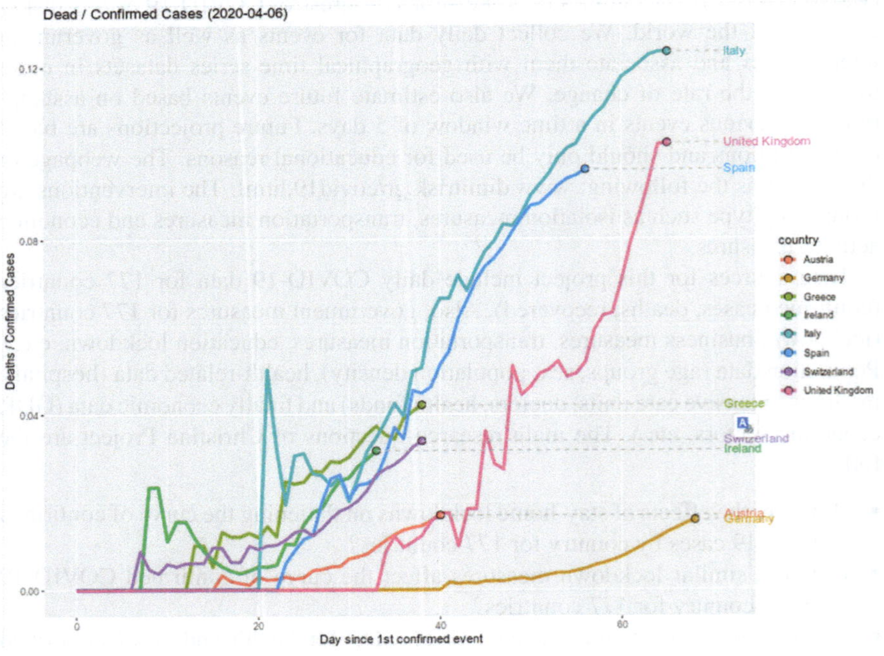

Fig. 1 Deaths over confirmed cases of COVID-19, since the day of 1st event. Germany is almost in the bottom of all other countries

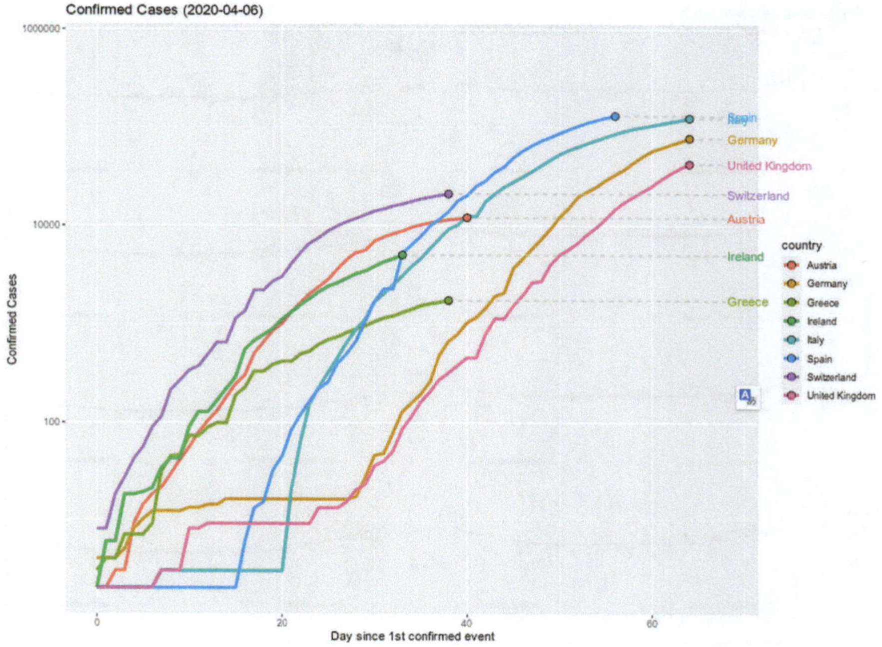

Fig. 2 Confirmed cases of COVID-19, since the day of 1st event. Germany is the 3rd country

Excessive Deaths by Age Group

Another very interesting topic from the CHRISTINA project is the analysis of excessive deaths by age group. This is the evaluation of death counts against previous non-COVID-19 years for each country. This approach can offer a more reliable way of comparison between countries especially when we consider temporal variations. Following Fig. 3 depicts excessive deaths be age group comparison between Greece (left) and Italy (right). Green and blue periods (bars in x axes) represent the two lockdown events in 2020. Black line indicates average deaths by week for the years 2005–2019. Red line is showing deaths by week for 2020. Finally, green and blue periods (x axes): the two lockdown events in 2020. It is more than clear that Italy (right column) has shown some excessive deaths across all age groups. The baseline for the excessive deaths calculations is the deaths of the last 10 years, for each age group. This measure is a comparative measure against the actual numbers of each country for the previous years. It is a consistent measure that is not influenced by temporal variations and is using a baseline of numbers before the COVID-19 pandemic.

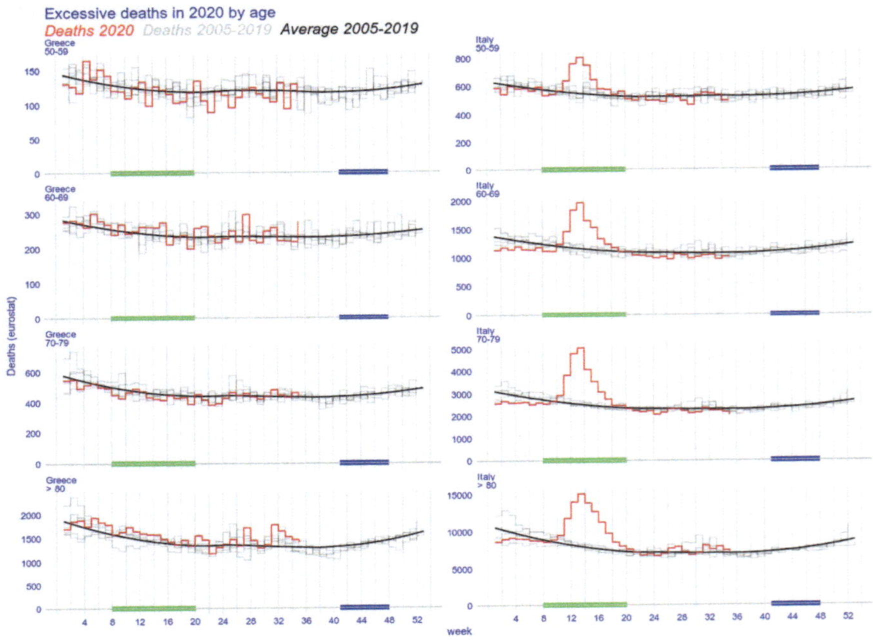

Fig. 3 Excessive deaths be age group comparison between Greece (*left*) and Italy (*right*). Green and blue periods (bars in x axes): the two lockdown events in 2020. *Gray lines*: deaths by week for the years 2005–2019. *Black line*: average deaths by week for the years 2005–2019. *Red line*: deaths by week for 2020

Logical Fallacies

Some of the most important logical fallacies when comparing COVID-19 data between countries are presented here in this part of the chapter. One of the most significant is *Cherry Picking*. It refers to selecting results that fit your claim and excluding those that don't. It is also related to selecting specific variables to present according to the argument we need to support. *Data Dredging* is another very interesting fallacy which refers to the act of repeatedly testing new hypotheses against the same set of data, failing to acknowledge that most correlations will be the result of chance. When comparing data between countries, a very common logical fallacy is *False Causality* (spurious correlation) which is falsely assuming when two events appear related that one must have caused the other. Also, when referring to COVID-19 sample data, *Sampling Bias* should also consider as one of the most essential logical fallacies. It is referring on drawing conclusions from a set of data that isn't representative of the population you're trying to understand.

Another significant misconception regarding probabilities of events is the *Gambler's Fallacy*, which is mistakenly believing that because something has happened more frequently than usual, it's now less likely to happen in future (and vice versa). This is quite relevant to COVID-19 comparisons when we falsely believe that because some rises and falls of the numbers (positive cases) are less likely to happen in the future. Also, *Simpson's Paradox* is very relevant in countries and regions comparisons. More specific when comparing intra-countries events, we may sometimes misunderstand the big picture for the total country. This paradox is referring to the case when a trend appears in different subsets of data but disappear or reverse when the groups combined. Finally, equally important is the *Publication Bias* which is also prominent in COVID-19-related studies. More specific, this is referring to the fact that interesting research findings are more likely to be published, distorting our impression of reality.

Cherry-Picking in Geography

Cherry-picking logical fallacy regarding COVID-19 geographic data can be a problematic practice as it can result in biased (or even incomplete) data that misrepresent the true state of the pandemic in a geographical area. Suppose a researcher wants to cherry-pick COVID-19 data for a specific city in the United States, such as, Los Angeles. The researcher can find a reliable source of COVID-19 data, such as the website of the Centers for Disease Control and Prevention[1] or the World Health Organization.[2] Once the researcher has located the data source, he can navigate to the page that provides data on COVID-19 cases in the United States and look for the data for the state of California. Next, he can locate the data for Los Angeles County, which is the largest county in California and includes the city of Los Angeles. The researcher can find data on the number of confirmed COVID-19 cases, deaths and other relevant metrics for Los Angeles County. However, cherry-picking this data for Los Angeles alone could give a misleading picture of the state of the pandemic in the city, as it does not take into account the wider context of the county, state, or even country. It is important to analyze the geographical data for the city in relation to the data for the broader geographic region and to consider factors such as population density, demographics and/or other social and economic factors that may affect the spread of the COVID-19 virus. It is therefore important to take into account the broader context of the pandemic to avoid misleading conclusions.

Data Dredging in Geography

Data dredging is the practice of selectively analyzing spatial data to find spatial patterns that appear significant, but are actually due to chance. This can lead to false conclusions and incorrect interpretations of the data. Suppose a researcher wants to

[1] https://www.cdc.gov.

[2] https://www.who.int.

determine if there is a correlation between temperature and the number of COVID-19 cases in Brazil. He starts by collecting data on the average temperature and the number of COVID-19 cases for each state in Brazil. The researcher then analyzes the data and finds that there appears to be a negative correlation between temperature and the number of COVID-19 cases, meaning that as the temperature increases, the number of cases decreases. However, this conclusion is based on a selective analysis of the spatial data and ignores many other factors that could be driving the spread of the virus, such as population density in Brazil, demographics, public health measures and other social and economic factors. It is also possible that the apparent correlation is simply due to chance, as correlations can appear by random chance in any dataset. In general, in order to avoid data dredging and draw valid conclusions, it is important to use a rigorous and systematic approach to spatial data analysis and to take into account all relevant factors that may affect the spread of the virus in a country. This can involve using multiple regression analysis to control for confounding variables and conducting sensitivity analyses to test the robustness of the results to different assumptions and model specifications.

False Causality in Geography

False causality, in geospatial sciences, is a type of error that occurs when a correlation between two spatial variables is assumed to indicate a causal relationship, even though there may be other spatial factors that are responsible for the observed relationship. Suppose we want to investigate the effect of air pollution on the number of COVID-19 cases in Athens, Greece. We collect data on air pollution levels and the number of COVID-19 cases for the city over a period of several months and analyze the data to find a positive correlation between air pollution levels and the number of cases. We may conclude that air pollution is causing an increase in COVID-19 cases in Athens, and suggest that reducing air pollution levels could help to mitigate the spread of the virus.

However, this conclusion is false, as there may be other factors that are responsible for the observed correlation, such as population density, demographics and/or public health measures. In order to avoid the pitfalls of false causality, it is important to use a rigorous and systematic approach to data analysis and to consider all relevant factors that may affect the relationship between variables. While it is important to investigate the relationship between COVID-19 and environmental factors such as air pollution, it is important to use caution when drawing causal inferences from observational data and to consider all relevant factors that may be responsible for the observed relationship.

Sampling Bias in Geography

Sampling bias is a type of error that occurs when samples are not representative of the population being studied. Sampling bias could affect the spatial analysis of COVID-19 data. For example, if we want to investigate the relationship between COVID-19 cases and income levels in a city, we first collect data on the number of COVID-19 cases and the median income for each neighborhood in the city. Then, we analyze the data to find that there is a negative correlation between "income levels" and COVID-19 cases. However, this conclusion may be biased, as the sample of neighborhoods we selected may not be quite representative of the entire city. It is possible that we may have selected only neighborhoods with higher income levels that have lower population densities and better access to healthcare facilities, which in turn may be responsible for the observed negative correlation. To avoid sampling bias, it is important to use a representative sample of data that accurately reflect the population being studied. This can be achieved using random sampling techniques or stratified sampling methods to ensure that all segments of the population in the city are represented in the sample.

Gambler's Fallacy in Geography

The gambler's fallacy is a type of cognitive bias that occurs when individuals assume previous random events will affect the outcome of future events, even though two events are statistically independent. For example, when analyzing the number of COVID-19 cases in a municipality over time, we may notice that there have been several consecutive days of increasing cases, and assume that this trend will continue into the future. However, this assumption is a form of the gambler's fallacy, as each day's COVID-19 case count is independent of the previous day's count and there is no statistical basis for assuming that the trend will continue. It is therefore important to use statistical methods to analyze data and account for the effects of randomness and variability. This can involve using time-series-analysis techniques to model trends and possible seasonal patterns in spatial data as well as conducting hypothesis tests to determine the statistical significance of observed patterns.

Simpson's Paradox

Simpson's paradox is a type of statistical paradox occurring when a trend appears in different groups of data, but disappears (or reverses) when groups are combined. For example, when investigating the relationship between COVID-19 cases and ethnicity in a city, initially we collect data on the number of cases and ethnic composition of each neighborhood. Then, we may find that in each neighborhood, the number of cases is higher among a specific ethnic group compared to residents of other ethnicities. However, when combining data across all neighborhoods, we may find that the opposite trend appears: the overall number of cases is higher among residents of other ethnicities compared to this specific ethnic group. This reversal of trend is a form of Simpson's paradox, as the relationship between ethnicity and COVID-19 cases changes when the data are aggregated at different spatial level. It is important to consider the underlying factors that may be driving the observed patterns in the data, and to use appropriate statistical methods to control for confounding variables. This can involve using regression analysis to model the relationship between ethnicity and COVID-19 cases while controlling for other variables such as age, income and access to health care.

Publication Bias

Publication bias is a type of bias occurring when the availability or publication of research results is influenced by their statistical significance. Suppose we conduct a systematic review of the relationship between COVID-19 cases and air pollution in Athens, Greece. We identify several studies that have investigated this relationship, but notice that most of the studies have reported significant positive associations between air pollution and COVID-19 cases, while only few studies have reported non-significant or negative associations. This pattern may be a form of publication bias, as studies with non-significant (or even negative) findings may be less likely to be published or included in the review, leading to an over-representation of only positive findings in the literature. To avoid publication bias, we should conduct a comprehensive search of the literature and include all relevant studies, regardless of their statistical significance. This may involve using search strategies that are not limited by language, geography or publication status, and using methods to assess the risk of bias in individual studies. It is important to use caution when interpreting the results of systematic reviews or meta-analyses, and to consider the potential impact of publication bias on the observed patterns in the spatial data.

Conclusions

Digital epidemiology is a cutting-edge field of epidemiology that is essential for improving people's health and quality of life. It is still in its early stages and may take some years to become fully realized. Currently, systems that use large amounts of data for digital epidemiology are being developed and can only detect and track the spread of contagious diseases; they have very limited forecasting abilities. Additionally, these systems typically rely on receiving confirmed data from other sources, so they cannot provide any warnings before the data are verified.

The continuous development of technology, the creation of faster computing systems as well as improvements in the field of artificial intelligence, within the next few years, for sure will offer new possibilities in the field of digital epidemiology. Thus, it will become strong the creation of new systems that can process and confirm information autonomously and faster than other epidemiological agencies. Also, it is possible to create forecasting systems that will be able to predict the advent of a new epidemic, even from the first cases, while there will also be possibility of creating spatial models that will predict its spread. By using such systems, health agencies will now be able to limit and deal with outbreaks of new communicable diseases in their very early stages, before they get out of control. An example which points to the need to create advanced systems in the field of digital epidemiology is the coronavirus pandemic, which it could be dealt with much faster and with better methods if we had a spatial model for predicting of its spread. The development of the field of spatial epidemiology and, in general, of the field of processing and analysis of "big" spatial data can lead to many improvements in the health and quality of life of the world's population. However, there are always some risks when collecting and processing data from electronic sources.

Logical fallacies can be found in almost any scientific field, including geography. Geographic research often involves complex spatial data, analysis and interpretation, which can make it vulnerable to a variety of cognitive biases and logical errors. This work illustrated some of the most prominent logical fallacies in geospatial science and more specific when dealing with COVID-19-related spatial data. It is important for geographers to recognize the limitations and uncertainties of spatial analysis of data and to be open to revising their conclusions as new evidence emerges.

References

Ali AH (2019) A survey on vertical and horizontal scaling platforms for big data analytics. Int J Integr Eng 11(6):138–150. Retrieved from https://publisher.uthm.edu.my/ojs/index.php/ijie/article/view/2892

Bansal S, Chowell G, Simonsen L, Vespignani A, Viboud C (2016) Big data for infectious disease surveillance and modeling. J Infect Dis 214:S375–S379. https://doi.org/10.1093/infdis/jiw400

CSSEGIS and Data (2022) COVID-19 data repository by the Center for Systems Science and Engineering (CSSE) at Johns Hopkins University.

ECDPC (2022) European Centre for Disease Prevention and Control COVID-19 [WWW Document]. European Centre for Disease Prevention and Control. https://www.ecdc.europa.eu/en/covid-19. Accessed 29 Nov 2022

EODY (2022) Greek National Public Health Organization Covid-19 [WWW Document]. https://eody.gov.gr/neos-koronaios-covid-19/. Accessed 29 Nov 2022

Hamet P, Tremblay J (2017) Artificial intelligence in medicine. Metabolism 69:S36–S40. https://doi.org/10.1016/j.metabol.2017.01.011

Holmes KV (2003) SARS coronavirus: a new challenge for prevention and therapy. J Clin Investig 111(11):1605–1609. https://doi.org/10.1172/JCI18819

Johns Hopkins University (2022) Web map of COVID-19 [WWW Document]. https://gisanddata.maps.arcgis.com/apps/dashboards/bda7594740fd40299423467b48e9ecf6. Accessed 29 Nov 2022.

Kavroudakis D (2022) CRISTINA project [WWW Document]. https://www.dimitrisk.gr/covid19.html. Accessed 29 Nov 2022

Locke S, Bashall A, Al-Adely S, Moore J, Wilson A, Kitchen GB (2021) Natural language processing in medicine: a review. Trends Anaesthesia Critical Care 38:4–9. https://doi.org/10.1016/j.tacc.2021.02.007

Mathieu E, Ritchie H, Rodés-Guirao L, Appel C, Giattino C, Hasell J, Macdonald B, Dattani S, Beltekian D, Ortiz-Ospina E, Roser M (2020) Coronavirus pandemic (COVID-19). Our world in data. Retrieve from: https://ourworldindata.org/coronavirus#citation

Park H-A, Jung H, On J, Park SK, Kang H (2018) Digital epidemiology: use of digital data collected for non-epidemiological purposes in epidemiological studies. Healthc Inform Res 24(4):253–262. https://doi.org/10.4258/hir.2018.24.4.253

Rajkomar A, Dean J, Kohane I (2019) Machine learning in medicine. New Engl J Med 380(14):1347–1358. https://doi.org/10.1056/NEJMra1814259

Saecker M, Markl V (2012) Big data analytics on modern hardware architectures: a technology survey. In: Aufaure MA, Zimányi E (eds) Business intelligence. eBISS 2012. Lecture Notes in Business Information Processing, 138. Springer, Berlin, Heidelberg, pp 125–149. https://doi.org/10.1007/978-3-642-36318-4_6

Salathé M (2018) Digital epidemiology: what is it, and where is it going? Life Sci Soc policy 14:1–5. https://doi.org/10.1186/s40504-017-0065-7

Salathe M, Bengtsson L, Bodnar TJ, Brewer DD, Brownstein JS, Buckee C, Campbell EM, Cattuto C, Khandelwal S, Mabry PL (2012) Digital epidemiology. PLoS Comput Biol 8(7):e1002616. https://doi.org/10.1371/journal.pcbi.1002616

Taubenberger JK, Morens DM (2006) 1918 influenza: the mother of all pandemics. Emerg Infect Dis 12(1):15–22. https://doi.org/10.3201/eid1201.050979

WHO (2022) Coronavirus disease (COVID-19)—World Health Organization [WWW Document]. https://www.who.int/emergencies/diseases/novel-coronavirus-2019. Accessed 29 Nov 2022

Zimmer SM, Burke DS (2009) Historical perspective—emergence of influenza A (H1N1) viruses. New Engl J Med 361(3):279–285. https://doi.org/10.1056/NEJMra0904322

Development of a Synthetic Index of Social Vulnerability to COVID-19 in the City of Zaragoza (Spain)

Julia de las Obras-Loscertales Sampériz, María Zúñiga Antón, Marcos Rodrigues Mimbrero, and Carmen Bentué Martínez

Abstract The impacts of COVID-19 are not uniformly distributed, but are closely related to the vulnerability of any given society. This study characterises social vulnerability to COVID-19 in the city of Zaragoza; in other words, the predisposition of certain groups to suffer greater impacts. A new vulnerability index (COVID-19 SVI) was created for this purpose, and its factors (population density, age, female population, foreign population, overcrowding, education level and income level) were weighted by the analytic hierarchy process (AHP) technique through a survey of 404 people from different professional profiles (health care, geography, social work and others). The most valued factors were population density, overcrowding and age, with no differences in criteria by professional profile. By mapping the results at different scales (district, basic health area, census section and block), spatial distribution patterns were identified and the suitability of the scales for analysis compared, revealing the census section as the best option. The use of COVID-19 SVI in social and public health policies can provide essential support to management and decision-making adapted to different situations in the short, medium and long term.

Keywords Geography · Social vulnerability · Decision-making · Public health · Socioeconomic characteristics · Cartography

---

J. de las Obras-Loscertales Sampériz
Instituto de Economía, Geografía y Demografía, CCHS-CSIC, C/Albasanz 26, Madrid, Spain
e-mail: julia.des.samperiz@cchs.csic.es

M. Zúñiga Antón (✉) · M. Rodrigues Mimbrero · C. Bentué Martínez
Departamento de Geografía y Ordenación del Territorio, Instituto Universitario de Investigación en Ciencias Ambientales de Aragón (IUCA), Universidad de Zaragoza, Pedro Cerbuna 12, Zaragoza, Spain
e-mail: mz@unizar.es

M. Rodrigues Mimbrero
e-mail: rmarcos@unizar.es

C. Bentué Martínez
e-mail: cbentue@unizar.es

A. Klonari et al. (eds.), *Re-visioning Geography*, Key Challenges in Geography,
https://doi.org/10.1007/978-3-031-40747-5_12

Introduction

The World Health Organisation (WHO) declared COVID-19 a pandemic on 11 March 2020. The spread of the new SARS-CoV-2 has been labelled the worst humanitarian crisis since World War II and the greatest health crisis since the twentieth century. Institutions such as the International Federation of Red Cross and Red Crescent Societies classified the pandemic as a natural disaster. As in any disaster, the effects of COVID-19 are not uniformly distributed, but have been closely related to the unequal level of vulnerability associated with welfare and development levels (Barrera et al. 2021; Seddighi 2020; Vincent 2004).

Since the outbreak of the pandemic, several studies have examined the relationship between vulnerability and COVID-19, with prominent ones focused on exposure. For instance, some works in the USA analyse the correlation at different scales between high vulnerability—measured through the Social Vulnerability Index (SVI)—and great likelihood of becoming an infection hotspot (Dasgupta et al. 2020; Karaye and Horney 2020); between high vulnerability and a high ratio of hospitalisation and mortality (Gaynor and Wilson 2020); or between higher vulnerability and a lower number of PCR tests, but a higher fraction of positive results (Borjas 2020). Similar studies conducted in Europe have established the relationship between variables such as age, wealth and ethnic origin and pandemic-induced mortality (Harris 2020).

This work takes a new approach to the study of vulnerability, viewing certain groups as predisposed to suffer greater effects due to socioeconomic factors, which are known as the Social Determinants of Health (SDHs) (World Health Organization 2003). This approach, linked to COVID-19, has not yet been fleshed out. However, some authors have already stressed the important role that SDHs and socioeconomic factors will play in intensifying the effects of the current pandemic.

Socioeconomic and urban spatial inequalities suggest that the pandemic's social impacts will differ according to intersectoral contrasts in variables such as overcrowding, age and income. As such, characterising, estimating and mapping vulnerability can provide essential support for management and decision-making, while laying the ground for improving future social, economic, housing and health conditions.

The main objective of this article is therefore to develop a new index of social vulnerability to COVID-19 (COVID-19 SVI) in the city of Zaragoza to characterise spatial distribution patterns, delving into the possible underlying causes and identifying the factors that experts from different fields consider the most significant in producing social vulnerability.

Methodology

This study was conducted in Zaragoza, the fifth-most populous city in Spain. Since its population of 681,877 is older than the national average, with an ageing ratio of 140.97% versus 125.75% (Instituto Nacional de Estadística), it has particular features that make it interesting to study, such as the coexistence of different urban and residential typologies. Zaragoza is characterised by an urban morphology that includes: (i) a historical centre with narrow and winding streets; (ii) working-class neighbourhoods with high population densities related to urban growth in the 1960s and 1970s; and (iii) suburbs with low population densities linked to the growing sprawl of recent years. This study will perform a multi-scale analysis at four different scales (Fig. 1):

District: territorial divisions with their own administrative bodies. Zaragoza is divided into 15 districts.

Basic health area (BHA): defined as the basic geographical and population framework of Primary Health Care, which may or may not coincide with other territorial demarcations (Fig. 1). There are 33 BHAs in Zaragoza and each has its own Health Council.

Census section: defined as a statistical partition consisting of population of between 500 and 2,000 inhabitants (Ley Orgánica 5/1985). Zaragoza has 491 sections, yet these demarcations vary over time (Villarín Clavería and Segura Calero 2015).

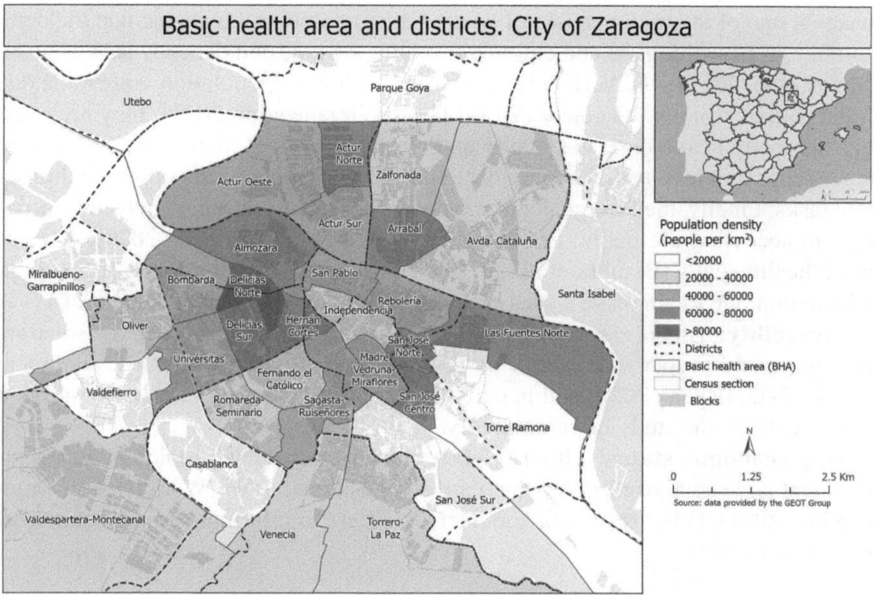

Fig. 1 Case study: City of Zaragoza

Block: defined as "the distance along a street between two intersections" (Cambridge Dictionary). Zaragoza has 2,815 inhabited blocks.

Since the outbreak of the pandemic, Zaragoza has had to deal with several waves of COVID-19 with different characteristics. As of 11 November 2022, the figures were 463,913 confirmed cases and 5,185 deaths (Gobierno de Aragón 2021).

Vulnerability Variables and Factors

Following a review of the literature, seven sociodemographic variables belonging to four dimensions were selected to construct the proposed vulnerability index (Table 1 and Fig. 2):

Urban space configuration: The layout of urban space can significantly influence the transmission of the disease (Sigler et al. 2021) and plays an essential role in producing stress, which has consequences on health (Dorantes Rodríguez 2002). The selected variable—population density per built-up area—was calculated by adding the surface area (in km^2) of the inhabited blocks for each scale of analysis and dividing the result by the total population in 2020. The aim of this operation was to overcome the limitations arising from the size of the demarcations (Escolano-Utrilla 2002).

Demography and population: The WHO considers the pandemic a geriatric emergency owing to the high death rates, the difficultly in diagnosing the disease (Pinazo-Hernandis 2020) and the clinical complexity (Boccardi et al. 2020), which makes it one of the major causes of death. Other related problems include the deterioration of psychological and emotional health, solitude and dependence (Alguacil Gómez et al. 2013). The WHO also considers the female population vulnerable due to the role of women in society (World Health Organization 2021); they are overrepresented in health care, hospitality and cleaning services, face greater economic dependence and less stable employment and shoulder more of the burden in household tasks. Finally, the foreign population was chosen, as foreigners may have problems in accessing the public health services (Gobierno de España 2020), a worse initial health status (Bambra et al. 2020) or impediments such as lack of access to information due to language issues or unstable employment (Liem et al. 2020).

Liveability: This is essential for complying with key policies such as isolation, quarantine and proper ventilation. Its relevant role in psychological development (Gobierno de España 2020) and in daily life (Alguacil Gómez et al. 2013) makes it fundamental in the study of vulnerability.

Socioeconomic status: This is closely linked to unstable employment, unemployment and social exclusion, which are all considered SDH (World Health Organization, 2021). Other related factors include access to information and technology (Cutter et al. 2003).

Table 1 Summary of selected variables

Dimension	Factor	Variable	Interpretation	Source
Urban space configuration	Population density	Population density per built-up area (inhabitants/km^2)	Higher density, higher vulnerability	Municipal Register February 2020
Demography and population	Age	Percentage of population over 60	Higher percentage of population over 60, higher vulnerability	Municipal Register February 2020
	Female population	Percentage of female population	Percentage of female population, higher vulnerability	Municipal Register February 2020
	Foreign population	Percentage of foreign population	Percentage of foreign population, higher vulnerability	Municipal Register February 2020
Liveability	Overcrowding	Percentage of people living with five or more people in less than 90 m^2	More people, higher vulnerability (higher overcrowding, higher vulnerability)	Cadastre February 2020
Socioeconomic status	Education level	Percentage of population that is illiterate or without primary education	Lower education level, higher vulnerability (higher percentage, higher vulnerability)	Municipal Register February 2020
	Income level	Per capita income (€)	Lower income level, higher vulnerability	National Statistics Institute (INE, acronym in Spanish), June 2017

Construction and Calculation of the Social Vulnerability Index

Applying weights to indexed variables is both essential and complex, and as all techniques have advantages and disadvantages. To avoid the dependence of statistical techniques, such as the principal components (Spielman et al. 2020), weights were obtained through a participatory process in which experts belonging to different fields

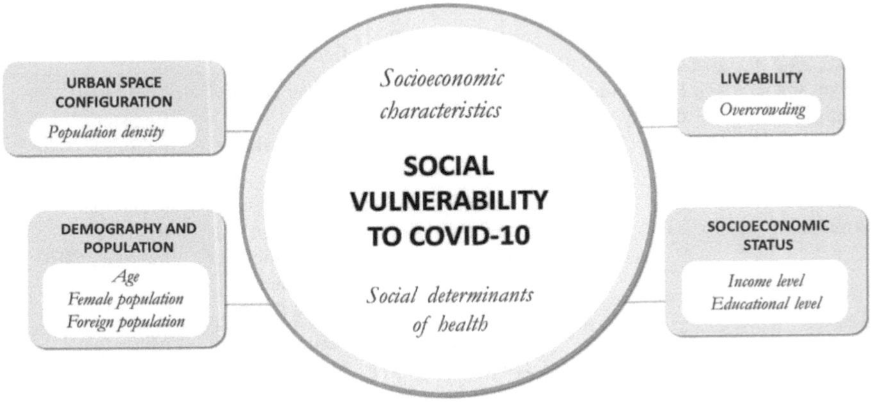

Fig. 2 Summary of selected variables. Diagram of how the selected variables interact with and influence social vulnerability. Created by the authors

(health care, social work and geography) and different sectors of the population took part in a survey. The survey was disseminated through different media and social networks both to specific research groups and to the general public. Although at first it was thought to be a random spatial sampling, in the end it was mainly focused on inhabitants and researchers of the city under study, Zaragoza.

The survey and methodology used to obtain weights were based on the analytic hierarchy process (AHP) technique, used and contrasted in many fields and applications. To get the weights, this method is based on conducting a pairwise comparison of the importance of each variable (Saaty 2008), in this case as vulnerability factors. A survey was executed for this purpose. The first section asked the respondents for general data that would later be needed for profile analysis (occupation, age or birthplace). The second section compared the factors. There were 21 questions in total in random order, so as not to put the last answers at a disadvantage. For each question, the survey respondent had to assess the relative importance of one variable compared to another on an ordinal scale: "very strong /strong /moderately strong / of equal importance /moderately low/low /very low". As this process adhered to the principle of reciprocity, the reciprocals were calculated with the answers obtained.

The results were processed with the AHP technique, using the "ahpsurvey" package in R software (Cho 2018). As a result, a matrix of weights by factor and professional profile was obtained. Variables were then normalised in a common range between 0 and 1 (Eq. 1) and the COVID-19 social vulnerability index (COVID-19 SVI) was calculated.

$$V_{n,v} = \frac{V_v - V_{v,\min}}{V_{v,\max} - V_{v,\min}} \tag{1}$$

where $V_{n,v}$ is the normalised value v; V_v is the original value; $V_{v,\min}$. is the minimum value; and $V_{v,\max}$ is the maximum value.

The final result of the COVID-19 SVI is a number between 0 (very low vulnerability) and 1 (very high vulnerability). The formula used to calculate the "index" is described below (Eq. 2).

$$\begin{aligned}\text{Covid19SVI} = {} & (\text{normPD} \times \text{weightPD}) + (\text{normA} \times \text{weightA}) \\ & + (\text{normFeP} \times \text{weightFeP}) + (\text{normFoP} \times \text{weightFoP}) \\ & + (\text{normOvc} \times \text{weightOvc}) + (\text{normEduc} \times \text{weightEduc}) \\ & + ((1 - \text{normInc}) \times \text{weightInc}) \end{aligned} \tag{2}$$

This process was repeated for every scale of analysis proposed: district, BHA, census section and block. The results were then mapped with a qualitative legend with five intervals (very high, high, moderate, low and very low vulnerability), discretised by the quintile method to compare both the different factors and the multi-scale analysis.

Results

Survey Results

The results of the survey, which was responded by 404 subjects, reached a significant level of $p < 0.05$ with a 95% confidence interval. Forty-five per cent of the respondents were aged 18–40, 34.4% were aged 41–60 and 20% were over 60. Broken down by sex, 59.4% were female and 40.6% male. By professional profile, 24.8% of the respondents came from the healthcare sector, 20.5% from the geography sector, 16.6% from the social work sector and 1.7% from the Public Health Administration (PHA), which was added to the social work profile. The "other" category accounted for the remaining 36.3%. Most respondents (72.8%) lived in the city of Zaragoza and 87.9% said they had knowledge about the city. In addition, 18% of the people surveyed suggested other factors to be included, 11.1% added other comments and 24.7% left their email address to receive information about the results (Fig. 3).

There are hardly differences among the profiles surveyed (Fig. 4). The three most valued factors are overcrowding (0.223), population over 60 years old (0.218) and population density (0.204). However, despite the apparent "agreement" among professional profiles, there is a certain dispersion in the individual ratings of each group. This is especially noticeable in factors with the lowest weights, such as female population (0.073), income level (0.103), education level (0.08) and foreign population (0.099).

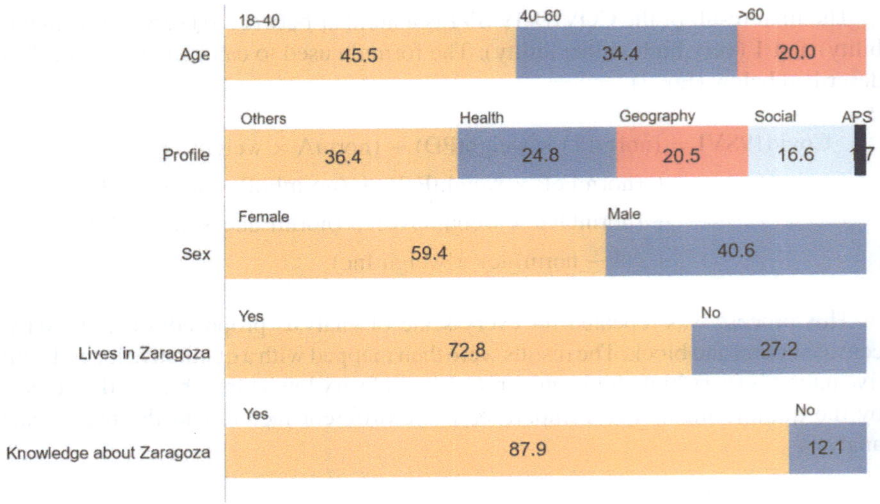

Fig. 3 General data of respondents

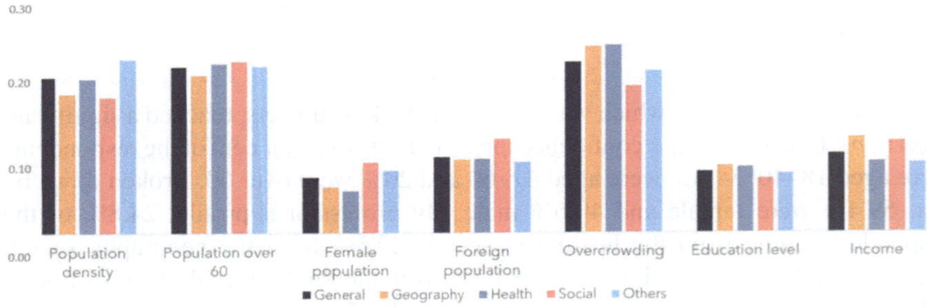

Fig. 4 Average weights assigned to vulnerability indicators per profile

Distribution and Characterisation of Social Vulnerability Factors at Different Scales

The spatial distribution of the variables shows a clear and identifiable pattern at all scales of analysis. However, the districts were seen as too large and varied to serve as a unit of analysis, as their mean values masked many internal differences. As such, this section will present the results of the BHA and census section for each factor and the index, while the latter will also be mapped at the block level.

BHAs in working-class districts have the highest vulnerability values of almost any variable, especially in the following factors: foreign population, overcrowding, education level and income level (Fig. 5). By contrast, BHAs in suburbs show values belonging to the first two quintiles for all variables except income level, which has average values compared to the rest of the city. Population density values here are much lower than in the rest of the city.

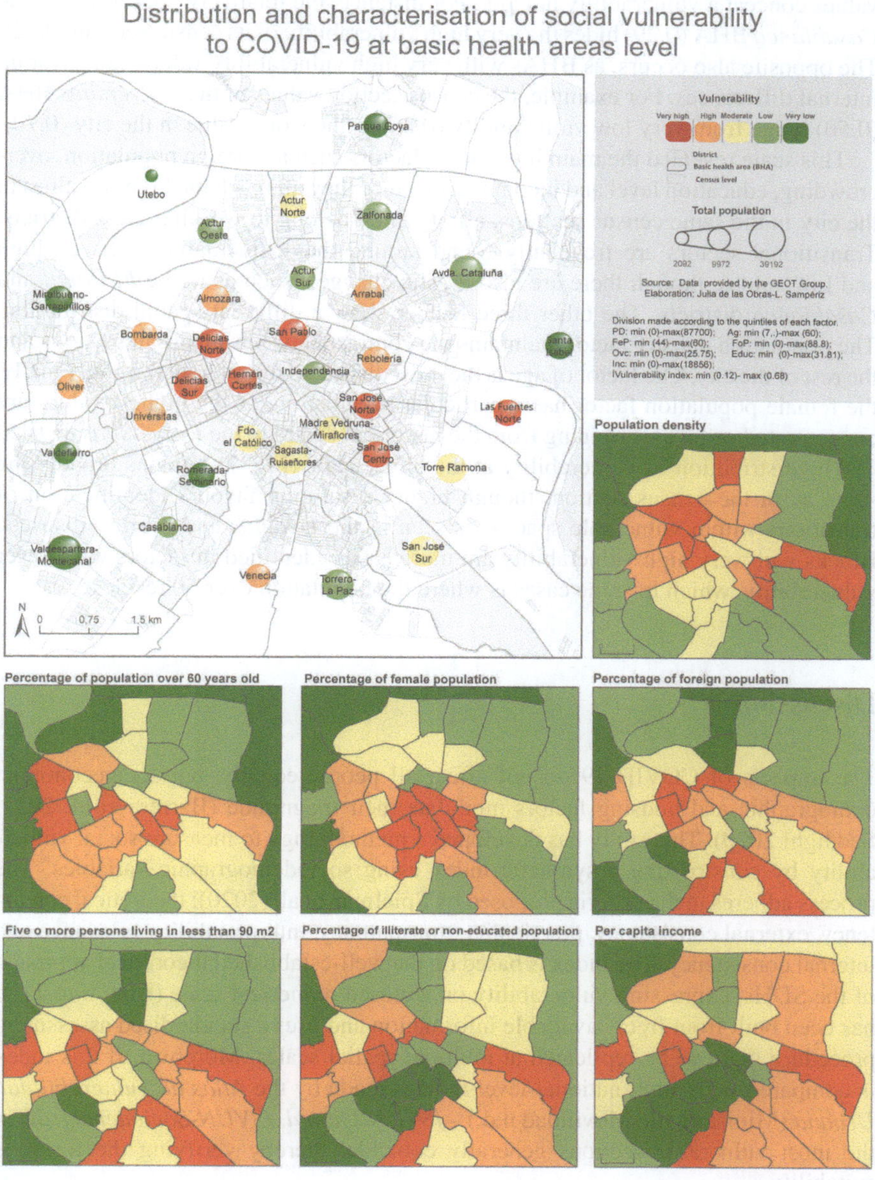

Fig. 5 Distribution and characterisation of social vulnerability to COVID-19 at the BHA scale by the GEOT Group

The pattern continues at the census section scale, but differences and nuances increase in the distribution (Fig. 6). Even though most sections with the highest vulnerability values belong to the BHAs, there are also cases where the average BHA values conceal a vulnerability hotspot. For instance, the relatively good value of the *Casablanca* BHA (0.29) hides the very high vulnerability of its census section (0.52). The opposite also occurs, as BHAs with very high vulnerability values can cover up internal differences. For example, the census section values of the *Univérsitas* BHA (0.50) range from very low vulnerability (0.26) to the worst value in the city (0.70).

This scale revealed the main hotspot for factors such as foreign population, overcrowding, education level and income level, since they all reach the highest values in the city in the same census sections, all of which belong to working-class districts. Transitional sectors are frequently found around them, fluctuating between high and low values, though there are also drastic changes, such as in the *El Rabal* and *Casablanca* districts. The other three factors show a different spatial distribution. The population density factor maintains the contrast between suburban sections and the rest of the city. The factor of age is the most dispersed throughout the city. Finally, the female population factor has a particular distribution, as the highest values are concentrated on an axis running from the historical centre to the *Univérsitas* district.

The distribution of vulnerability at the block scale (Fig. 7) remains broadly the same as in the census section, though higher resolution favours identification of transitions within vulnerable spaces, as seen in the *Delicias* and *Torrero* districts. Blocks with very high vulnerability are more easily identified in sectors with lower vulnerability, which in some cases is where the population over 60 resides.

Discussion

The impacts of COVID-19 do not affect all people equally, but socioeconomic, demographic and housing factors modulate their magnitude (Barrera et al. 2021; Seddighi 2020). This study has developed a methodology to measure social vulnerability by constructing a synthetic index using sociodemographic variables. The process adheres to the criteria proposed by Spielman et al. (2020): theoretical consistency, external consistency, practicality, transparency, interpretability, relevance and internal consistency. The index is based on the well-established theoretical approach of the SDHs, expressing vulnerability on a closed numerical scale (from 0 to 1). It has been built from freely available information and uses a standardised assessment procedure that can be replicated at multiple spatial scales. Moreover, if this index is compared with other national-level indices made by the *Atlas de Vulnerabilidad Urbana* (Ministerio de Movilidad n.d.)—*ISVUN-Global, ISVUN-Sociodemográfico*, the most vulnerable sections generally coincide, thereby verifying the former's suitability.

An essential resource resulting from this study is mapping, which can be key for COVID-19 management (Rose-Redwood et al. 2020). The spatial knowledge produced helps to understand complex phenomena in origin involving different

Fig. 6 Distribution and characterisation of social vulnerability to COVID-19 at census section scale by the GEOT Group

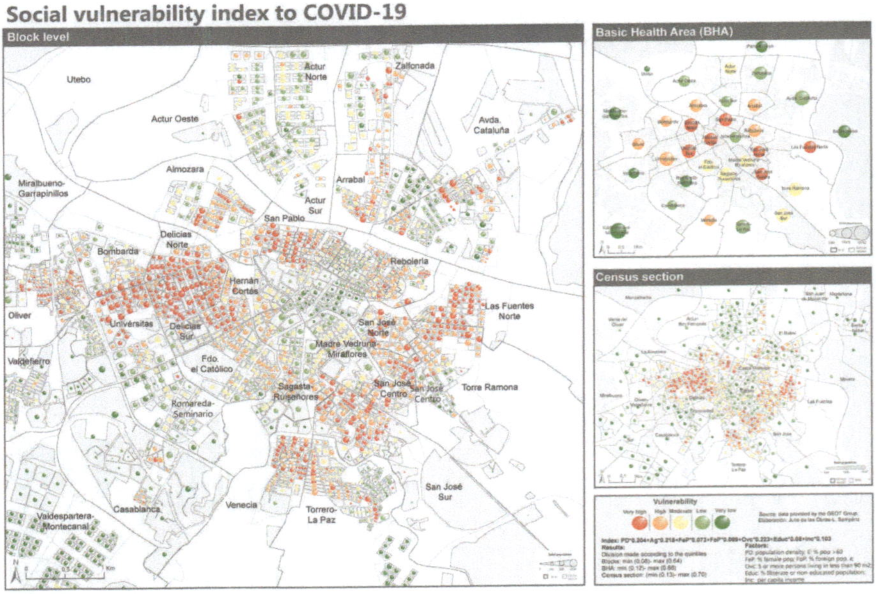

Fig. 7 Distribution of social vulnerability index to COVID-19

factors (Pueyo-Campos et al. 2016). For instance, this analysis reveals distribution patterns, which are fundamental for understanding the explanatory factors and therefore require action. In Zaragoza specifically, the distribution of vulnerability is closely related to the effects of urban transformation during the twentieth and twenty-first centuries (Escolano-Utrilla et al. 2018; de Miguel González 2014; Monclús 2013; Pérez-Moreno and Kurtz 2019). As such, the spatial pattern is characterised by alternating concentric rings reflecting socioeconomic and urban differences, thereby confirming the strong social component of COVID-19. At the centre of these rings is the *Independencia* sector, which has "low vulnerability". The commercial and service orientation of this part of the city, as well as the high price of land, as the latest data show (B.A. 2020), attracts upper-class residents to settle in large homes in a space with low population density. These characteristics are extended along small axes starting from the centre, such as *Sagasta* and *Gran Vía* (Escolano-Utrilla et al. 2018).

The second ring includes the city's working-class districts: *Delicias, Almozara, Las Fuentes, San José, Torrero* and *El Rabal*, which exhibit higher levels of vulnerability. This can be explained by the lack of planning and facilities in these districts during the population explosion in the twentieth century. The boom was particularly intense in the 1960s, when 5,000 social housing units (Pérez-Moreno and Kurtz 2019) and 60,000 new housing units are estimated to have been built without the necessary planning (de Miguel González 2014), with consequences that are still felt today. These districts are characterised by high population densities and poor-quality housing, with a high ageing rate that is only reversed in some sectors by the arrival of foreign populations attracted by lower land prices (Escolano-Utrilla et al. 2018).

The last ring includes recently built districts located between the belts formed by the Z-30 and Z-40 roads. In contrast to the second ring, these neighbourhoods are characterised by lower population densities and a higher rate of young people who have left the city centre in recent years (Escolano-Utrilla et al. 2018). Thus, this ring has low levels of vulnerability.

Nevertheless, these rings show significant differences in terms of urban morphology, housing quality and urban space, which are decisive in population density and land prices, and therefore in other factors analysed. This leads to differences between and within districts and creates vulnerability hotspots due to the overlap of several factors: foreign population, overcrowding, income level and education level.

This study reveals that selecting the right unit of analysis is key to pandemic management. In the light of the results, the BHA appears too varied to serve as a unit of analysis, as its average values may mask a vulnerability hotspot and significant internal differences due to its large size. Instead, the census section seems to be the most appropriate unit of analysis due to its similarity to the block-level distributions and because it is a consistent and valuable source of information (Villarín Clavería and Segura Calero 2015), despite some problems such as it susceptibility to change.

Finally, it is necessary to consider what action should be taken to improve the situation. While taking this into account in short-term measures to prevent an increase in socioeconomic and health inequalities (World Health Organization 2021), it can also inform mid- and long-term strategies. For example, in the medium term, social policies are needed to target, monitor and support the most vulnerable groups, while long-term policies should aim to lower vulnerability levels, especially as the effects of COVID-19 are likely to exacerbate social inequalities (Ocaña et al. 2020). As such, this article argues for the need to improve social policy management and coordination, especially with regard to providing health services. It also calls for the implementation of strategies targeting factors that can serve as a driving force to improve the rest, such as support for education and inclusion. While urging continuous progress in gathering knowledge about the SDHs, it also proposes future lines of research, such as cross-referencing levels of vulnerability with incidence and fatality rate data to check if there is a correlation and monitoring vulnerability in these spaces to quantify the consequences of the pandemic.

Conclusions

The impacts of the COVID-19 pandemic will not affect all groups equally. Therefore, the creation and mapping of a vulnerability index can be crucial for pandemic-related decision-making and for identifying spatial distribution patterns. While constructing and applying this index of social vulnerability to COVID-19 (COVID-19 SVI), some conclusions of this research can be extrapolated to other case studies.

The factors with the highest values were overcrowding, age and population density, while the factors with the lowest values were female population, education level, income level and foreign population. However, there were a range of opinions when assessing the factors, particularly those with lower values. Nevertheless, this diversity does not reflect the different fields of knowledge so much as each respondent's individual opinion.

Once the multi-scale analysis was carried out, the importance of selecting the right scale for the analysis became clear. The conclusion was that the census section provides the optimal scale in this case, since a smaller scale would conceal internal differences in the neighbourhoods and preclude the proper detection of vulnerability hotspots.

Finally, social vulnerability to COVID-19 in Zaragoza is reflected in a spatial pattern of alternating concentric rings closely related to urban morphologies and the quality of housing and urban space. Areas dominated by the commercial sector show lower vulnerability, while working-class districts, which are more dense and populous, present high levels of vulnerability. Factors such as overcrowding, foreign population, education level and income level overlap in certain working-class areas, producing significant vulnerability hotspots.

In conclusion, this article advocates the measurement and spatialisation of vulnerability as a key management and decision-making tool. It also stresses the need to monitor socioeconomic developments and the social determinants of health and to implement policies to improve these spaces and prevent the situation from worsening. For all these reasons, further work on this subject is needed, as it has received too little study to date.

References

Alguacil Gómez J, Camacho Gutiérrez J, Hernández Ajá A (2013) La vulnerabilidad urbana en España. Identificación y evolución de los barrios vulnerables. Empiria. Revista de Metodología de Ciencias Sociales 27:73–94. https://doi.org/10.5944/empiria.27.2014.10863

B.A (2020) El mapa de los precios de la vivienda en zaragoza por códigos postales. Heraldo de Aragón. https://www.heraldo.es/noticias/economia/2020/12/21/mapa-de-los-precios-de-venta-de-la-vivienda-en-zaragoza-por-codigos-postales-1410984.html. Accessed 7 Dec 2022

Bambra C, Riordan R, Ford J (2020) The COVID-19 pandemic and health inequalities. J Epidemiol Community Health 74:964–968. https://doi.org/10.1136/jech-2020-214401

Barrera A, Bonilla A, Espinosa S et al (2021) Índice de vulnerabilidad y trayectorias espaciales del covid-19 en el distrito metropolitano de Quito. Geopolítica(s). Revista de Estudios sobre Espacio y Poder 12(1):51–76. https://doi.org/10.5209/geop.70908

Boccardi V, Ruggiero C, Mecocci P (2020) COVID-19: a geriatric emergency. Geriatrics 5(2). https://doi.org/10.3390/geriatrics5020024

BOE. Legislación consolidada. Ley Orgánica 5/1985, de 19 de Junio, del Régimen Electoral General. https://www.boe.es/eli/es/lo/1985/06/19/5/con. Accessed 2 Dec 2022

Borjas GJ (2020) Demographic determinants of testing incidence and COVID-19 infections in New York City neighborhoods. SSRN Electron J. https://doi.org/10.2139/ssrn.3572329

Cambridge Dictionary. https://dictionary.cambridge.org/. Accessed 5 Dec 2022

Cho F (2018) Analytic hierarchy process for survey data in R. https://cran.r-project.org/web/packages/ahpsurvey/vignettes/my-vignette.html
Cutter SL, Boruff BJ, Shirley WL (2003) Social vulnerability to environmental hazards. Soc Sci Q 84(2):242–261. https://doi.org/10.1111/1540-6237.8402002
Dasgupta S, Bowen VB, Leidner A et al (2020) Association between social vulnerability and a county's risk for becoming a COVID-19 hotspot—United States, June 1–July 25, 2020. MMWR. Morb Mortal Weekly Rep 69(42):1535–1541. https://doi.org/10.15585/mmwr.mm6942a3
Decreto Legislativo 2/2004, de 30 de Diciembre, Del Gobierno de Aragón, por el que se aprueba el texto refundido de la Ley del Servicio Aragonés de Salud. https://www.boe.es/buscar/act.php?id=BOA-d-2005-90000. Accessed 7 Dec 2022
de Miguel González R (2014) Metamorfosis urbana en Zaragoza. Biblio 3W. Revista Biblio-gráfica de Geografía y CIencias Sociales XIX(1083). https://www.ub.edu/geocrit/b3w-1083.htm
Dorantes Rodríguez C (2002) El estrés y la ciudad. Revista del Centro de Investigación de la Universidad La Salle 5(17/18). https://doi.org/10.26457/recein.v5i17/18.620
Escolano-Utrilla S (2002) Densidad de población y sustentabilidad en la ciudad de Zaragoza. In: Peña JL, Longares A (eds) Aportaciones geográficas en memoria del Profesor Miguel Yetano. Universidad de Zaragoza, Zaragoza, pp 173–182
Escolano-Utrilla S, López-Escolano C, Pueyo-Campos A (2018) Urbanismo neoliberal y fragmentación urbana: el caso de Zaragoza (España) en los primeros quince años del siglo XXI. EURE (Santiago) 44(132):185–212. https://doi.org/10.4067/s0250-71612018000200185
Gaynor TS, Wilson M (2020) Social vulnerability and equity: the disproportionate impact of COVID-19. Public Adm Rev 80(5):832–839. https://doi.org/10.1111/puar.13264
Gobierno de Aragón (2021) Mapa de casos COVID-19 en Aragón. https://datacovid.salud.aragon.es/covid/. Accessed 6 Dec 2022
Gobierno de España (2020) Equidad en salud y COVID-19. https://www.sanidad.gob.es/profesionales/saludPublica/ccayes/alertasActual/nCov/documentos/COVID19_Equidad_en_salud_y_COVID-19.pdf. Accessed 2 Dec 2022
Harris R (2020) Exploring the neighbourhood-level correlates of Covid-19 deaths in London using a difference across spatial boundaries method. Health & Place 66(102446). https://doi.org/10.1016/j.healthplace.2020.102446
Instituto Nacional de Estadística (2020) Indicadores Demográficos Básicos https://www.ine.es/dyngs/INEbase/es/operacion.htm?c=Estadistica_C&cid=125473617%0A7003&menu=resultados&idp=1254735573002#!tabs-1254736195380. Accessed 6 Dec 2022
Karaye IM, Horney JA (2020) The impact of social vulnerability on COVID-19 in the U.S.: an analysis of spatially varying relationships. Am J Prev Med 59(3):317–325. https://doi.org/10.1016/j.amepre.2020.06.006
Liem A, Cheng W, Wariyanti Y (2020) The neglected health of international migrant workers in the COVID-19 epidemic. Lancet Psychiatry 7(4). https://doi.org/10.1016/S2215-0366(20)30076-6
Ministerio de Movilidad, Transporte y Agenda Urbana. Atlas de Movilidad Urbana. https://atlasvulnerabilidadurbana.mitma.es/#c=home. Accessed 6 Dec 2022
Monclús FJ (2013) Paisajes urbanos residenciales en la Zaragoza contemporánea. Prensas Universitarias, Zaragoza
Ocaña C, Bandrés E, Chuliá E et al (2020) Impacto social de la pandemia en España. una evaluación preliminar. Funcas, Madrid
Pérez-Moreno L, Kurtz F (2019) La vivienda social en la ciudad de Zaragoza bajo las políticas del régimen franquista. In: López-Mesa B (ed) Nuevos enfoques en la rehabilitación energética de la vivienda hacia la convergencia europea La vivienda social en Zaragoza, 1939–1979. Prensas de la Universidad de Zaragoza, pp 77–94
Pinazo-Hernandis S (2020) Impacto psicosocial de la COVID-19 en las personas mayores: problemas y retos. Revista Española de Geriatría y Gerontología 55(5): 249–252. https://doi.org/10.1016/j.regg.2020.05.006
Pueyo-Campos A, Postigo-Vidal R, Arranz-López A et al (2016) La cartografía temática: una herramienta para la gobernanza de las ciudades. Aportaciones de la semiología gráfica clásica en

el contexto de los nuevos paradigmas geográficos. Revista de Estudios Andaluces 33(1):84–110. https://doi.org/10.12795/rea.2016.i33.05

Rose-Redwood R, Kitchin R, Apostolopoulou E et al (2020) Geographies of the COVID-19 pandemic. Dialogues Hum Geogr 10(2):97–106. https://doi.org/10.1177/2043820620936050

Saaty TL (2008) Decision making with the analytic hierarchy process. Int J Serv Sci 1(1):83. https://doi.org/10.1504/IJSSCI.2008.017590

Seddighi H (2020) COVID-19 as a natural disaster: focusing on exposure and vulnerability for response. Disaster Med Public Health Prep 14(4):42–43. https://doi.org/10.1017/dmp.2020.279

Sigler T, Mahmuda S, Kimpton A et al (2021) The socio-spatial determinants of COVID-19 diffusion: the impact of globalisation, settlement characteristics and population. Glob Health 17(1):56. https://doi.org/10.1186/s12992-021-00707-2

Spielman SE, Tuccillo J, Folch DC et al (2020) Evaluating social vulnerability indicators: criteria and their application to the social vulnerability index. Nat Hazards 100(1):417–436. https://doi.org/10.1007/s11069-019-03820-z

Villarín Clavería MC, Segura Calero S (2015) Relevancia de La Sección Censal Como Escala de Estudio. Caso Práctico Del Consumo Doméstico de Agua En El Municipio de Sevilla. In: de la Riva J, Ibarra P, Montorio R, Rodrigues M (eds) Análisis espacial y representación geográfica: innovación y aplicación. Universidad de Zaragoza-AGE, pp 433–441

Vincent K (2004) Creating an index of social vulnerability to climate change in Africa. Tyndall Centre Working Paper No. 56

World Health Organization (2003) Social determinants of health: the solid facts. World Health Organization

World Health Organization. Regional Office for Europe (2021) Strengthening population health surveillance: a tool for selecting indicators to signal and monitor the wider effects of the COVID-19 pandemic. World Health Organization. Regional Office for Europe

Students' Satisfaction with Synchronous Online Learning in Times of COVID-19: A Case Study of Greek Geography Students

Aikaterini I. Klonari, Anthoula-Styliani Passadelli, and Ourania Rizou

Abstract The COVID-19 pandemic has had a great impact in education, and as an emergency response, education implemented some form of remote teaching and learning. This study aimed at describing the satisfaction level of university students in synchronous online teaching and learning at the Geography Department, University of the Aegean, Greece, during the COVID-19 pandemic. The sample of the research was 60 undergraduate students, who participated in online synchronous teaching geography courses in 2020–2021 (Semesters: Spring - Fall 2020; Spring 2021). Data collection was performed via a Google Forms questionnaire which the researchers distributed via e-mail; data analysis was done quantitatively, aimed at measuring students' satisfaction on the quality of applying the online teaching methods. The result of students' responses showed that 35 students (65.0%) were satisfied with the implementation of online learning organized by the department, and 13 other students (21.7%) stated that they were very satisfied, although there were also a small number namely 8 students (13.3%) expressed dissatisfaction. Most students were satisfied with the good communication with the professors and the students believe that the professors were facilitators. However, the students stated that they were dissatisfied with their effectiveness because they were bored with the independent assignments.

Keywords COVID-19 pandemic · Synchronous online learning · Satisfaction level · University students

A. I. Klonari · A.-S. Passadelli · O. Rizou (✉)
Department of Geography, University of the Aegean, Mytilene, Greece

A. I. Klonari
e-mail: aklonari@geo.aegean.gr

A.-S. Passadelli
e-mail: stellapass@geo.aegean.gr

A. Klonari et al. (eds.), *Re-visioning Geography*, Key Challenges in Geography,
https://doi.org/10.1007/978-3-031-40747-5_13

Introduction

According to Rizou et al. (2021), the past two years were undoubtedly marked by the global spread of the COVID-19 virus, to such an extent that it has since become an integral part of our daily routine. Moreover, this co-existence has tipped the scales on our personal and socio-economical lives in many aspects, including education (Yan et al. 2020). It has ushered many schools around the globe to switch to online synchronous teaching methods, as a means of continuing their educational efforts despite the widespread closures of all levels of schools and institutions. A survey showed that more than 1,723 billion students (or 80%) of compulsory and university education had been lead away from school facilities (Crawford et al. 2020). Approximately 191 countries have implemented nationwide closures, affecting approximately 99.4 percent of the world's student population (World Health Organization 2020a; World Health Organization 2020b). School shutdown has affected several stakeholders such as students, teachers, and families, besides bearing other economic and social consequences (UNESCO 2020). To facilitate distant learning, such tools as online learning programs, apps, and web-based platforms were made available to the public, both students and faculty, in order to reach remote learners and reduce disruption to education (Steffens 2020; World Health Organization 2020b).

In 2020, Greece has undergone three waves of COVID-19 outbreak. On February 26, 2020, Greece recorded the first case of COVID-19. After that, there were new cases consecutively in various cities and provinces across the country. On March 23, 2020, the Prime Minister issued a directive on the implementation of urgent measures to prevent and control the COVID-19 pandemic, and in April 2020, the whole country implemented social isolation and a lockdown nationwide. In response to the emergency, most of educational institutions in the country immediately thought of ways to ensure the teaching and learning activities would take place (Government Official Journal 42/A/25.02.2020; Government Official Journal 956/B/21.03.2020; Join Ministerial Decision (JMD): No D1a/Under General Register No 20021; JMD: No. D1a/ Under General Register No 24343). Therefore, synchronous online teaching and learning[1] was applied as the main measure to deal with the situation. However, some challenges have been identified since not every institution has previous experience with e-learning. On May 11, it issued a new directive on gradual reopening shops and 3rd grade of upper secondary schools with certain personal protection measures (use of masks, plastic gloves and keeping distances). On May 18, the face-to-face teaching resumed for all courses and all secondary school students. On the contrary, online teaching continued for all higher education as well as the final exams (JMD: 63,314/General Decree (G.D.) 4/26–05-2020).

[1] Synchronous online learning means that students are required to log in and participate in class at a specific time each week. The class is a firm, weekly time commitment that cannot be rescheduled. Much like an on-campus class, students will have readings and assignments to complete outside of class time to help prepare them to participate in the discussion. This kind of preparation from students, along with a dedicated agenda, set by the instructor (Schneider 2020).

On November 14, 2020, primary schools and kindergartens closed, initially for two weeks, and from November 18, 2020, they switched to distance learning. The schools remained closed until Monday, January 11, 2021, when the primary schools and kindergartens opened while the lower secondary schools opened on Monday February 1, 2021, with the announcement of the Minister of Education, while in the "red" areas such as Attica the upper secondary schools continued the lessons with distance synchronous teaching.

On September 13, 2021, all schools started face-to-face teaching until the end of the academic year 2021–2022. Additionally, according to no. 111525/G.D. 4/10–9-2021 (Government Official Journal 4188/B) decision of the Ministry of Education, the school units of the primary and of secondary education are obliged, exceptionally, to provide distance education to students if the risk of spreading the coronavirus COVID-19 remains, or to those who are unable to attend the educational process in person.

According to the ministry of education in Greece, the universities had to do online courses. Universities can use online learning more easily than secondary education. But many universities' students had difficulties to connect on the internet network or they have not facilities. The universities' campus has the appropriate facilities, but the students were not allowed to enter in the universities during the COVID-19 era. So, it was of great importance to conduct an in-depth research on the level of satisfaction of the students regarding online learning and whether the particular method met their requirements and expectations as it is basically distance learning (Taufik and Fiptar 2022).

E-learning (e-Learning) is a form of teaching that uses the Web as its main tool. According to Nishimura, 2021, there are three main reasons why this type of learning can contribute to improving educational issues. The first one is that students can study theoretically anywhere, although an internet connection is a prerequisite. The second reason is that the quality of education provided is uniform, since the same learning materials are used by all students. Thirdly, it can be introduced at a reduced cost, in comparison. As the same teaching material can be delivered to many learners and since there is no need to secure appropriate teachers or instructors or dispatch them to the areas where the learners are located, the total cost is significantly reduced. E-learning, as a flexible learning pathway, enables the inclusion of diverse learners and improves the concept of lifelong learning. Lately, e-learning is becoming a crucial strategy that increases accessibility opportunities to higher education for all people. Several research studies have been published on the benefits of e-learning in higher education. With these benefits in mind, we consider e-learning as an opportunity to achieve the sustainability of education, given the current huge shift to digital education due to the COVID-19 pandemic (Barclay et al. 2018; Andrade and Alden-Rivers 2019; Basilaia, et al. 2020). In this chapter, we explore the impacts of the COVID-19 pandemic and e-learning on achieving one of the Sustainable Development Goals (SDG). In particular, we are referring to SDG Number Four, one of the most prominent ones, which is to provide learners with high-quality education and ensure inclusive and equitable quality education and promote lifelong learning opportunities for all.

Literature Review

The COVID-19 outbreak affected education in such manners as reducing school's productivity and teaching quality, mainly by diminishing accessibility to educational services and thus causing frustration and disorientation to students and educators alike. Limited accessibility, combined with deficiency in home teaching material and parents' preparedness for distant e-learning, further caused social isolation and emotional imbalance and in many cases, school dropouts (World Health Organization 2020a). Universities were faced with a major challenge: to protect their staff and students, while still providing quality services and high-level academic education. To counter these impacts, governments were quick to take measures to ensure education continuity through an urgent approach to online teaching and distance learning with many ongoing e-learning solutions (World Health Organization 2020a). According to Hodges et al. (2020), remote teaching and learning as a response to the emergency is a temporary change of the currently applied teaching method, to the use of fully online teaching solutions and to flexible adoption or alternating between online and traditional teaching or combined instruction when the disease is well controlled to minimize the risks and damages. The main goal of adopting online teaching methods is to provide a potentially lasting teaching method that supports the learning process and to avoid the interruption of education during the COVID-19 crisis (Hodges et al. 2020).

Synchronous online teaching is not a new concept (Singh & Thurman 2019); however, for many of academics, it is a shift away from the traditional classroom teaching model (Schneider and Council 2021). Technology is quite naturally the cornerstone of online teaching (via discussion boards or learning management systems, etc.) as geographically separated students and departments interact through two-way communication technologies (Schneider 2020) (Fig. 1). Online teaching tools that are mainly used in Greece are WebEx, Zoom, BBB (BigBlueButton) platforms, etc., and they are used in all levels of primary to tertiary education. Teaching with their aid of them can go both ways supporting synchronous and asynchronous learning.

With online learning, distance is eliminated; both teachers and students interact by seeing and chatting (with) each other while students can still perform group activities. Undoubtedly, online teaching reduces direct (i.e., physical) interaction which becomes an advantage of course in the fight against the COVID-19 spread, effectively minimizing risks and sources for it (Molina Higgins et al. 2022). However, online courses come with hidden challenges for both educators and students, making interaction and participation for students harder than it is in a conventional classroom (Lore 2000; Purcell-Robertson and Purcell 2000). In addition, during COVID-19 remote teaching practical exercises in geography courses, practice in school and fieldwork including laboratory work was not implemented. It is a well-known fact that geography education courses aid in students' developing motor skills, teamwork, and promoting social activities, through group or competitive activities—and this isn't something that can be carried out due to the circumstances enforced upon us

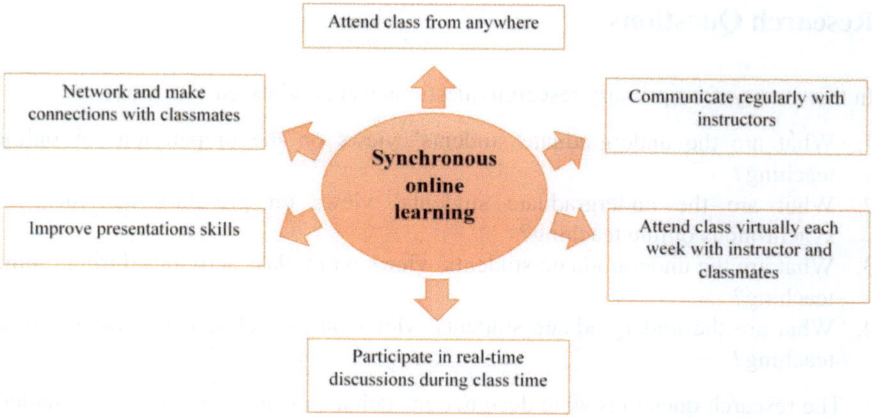

Fig. 1 Synchronous online learning (Scheiderer 2020)

by the epidemic, since none of the online geography classes can offer what they would normally. Even through online learning, where educators provide lectures via virtual classes that can be accessed anywhere and anytime, unbound by time and space (Hastuti et al. 2021), they still are unable to develop the above-mentioned competencies.

Nonetheless, during the COVID-19 outbreak, geography courses were carried out online in order to obtain learning progress. It is worth mentioning some of the advantages of such applications as online learning for students: (1) versatility (appropriate time and place); (2) enhance knowledge (accessing a large amount of data through the Web); (3) opportunities for interaction (via online discussion forums); (4) cost-effective (no need traveling); (5) compensates the lack of academic staff and facilities; (6) learning in their own pace, (asynchronous or synchronous way) (Hastuti et al. 2021). Thus, while it's not wrong to say that online teaching has more advantages than disadvantages and can guarantee, to some extent, uninterrupted learning progress against COVID-19 circumstances, we must still examine if it has a positive impact on the quality of teaching for students or not. We are hence faced with two important factors: student learning outcomes (expressed through mid-term assessment or final assessment) as well as student satisfaction with the aforementioned method. In this study, the authors focus on surveying and analyzing students' satisfaction with the applied online teaching method, which is a limitation of the study. There are a small number of studies globally, concerning the student's level of satisfaction with the applied teaching methods, and none about the application of online teaching in geography education in Greece is found. This study is done to bridge the gap, to some degree.

Research Questions

In this survey, four primary research questions were addressed:

1. What are the undergraduate students' views on the preparation for online teaching?
2. What are the undergraduate students' views on pre-class activities for synchronous online teaching?
3. What are the undergraduate students' views on in-class activities during online teaching?
4. What are the undergraduate students' views on post-class activities of online teaching?

The research questions were designed in such a way as to investigate the undergraduate students' satisfaction level concerning the preparation stage and the pre-class, in-class, and post-class activities of online teaching. Preparation steps included providing all undergraduate students with instructions regarding synchronous online learning, as well as the use of the platforms and applications they would come in contact with during the semester. Next, they were added to their respective e-class(es) by their teachers, so they could be able to upload exercises and activities or download the available teaching material. Additionally, through their e-class dashboard, they could send and receive messages and receive degrees. It was required of them to be in the possession of a digital device (laptop or PC) at home, complete with (broadband) internet connection.

With regard to the pre-class activities, teachers sent teaching material to their students to be prepared in advance for the online class, while posting messages in their e-class(es) boards providing instructions about the preparation and the tasks that students had to carry out (Gay 2016; Kay 2022). During the in-class activities, all students had to have their camera turned on and teachers facilitated discussion among them during the online course (Kay 2022). It was important for students to interact with each other because they were physically not present in the same room, and complete the given tasks on time. Lastly, the post-activities involved teachers being consistently available to correct mistakes and comment on the assignments so that their students would hand in a completed assignment or essay in the end; feedback was immediate and on-point, in order to answer questions or resolve any difficulties in the part of the students.

Research Methodology

The research carried out during the turmoil of the COVID-19 pandemic took 3 weeks to conduct (from May to June 2022). A total of 120 undergraduate students from the Geography Department of the University of the Aegean was our population, and the final sample comprised 60 (32 males or 53.30% and 28 females or 46.70%) of

Table 1 Description on the sections of the questionnaire instrument

Section	Description	Number of items
A	Demographic data	4
B	Preparation for online teaching	6
C	Pre-class activities in the online teaching	3
D	In-class activities in the online teaching	9
E	Post-class activities in the online teaching	3
	TOTAL	25

them, who have participated in online synchronous learning geography courses in 2020–2021 (Semesters: Spring - Fall 2020; Spring 2021).

An online questionnaire was created for this purpose (with the aid of Google Forms) and was sent to the selected participants. It consisted solely of closed-type questions (25 in total), and it was estimated that 10' minutes were adequate for completing and submitting it. With the exception of the demographic data section, we used Likert Scale with five (5) possible answers. The scale started (low point) with "Strongly Disagree"/ "Nothing" and went up (high point) to "Strongly Agree"/ "More than expected." The responses were organized, coded, and analyzed using IBM SPSS v25. It must be noted that most of the closed-type questions were taken (verbatim) from the article (Thi and Thi 2021) since they reflect precisely the topics we need to examine as well.

The questionnaire consisted of four main sections, pertaining to students' satisfaction with the preparation for the online teaching, with pre-class, in-class, and post-class activities in the online teaching plus one more for the demographic information. The details for each section are summarized in Table 1.

Findings and Discussion

Students responded by submitting their completed questionnaire, which had been previously e-mailed to their own academic e-mail. The composition of the sample is as follows (Table 2):

Students' Satisfaction with the Preparation for the Online Teaching

We can acknowledge that with the aid of digital and mobile devices such as smartphones, laptops, tablets, and personal computers, supporting e-learning through the internet becomes an easy task anytime and from anywhere (Gikas and Grant 2013).

Table 2 Sample demographic data

Sex	Male	Female			
	53.30%	46.70%			
Age (years)	19	20	21	22	>22
	15.00%	16.70%	15.00%	26.70%	23.30%
Studies (year)	2nd	3rd	4th	>4th	
	28.30%	33.30%	30.00%	8.30%	
Scientific Field	None	Physical Geography and Environment	Geo-informatics	Human Geography	Spatial Development and Planning
	11.70%	36.70%	25.00%	21.70%	7.00%

Zoom is a platform or software application for online learning activities; this platform is used by students and teachers of Geography Department. However, internet connection was very necessary for e-learning. Based on the results of student's questionnaires (Table 3), most of students (81.60%) had already desktop and internet connection, so this was not barrier for online learning. Although most of the students are very familiar with computers, internet, and online platforms, most of them (78.45%) said that they were informed about the platforms and applications used in geography courses during the COVID-19 and they were added to open e-class by the instructors to have access to educational material, to upload learning activities, receive and send messages as well as the grades, during the COVID-19. Another problem that some students had faced was that some of them (43.40%) they hadn't a quiet place for studying online during the COVID-19.

Students' Satisfaction with Pre-Class Activities for the Online Teaching

Students' satisfaction is an essential element of assessing the effectiveness of teaching and learning (O' Flaherty & Phillips 2015) as it can influence students' motivation, performance, and success (Thi & Thi 2021). According to the results (Table 4), students were satisfied with online learning because most of them said that educators send the learning materials to be read before the online session (41.70%) and they post messages in the class group to remind the lesson preparation before each online session (56.70%).

Table 3 Data for students' satisfaction with the preparation for online teaching

Questions	Strongly disagree	Disagree	Neutral	Agree	Strongly Agree	Mean value	Standard Deviation
I received instruction on how to learn online during the COVID-19	1(1.7%)	9(15.0%)	19(31.7%)	22(36.7%)	9(15.0%)	3.48	0.983
I was informed about the platforms and applications used in my geography courses during the COYID-19	0(0.0%)	3(5.0%)	9(15.0%)	34(56.7%)	13(21.7%)	3.97	0.765
I was added to open e-class by the instructors to have access to educational material, to upload learning activities, receive and send messages as well as the grades, during the COYID-19	0(0.0%)	2(3.3%)	5(8.3%)	26(43.3%)	25(41.7%)	4.28	0.768
I have a desktop and internet available to learn online during the COYID-19	1(1.7%)	3(5.0%)	7(11.7%)	20(33.3%)	29(48.3%)	4.22	0.958

(continued)

Table 3 (continued)

Questions	Strongly disagree	Disagree	Neutral	Agree	Strongly Agree	Mean value	Standard Deviation
I have a laptop and internet available to learn online during the COYID-19	1(1.7%)	1(1.7%)	5(8.3%)	23(38.3%)	29(48.3%)	4.32	0.840
My place is quiet and there is a suitable space for studying online during the COYID-19	3(5.0%)	7(11.7%)	16(26.7%)	17(28.3%)	16(26.7%)	3.61	1.160

Table 4 Data for students' satisfaction with pre-class activities in the online teaching

Questions	Strongly disagree	Disagree	Neutral	Agree	Strongly Agree	Mean Value	Standard Deviation
Lecturers send the learning materials for students to read before the online session	2(3.3%)	12(20.0%)	21(35.0%)	18(30.0%)	7(11.7%)	3.27	1.023
Instructors post messages in the class group to remind the lesson preparation before each online session	0(0.0%)	6(10.0%)	20(33.3%)	25(41.7%)	9(15.0%)	3.62	0.865
Instructors assign tasks and ask students to practice before the online session	2(3.3%)	8(13.3%)	16(26.7%)	18(30.0%)	15(25.0%)	3.61	1.114

Students' Satisfaction with In-Class Activities During the Online Teaching

Given the fact that neither teachers nor students need to travel to school, lack of punctuality—if any—should be attributed to different reasons rather than the usual ones (e.g., technical issues instead of traffic, etc.). Furthermore, the absence of physical presence, which normally contributes to students' increased participation and higher focus, must be compensated for with other methods—such as visual contact (through camera) and higher interactivity (via individual Q&A), while keeping the class in order and everyone on the same page all along.

Data analysis (Table 5) showed that students found their teachers to be "averagely" punctual (a combined total of 60.00% "Agree" or "Strongly agree," Mn = 3.65) while at the same time consider themselves to be "mostly" punctual (equivalent total of 66.70%, Mn = 3.88). On the matter of making up for the lack of physical presence via visual contact, we found that only about 5% of the students chose to use their camera. However, this could be owed to a number of reasons: absence of camera, lack of confidence or high self-consciousness, tendency to refuse to comply, etc. They did feel that the instructors had moderate control over the discussion and upkeep of focus, with their answers being either "Neutral" (35.00% and 33.30%, respectively) or "Agree," at best (33.30% and 30.00%, respectively). Their opinions regarding the opportunity to interact with each other were more evenly distributed around the "Neutral" value (Mn = 3.12). Nevertheless, they replied that instructors spent time to provide comments and correct their mistakes throughout the lesson (combined total of 66.7% "Agree" & "Strongly agree," Mn = 3.62). Additionally, instructors were thought to spend an equal amount of time providing guidance (Mn = 3.36) as well as correcting mistakes and giving feedback to students (Mn = 3.62). Finally, it is worth pointing out that half of the students claimed to encounter technical issues during their online class (50.00%, Mn = 2.72) with an extra 20.00% of them being on the "Neutral' mark.

Student Satisfaction with Post-Class Activities of the Online Teaching

A combined percentage of 63.30% of the students declared themselves satisfied ("Agree" & "Strongly agree") with the immediacy of the teachers' replies to the questions they submitted. At the same time, 2/3 of them consider that teachers promptly and at regular intervals posted announcements about corrections that must be made or warnings about task deadlines (Table 6). Additionally, 73.30% of them found that the lecturers were up to the task of assisting them in their activities by providing targeted advice and on-point suggestions. It must be noted here that this final question gathered the most positive responses, as it features the highest average compared to the other two (Mn = 3.82 and Mn = 3.68, respectively).

Table 5 Data for students' satisfaction with in-class activities during the online teaching

Questions	Strongly disagree	Disagree	Neutral	Agree	Strongly Agree	Mean Value	Standard Deviation
Instructors always join the online class on time	3(5.0%)	4(6.7%)	17(28.3%)	23(38.3%)	13(21.7%)	3.65	1.055
I always join the online class on time	2(3.3%)	3(5.0%)	14(23.3%)	21(35.0%)	19(31.7%)	3.88	1.035
All the students in my class turn on the camera during the study	12(20.0%)	25(41.7%)	20(33.3%)	2(3.3%)	1(1.7%)	2.25	0.876
The instructor facilitates student discussion during online lessons	2(3.3%)	11(18.3%)	21(35.0%)	20(33.3%)	5(8.3%)	3.25	0.975
Lecturers have good control over students' attendance during online lessons	6(10.0%)	10(16.7%)	20(33.3%)	18(30.0%)	6(10.0%)	3.13	1.127
Students have the opportunity to interact with each other during online lessons	6(10.0%)	12(20.0%)	18(30.0%)	17(28.3%)	7(11.7%)	3.12	1.166

(continued)

Table 5 (continued)

Questions	Strongly disagree	Disagree	Neutral	Agree	Strongly Agree	Mean Value	Standard Deviation
Instructors spend most of their time guiding students through their activities	3(5.0%)	5(8.3%)	23(38.3%)	24(40.0%)	4(6.7%)	3.36	0.924
I have no technical problems in the online learning process (broken computer, poor network, power failure ...)	10(16.7%)	20(33.3%)	12(20.0%)	13(21.7%)	5(8.3%)	2.72	1.223
The instructor conects mistakes and gives comments to each of student while practicing	2(3.3%)	9(15.0%)	9(15.0%)	30(50.0%)	10(16.7%)	3.62	1.043

The students' responses on the overall evaluation of online synchronous learning show a rather neutral attitude. Just 8 of them (13.40%) chose not to answer or expressed negative impressions, while most (35 – 65.00%) appear to consider that e-learning offered them what they expected. An additional 13 of them (21.70%) said it even exceeded their expectations (Table 7).

Conclusions

The pandemic brought school closures at an unforeseen scale. Based on the findings of this research, the researchers conclude that most students of the Geography Undergraduate Study Program at the University of the Aegean are mostly satisfied with the implementation of online learning organized by the department, while only a small

Table 6 Data for students' satisfaction with post-class activities of the online teaching

Questions	Strongly disagree	Disagree	Neutral	Agree	Strongly Agree	Mean Value	Standard Deviation
The instructor promptly responded to my submissions in the group	1(1.7%)	6(10.0)	15(25.0%)	27(45.0%)	11(18.3%)	3.68	0.948
The instructor regularly published notices reminding students to review the learned content and complete the assigned assignments	0(0.0%)	7(11.7%)	13(21.7%)	32(53.3%)	8(13.3%)	3.68	0.854
Teachers are ready to assist if students have difficulty with homework	1(1.7%)	2(3.3%)	13(21.7%)	35(58.3%)	9(15.0%)	3.82	0.792

Table 7 Data for students' satisfaction of the distance synchronous teaching (overall)

Overall Distance synchronous learning	Nothing	Less than expected	As expected	Neutral	More than expected	Mean value	Standard Deviation
Level of Satisfaction	1(1.7%)	7(11.7%)	21(35.0%)	18(30.0%)	13(31.7%)	3.58	1.013

part of them are either very satisfied or dissatisfied. E-learning has some advantages for students: (a) the students can choose the place where they attend lectures and they don't need to spend money for journeys; (b) the students can learn at their pace; (c) the students can communicate with each other in online forums and it helps someone with fear of speaking. The level of students' satisfaction with online learning depends on the communication they had with the professor and how well he coordinated the course. Research by Sun et al. (2008) informs that the flexibility of time, learning methods, and place in online learning influences student satisfaction with learning. On the other hand, the students were dissatisfied because they thought they were not very efficient and they felt bored with the independent assignments. Also, e-learning has the danger of disengaged participation in class (e.g., passive listening or watching the teacher's lecture, silently reading peer statements in the chat) (Smith and Smith, 2014). The use of online synchronous learning can increase learning independence (Al Rawashdeh et al. 2021; Nielsen 2012). In general, students show a positive attitude toward the implementation of online teaching. They believe that the overall design and implementation of this type of teaching overcame existing obstacles and offered—under the current conditions—an alternative form of learning. Their positive feelings, as analyzed above, were plentiful, albeit skepticism remains; the way this course was implemented demands further investigation, mainly from the aspects of pedagogical application of online learning and the degree of involvement and autonomy of the students (Díaz-Noguera et al. 2022) concerning also the availability of internet services and additional costs that must be borne by students (Hastuti et al. 2021).

References

Al Rawashdeh AZ, Mohammed EY, Al Arab AR, Alara M, Al-Rawashdeh B (2021) Advantages and disadvantages of using e-Learning in University Education: analyzing students' perspectives. Electron J E-Learn 19:107–117. https://files.eric.ed.gov/fulltext/EJ1296879.pdf (Accessed on 3 Dec 2022)

Andrade MS, Alden-Rivers B (2019) Developing a framework for sustainable growth of flexible learning opportunities. High Educ Pedagog 4(1):1–16. https://doi.org/10.1080/23752696.2018.1564879

Barclay C, Donalds C, Osei-Bryson KM (2018) Investigating critical success factors in online learning environments in higher education systems in the Caribbean. Inf Technol Dev 24(3):582–611. https://doi.org/10.1080/02681102.2018.1476831

Basilaia G, Dgebuadze M, Kantaria M, Chokhonelidze G (2020) Replacing the classic learning form at Universities as an immediate response to the COVID-19 virus infection in Georgia. Int J Res Appl Sci & Eng Technol (IJRASET) 8:101–108. https://doi.org/10.22214/ijraset.2020.3021

Crawford J, Butler-Henderson K, Rudolph J, Malkawi B, Glowatz M, Burton R, Magni P A, Lam S (2020) COVID-19: 20 countries' higher education intraperiod digital pedagogy responses. J Appl Learn Teach 3:1–20. https://doi.org/10.37074/jalt.2020.3.1.7

Díaz-Noguera MD, Hervás-Gómez C, De la Calle-Cabrera AM, López-Meneses E (2022) Autonomy, Motivation, and digital pedagogy are key factors in the perceptions of spanish higher-education students toward online learning during the COVID-19 Pandemic. Int J Environ Res Public Health 19:1–14. https://doi.org/10.3390/ijerph19020654

Gay GH (2016) An assessment of online instructor e-learning readiness before, during, and after course delivery. J Comput High Educ 28(2):199–220. https://doi.org/10.1007/s12528-016-9115-z

Gikas J, Grant MM (2013) Mobile computing devices in higher education: student perspectives on learning with cellphones, smartphones & social media. Internet High Educ 19:18–26. https://doi.org/10.1016/j.iheduc.2013.06.002

Government Official Journal (42/A/25.02.2020) Legislative Content Act of 25–2–2020 (Government Official Journal 42/A/25.02.2020): urgent measures to avoid and limit the spread of coronavirus (in Greek)

Government Official Journal (956/B/21.03.2020) JMD No. D1a/ Under General Register No 20021 (Government Official Journal 956/B/21.03.2020). Enforcement of the measure of temporary prohibition of operation of nurseries, kindergartens, schools, higher education institutions, foreign language centers, tuition centers and all kinds of educational structures, institutions, public and private, of any type and degree of prohibition of the country, temporary educational trips inside and outside the country and a temporary ban on travel and visits of pupils, students and teachers, as well as a temporary ban on the operation of Greek-language schools and all kinds of educational structures abroad for the period from 21.3.2020 to 10.4.2020 (in Greek)

Hastuti KP, Angriani P, Alviawati E, Arisanty D (2021) The perspective of geography education students on the implementation of online learning during covid-19 pandemic. IOP Conf Ser: Earth Environ Sci 747(2021):012012. https://doi.org/10.1088/1755-1315/747/1/012012

Hodges C, Moore S, Lockee B, Trust T, Bond A (2020) The difference between emergency remote teaching and online learning. Educause 2020(3). https://er.educause.edu/articles/2020/3/the-dif ference-between-emergencyremoteteaching-and-online-learnin (Accessed on 3 Dec 2022)

Join Ministerial Decision (JMD): No. D1a/Under General Register No 20021; JMD: No. D1a/Under General Register No 24343) (Government Official Journal 1293/B/10.04.2020). Extension of the Joint Ministerial Decision of Ministry of Development and Investment, Ministry of Citizen Protection, Ministry of Education and Religious Affairs, Ministry of Labour and Social Affairs, Ministry of Health, Ministry of Culture and Sports and Ministry of the Interior, Decentralization and Electronic Governance No D1a/ Under General Register No 20021/21.3.2020 (B΄ 956) till 10.5.2020 (in Greek)

Join Ministerial Decision: 63314/G.D. 4/26–05–2020. JMD: No. 63314/G.D. 4/2020 (Government Official Journal 2026/B/27.5.2020). How to reopen primary education schools and special education schools after the end of their temporary closure (in Greek)

Kay RH (2022) Pre-Course Activities for Online Learning. In Kay R & Hunter B (eds) Thriving online: a guide for busy educators. Ontario Tech University, 58–64. https://doi.org/10.51357/TMSM9420

Lore M-P (2000) Elements of a successful distributed learning program. In: Lau L (ed) Distance learning technologies: issues, trends and opportunities. Idea Group Publishing, Hershey, pp 82–90

Molina Higgins MC, Taylor AH, Rocco SA, Kimel RA, Sinnott SB (2022) COVID-19 pandemic student engagement strategies for materials science and engineering courses. MRS Bull 47:5–9. https://doi.org/10.1557/s43577-021-00252-8

Nielsen HL (2012) E-learning and the dilemma of learner autonomy: a case study of first year university students of Arabic. Orientalia Suecana 61:89–106. https://www.semanticscholar.org/paper/E-learning-and-the-dilemma-of-learner-autonomy%3A-A-Nielsen/72b1156a23e1ffdf0da75e483044000bcebe3da9 (Accessed on 3 Dec 2022)

Nishimura Y (2021) Three reasons why e-learning can contribute to the SDGs education goals. https://learningbox.online/en/2021/06/29/blog_sdgs-elearning/ ((Accessed on 3 Dec 2022)

O' Flaherty J, Phillips C, (2015) The use of flipped classrooms in higher education: a scoping review. Internet High Educ 25:85–95. https://doi.org/10.1016/j.iheduc.2015.02.002

Purcell-Robertson RM, Purcell DF (2000) Interactive distance learning. In: Lau L (ed) Distance learning technologies: issues, trends and opportunities. Idea Group Publishing, Hershey, pp 16–21

Rizou O, Klonari A, Kavroudakis D (2021) Supporting statistical literacy with ICT-based teaching scenario. Int J Educ (IJE) 9:59–78. https://doi.org/10.5121/ije.2021.9405

Scheiderer J (2020) What's the difference between asynchronous and synchronous learning? https://online.osu.edu/resources/learn/whats-difference-between-asynchronous-and-synchronous-learning (Accessed on 3 Dec 2022)

Schneider SL, Council ML (2021) Distance learning in the era of COVID-19. Arch Dermatol Res 313:389–390. https://doi.org/10.2807/1560-7917.ES.2020.25.14.2000550

Singh V, Thurman A (2019) How many ways can we define online learning? A systematic literature review of definitions of online learning (1988–2018). Am J Distance Educ 33(4):289–306. https://doi.org/10.1080/08923647.2019.1663082

Smith D, Smith K (2014) Case for 'passive' learning—the 'silent' community of online learners. Euro. J. Open Distance E-Learn. 17:86–99. https://doi.org/10.2478/eurodl-2014-0021

Steffens I (2020) A hundred days into the coronavirus disease (COVID-19) pandemic. Euro Surveill 25:2000550. https://doi.org/10.2807/1560-7917.ES.2020.25.14.2000550

Sun PC, Tsai RJ, Finger G, Chen YY, Yeh D (2008) What drives a successful e-Learning? An empirical investigation of the critical factors influencing learner satisfaction Comput & Educ 50:1183–1202. https://doi.org/10.1016/j.compedu.2006.11.007

Taufik, and Fiptar Abdi A (2022) The survey on students' satisfaction degree towards online learning during COVID-19 pandemic condition. J Engl Lang Teach Lit (JELITA) 3:44–56. https://doi.org/10.56185/jelita.v3i1.94

Le TKT, Trinh TTT (2021) An investigation into student satisfaction with the implementation of online learning in physical education during the COVID-19 at a vocational college in Vietnam. In Proceedings ICEMT 2021: 2021 5th International Conference on Education and Multimedia Technology, July 2021, pp 213–219. https://doi.org/10.1145/3481056.3481072

UNESCO (2020) Adverse consequences of school closures. https://en.unesco.org/covid19/educationresponse/consequences (Accessed on 3 Dec 2022)

World Health Organization (2020a) Director-General's remarks at the media briefing on 2019-nCoV. https://www.who.int/dg/speeches/detail/who-director-general-s-remarks-at-the-media-briefing-on-20 19-ncov-on-11-february-2020 (Accessed on 3 Dec 2022)

World Health Organization (2020b) Coronavirus disease (COVID-2019) situation reports. Situation report—64 (29 March 2020). https://www.who.int/emergencies/diseases/novel-coronavirus-2019/situation-reports (Accessed on 3 Dec 2022)

Yan Y, Shin WI, Pang YX, Meng Y, Lai J, You C, Zhao H, Lester E, Wu T, Pang CH (2020) The First 75 Days of Novel Coronavirus (SARS-CoV-2) Outbreak: recent advances, prevention, and treatment. Int J Environ Res Public Health 17:2323. https://doi.org/10.3390/ijerph17072323

Evaluation and Analysis of Literature for COVID-19 and Climate Change

Eirini Chatzara, Apostolia Galani, George Arhonditsis, and Evangelia Mavrikaki

Abstract The COVID-19 pandemic really went viral (pun intended) and dominated the media and public discourse, often overshadowing coverage of the equally urgent climate crisis. Despite their distinct features, both crises share important characteristics, such as global impact, economic repercussions, and associations with biodiversity loss, which disproportionately affect marginalized communities. Based on the bibliometric analysis, a significant body of literature exists on the impact of climate change on virus transmission and the environmental consequences of the pandemic. This work focuses on articles that explore the COVID-19 crisis as a learning experiment for the climate crisis, suggesting that in both cases, mitigation policies are less costly than adaptation policies, a collective global response should be agreed upon through international agreements, support for vulnerable communities is crucial, and effective communication to policymakers and the public is paramount.

Keywords Climate crisis communication · COVID-19 · Bibliometric analysis · Future climate actions

Introduction

The COVID-19 pandemic has had a significant impact on global society, with far-reaching effects on daily life and the global economy. However, it has also underscored the interdependence of planetary-scale crises, particularly the climate crisis. Both crises share specific characteristics, such as being connected to biodiversity loss, having a major economic impact, and impacting communities worldwide, particularly low-income and marginalized communities (IPCC 2014; WHO 2020). Studies have demonstrated that the COVID-19 pandemic led to a temporary decrease in

E. Chatzara (✉) · A. Galani · E. Mavrikaki
National and Kapodistian University of Athens, Athens, Greece
e-mail: eirchatzara@primedu.uoa.gr

G. Arhonditsis
University of Toronto, Toronto, ON, Canada

A. Klonari et al. (eds.), *Re-visioning Geography*, Key Challenges in Geography,
https://doi.org/10.1007/978-3-031-40747-5_14

greenhouse gas emissions, providing a glimpse of the possible effects of collective action on climate change (Le Quéré et al. 2020). Moreover, recent research has emphasized the connection between climate and virus transmission, suggesting that variations in temperature and humidity can influence the survival and transmission of specific viruses (Wu et al. 2020). The COVID-19 pandemic has also led to substantial changes in human behavior and policy responses, which may have significant implications for efforts to mitigate and adapt to climate change.

Furthermore, addressing the COVID-19 pandemic and the climate crisis is essential for achieving the Sustainable Development Goals (SDGs), a set of 17 interrelated goals aimed at achieving a more equitable, sustainable, and just global society by 2030, developed by the United Nations (United Nations 2015). Given the importance of both crises and the potential for interconnected effects, it is critical to investigate the possible links and overlaps between the COVID-19 pandemic and the climate crisis. As a result, investigating the lessons that can be gained from the COVID-19 crisis may provide valuable insights into how to tackle the climate crisis, especially in terms of mitigation and adaptation strategies. As Fuentes et al. (2020) wondered, what should the world have done in November 2019, a month before the COVID-19 outbreak, to avoid what happened next? Does this question provide insight into how the world should prepare for something that might occur in the next 30 to 50 years?

By carrying out this study, we aimed to examine that question, guided by the following research questions:

RQ1: How has COVID-19 communication affected climate crisis public discourse?
RQ2: What are the themes that arise in the literature's intersection of the two crises?
RQ3: What are the lessons from the COVID-19 crisis response?

Methodology

The COVID-19 pandemic and the climate crisis have attracted significant interest from both the academic and media communities, with many researchers looking into the overlaps and connections between the two global crises. However, to fully comprehend the magnitude and implications of these crises, a rigorous methodology is required. In the present study, we used a mixed-methods approach to answer our research questions, as such an approach allows for a nuanced understanding of the connections and overlaps between the COVID-19 pandemic and the climate crisis. In particular, the methodology we applied to the research questions is the following:

Public Discourse on Climate Crisis and COVID-19 Pandemic (RQ1)

To answer the first research question (RQ1) about public discourse on climate crisis and COVID-19, we utilized both academic and media sources. For the academic discourse, we collected data from the Web of Science database and created a chronological distribution of climate and coronavirus-related publications. To compare the volume of coronavirus-related literature to that of climate-related literature, we obtained the number of publications for each subject from the Web of Science database. We chose Web of Science (hereinafter "WoS"), because a) it is widely used and contains a large number of publications, and b) the bibliographic data can be exported in a VOSviewer-compatible file format, useful for answering RQ2.

The search queries were:

TI = CLIMAT*[1]

TI = COVID* OR CORONAVIRUS*[2]

We included publications spanning a decade (2012–2021). This timeframe was selected to encompass the most recent decade and include the emergence of COVID-19 as a global pandemic. It also allows for an analysis of how the volume of literature on climate change and COVID-19 has evolved over time, particularly in the context of the Sustainable Development Goals established by the United Nations in 2015. We excluded publications from 2022 since we were only halfway through that year at the time of writing and did not have the full number of publications. To examine the chronological distribution of climate-related publications, we used linear regression analysis. This analysis provides insights into how academic discourse surrounding the climate crisis has been affected by the COVID-19 pandemic.

For the media discourse, in order to examine the impact of COVID-19 on Internet users' search queries about climate change, we collected data from the Google Trends tool. Google Trends (hereinafter referred to as "GT") provides normalized values ranging from 0 to 100, representing the relative search volume for a search term during a given period and location (Google 2015). Increasingly, researchers across diverse fields have been utilizing search-trend analysis tools like GT to gain insights into Internet users' interests (Ficetola 2013; Schaub et al. 2020).

To compare interest in the topics of "climate change" and "COVID-19," we queried GT from January 2019 until December 2021 using topics instead of keywords. This allowed for the comparison of interest across all relevant terms searched by users in all possible languages and areas (Kamiński et al. 2019).

Emerging Themes in the Intersection of Climate and Pandemic Crisis (RQ2)

To examine the themes that pervade the selected literature (RQ2), we employed bibliometric analysis, which provides a better understanding of the relationship between COVID-19 and climate change in terms of themes and research topics.

[1] To include variations like *climatic*.

[2] To include variations such as *covid*, *covid-19*, etc.

We chose the Web of Science database to obtain bibliographic data because it is widely used and contains a vast amount of publications, and its bibliographic data can be exported in a VOSviewer-compatible file format. We conducted a keyword co-occurrence analysis to provide an overview of the underlying topical themes of the literature and to identify which themes pervade the entire collection of documents and how they relate to each other. We used VOSviewer software for this analysis (McAllister et al. 2022).

Our query for the Web of Science database was:

(TS = ((("climate change") OR ("climate crisis"))

And

(covid* OR coronavirus*))).

And

((DT = = ("ARTICLE" OR "REVIEW") AND DT = = ("ARTICLE" OR "REVIEW") AND LA = = ("ENGLISH")).

We limited our search to the years 2020, 2021, and 2022, which yielded 1,281 results, and then refined the results to exclude subjects that were irrelevant to our study by using the WoS filter "Web of Science categories," resulting in 1,047 articles.

To avoid redundancy of similar terms (e.g., COVID-19/COVID 19, carbon dioxide/CO2), we created a thesaurus Excel file with 37 terms, which was uploaded to VOSviewer for analysis (Appendix A). After some testing, we set the minimum number of articles in which a keyword appears to eight, resulting in 129 keywords when combined with the thesaurus file. We included titles, abstracts, and keywords in our analysis, which generated a map illustrating the main themes related to the issue of the climate crisis in relation to the COVID-19 crisis.

Lessons Learned from COVID-19 Crisis Response (RQ3)

To answer RQ3, we conducted a narrative literature review after shortlisting the articles used in answering RQ2, to those pertaining to public understanding of the pandemic and climate crisis.

The 1,047 articles included in our bibliometric analysis served as our starting point. Based on the theme clusters found in RQ2, we defined criteria for excluding articles so that only those that were relevant to RQ3 were selected (Table 1).

Following that, we performed first a title filtering and then a title/abstract filtering, which resulted in excluding 868 articles that did not meet the specified criteria. The 179 remaining articles were sorted according to their citations, and lastly, the ones considered most relevant for answering RQ3 were taken into account.

Table 1 Inclusion and exclusion criteria

Inclusion criteria	Exclusion criteria
Contains relevant research keywords and the examined topic is discussed in the abstract	Contains relevant research keywords, but the examined topic is not discussed in the abstract
	(Inter)national policies
	Environmental impacts of COVID-19 to climate
	Association of climatic indicators with COVID-19
	Food supply chain

Results

Public Discourse on Climate Crisis and COVID-19 Pandemic (RQ1)

Academic Discourse

In this section, we present the results of our analysis for RQ1, which aimed to investigate the comparative volume of climate-related and coronavirus-related research in the Web of Science database. To this end, we created a chart comparing the number of publications for each topic over time (Fig. 1).

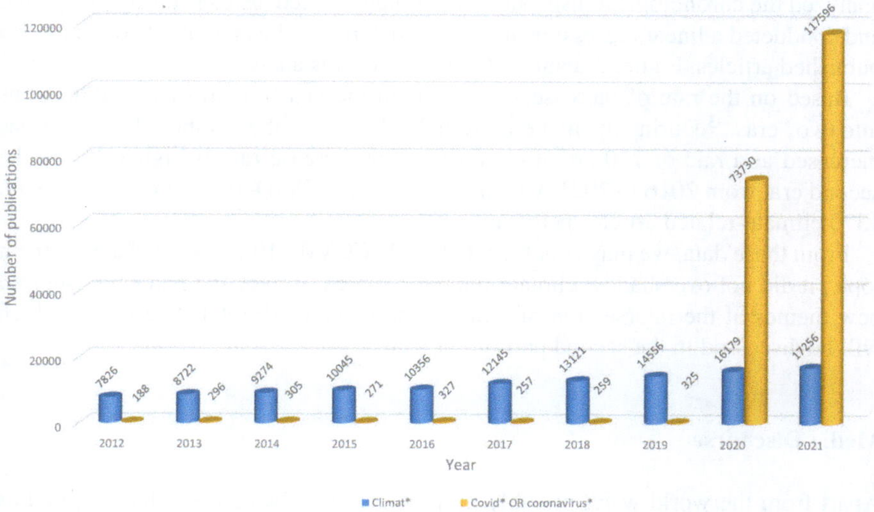

Fig. 1 Chronological distribution of climate versus coronavirus research volume

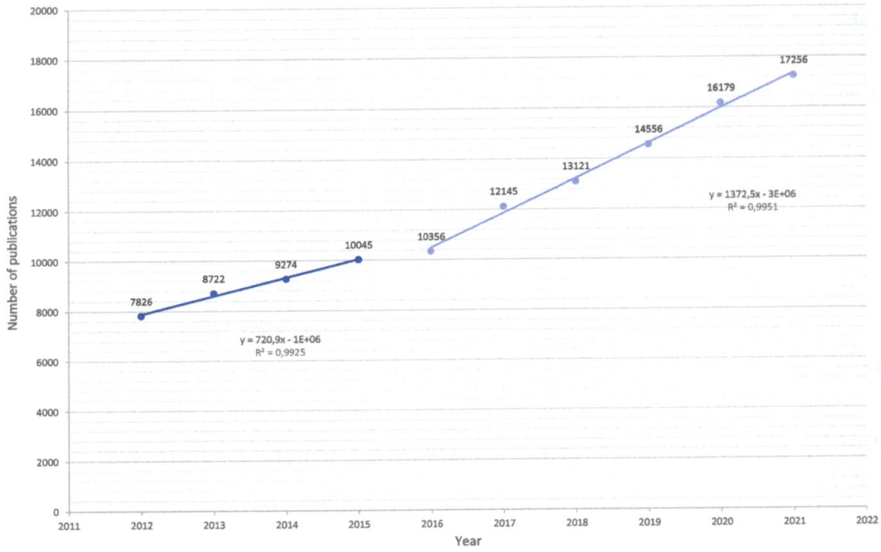

Fig. 2 Chronological distribution of climate-related publications

Research on coronaviruses has been limited prior to 2019, whereas climate-related research has shown a steady increase over time. While the sheer number of publications on coronavirus in 2020 and 2021 appears to be extensive compared to climate research, it is important to determine if this is truly the case. To investigate this, we analyzed the chronological distribution of climate-related articles during this period and conducted a linear regression analysis to determine the annual rate of change in published articles. Figure 2 displays the results of this analysis.

Based on the rate of increase, we classified the yearly number of publications into two "eras."[3] During the first era, from 2012 to 2015, the number of publications increased at a rate of 720 articles per year. The increase rate is higher during the second era, from 2016 to 2021, which includes the COVID-19 article "boom," with 1370 climate-related articles per year.

From these data, we may conclude that while COVID-19 was indeed a prominent topic, it did not overshadow climate-related research. In fact, the pandemic spurred new themes at the intersection of climate and COVID-19-related research, which will be discussed in the second part of our study.

Media Discourse

Apart from the world wars, possibly no previous worldwide event has inspired as much social concern about the well-being of people affected or occupied as much

[3] We consider the "era" as a period of time with a certain rate of increase in publications.

media coverage as the pandemic. However, global public attention to climate crisis coverage fell by 59% from January to April 2020 (Mocatta and Hawley 2020). According to the findings of Loureiro and Alló (2021), COVID-19 had a negative impact on the public discussion of climate change on Twitter in the majority of countries they investigated. Rauchfleisch et al. (2021) investigated COVID-19's coverage in Swiss media and Twitter between April 2019 and October 2020 and estimated a substantial negative impact of COVID-19 on media attention as well as on tweets about climate. Another study by Spisak et al. (2022) utilized data from Facebook and ascertained a drop in climate-related posts in the United States from August 2019 to December 2020.

Figure 3 is a time series created by the values we acquired for the said period. The obtained data are in weekly intervals.

According to the time series, the first peak in Internet users' interest in COVID-19 was in January 2020. We cross-checked it with the World Health Organization's (WHO) timeline. On January 10, 2020, *"WHO issued its first guidance on the novel coronavirus,"* and on January 13, *"the first case of the novel coronavirus outside of China was confirmed,"* so it was bound to catch the public's attention. The second and highest peak was in March 2020. Again, according to WHO's timeline, on March 12, *"WHO characterizes COVID-19 as a pandemic"* (WHO 2020).

At the same time, the climate change time series reveals that public interest in climate change is substantially lower than for COVID-19, with a relative interest (number of climate change queries/maximum number of searches) of less than one.

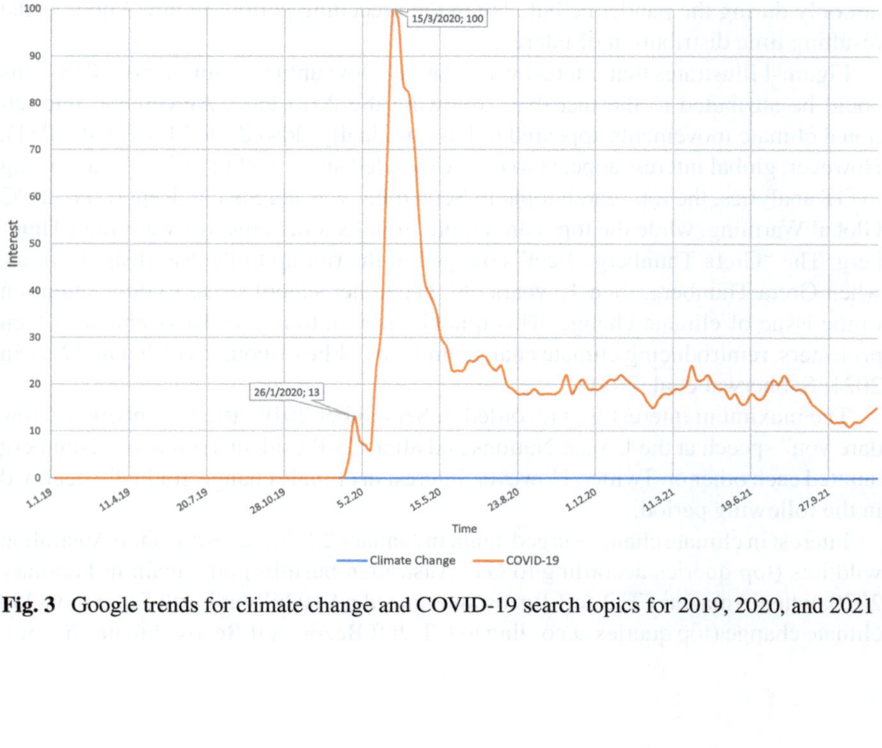

Fig. 3 Google trends for climate change and COVID-19 search topics for 2019, 2020, and 2021

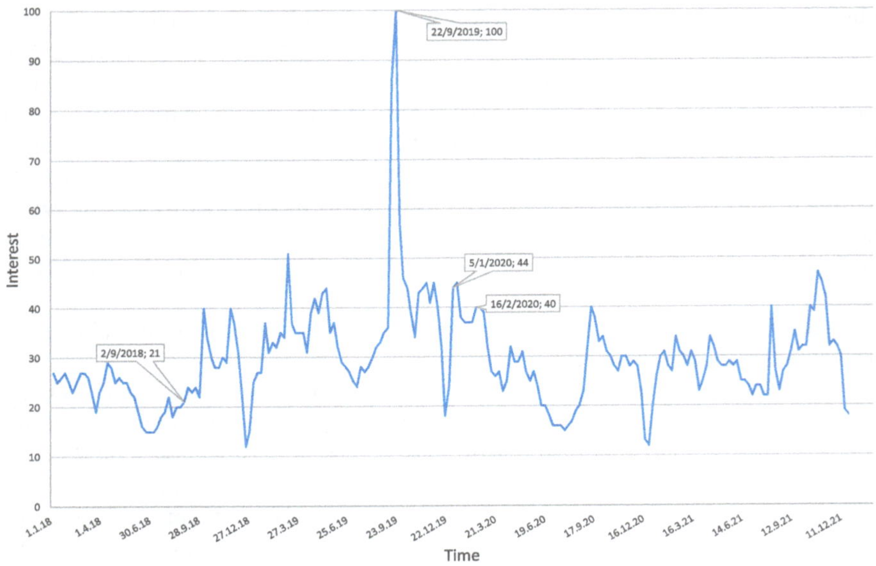

Fig. 4 Worldwide interest in Climate Change versus time

To determine whether and to what extent COVID-19 overshadowed climate change interest, we examined trends concerning the search queries for the latter, not only during the pandemic but also in the preceding period. Figure 4 depicts the resulting time distribution of interest.

Figure 4 illustrates that interest was relatively low until the summer of 2018. This could be attributed to the fact that, following the 2015 Paris Agreement, international climate movements appeared to have gradually slowed (de Moor et al. 2021). However, global interest appears to have expanded since the fall of 2018. According to GT analytics, the top search topic in September was the Special Report on 1.5 °C Global Warming, while the top search topic from October onwards was Greta Thunberg. The "Greta Thunberg effect" emerged in the run-up to the Swedish elections, when Greta Thunberg, then 16 years old, began her school strike to draw attention to the issue of climate change. This quickly grew into a global movement of teen protesters, reintroducing climate change into the public discourse (Fisher and Nasrin 2021; Sabherwal et al. 2021).

The maximum interest was recorded in September 2019, after Thunberg's "How dare you" speech at the United Nations and after US President Trump and Thunberg taunted each other on Twitter. However, interest in climate change gradually declined in the following period.

Interest in climate change surged again in January 2020 as a result of the Australian wildfires (top queries according to GT: Australian bushfire) and again in February 2020, when Amazon CEO Jeff Bezos announced a $10 billion Earth Fund to tackle climate change (top queries according to GT: Jeff Bezos, Jeff Bezos climate change).

Despite this renewed interest, public interest in climate change diminishes after March 2020 (when the WHO declared COVID-19 a pandemic) and continues until September 2020 (WHO 2020).

In conclusion, it can be inferred that COVID-19 had a negative impact on global public discourse. Some researchers attribute this effect to the "finite pool of worry" hypothesis, which suggests that when confronted with another, more immediate threat, people become less concerned about the previous one.

Emerging Themes in the Intersection of Climate and Pandemic Crisis (RQ2)

In the second research question, the focus is on identifying the keyword clusters and their interrelationships in the literature related to climate change and COVID-19. The resulting keyword network is presented in Fig. 5.

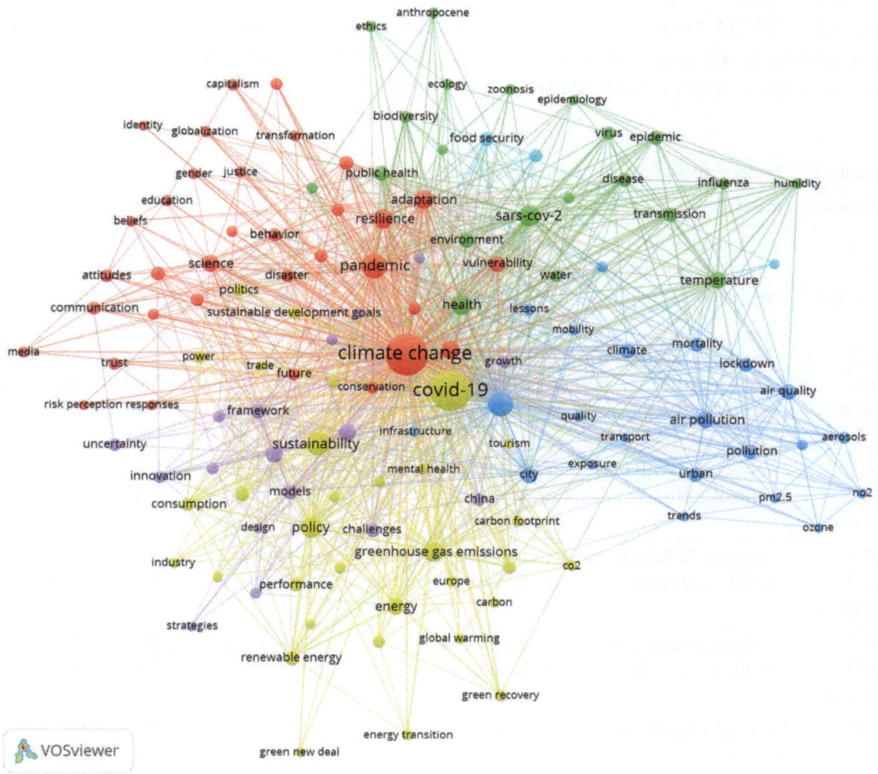

Fig. 5 Co-occurrence of keywords appearing at least in 8 articles

The size of the nodes indicates the focus on each of the keywords. The co-occurrence of the nodes (keywords) in the same publication is represented by the curves connecting them. The smaller the distance between two nodes, the more frequently the two keywords co-occur. In addition, different colors represent different general research clusters (van Eck and Waltman 2020).

According to Fig. 5, there are six keyword clusters forming. In Table 1, we included the 10 most central keywords of each cluster and in Table 2 the cluster's general theme).

Based on Table 1, the answer to RQ2 can be found in the themes that emerged from the intersection of the climate crisis and the COVID-19 crisis:

1. Public response and understanding of crises, and the role of media and education
2. Policies toward a "green" transition
3. The effect of lockdowns and transportation restrictions on air quality
4. Climate/environmental factors that contribute to virus transmission
5. Future challenges related to the economy and technology
6. Impacts of/on the food supply chain.

In conclusion, the themes that emerged from the intersection of the climate crisis and the COVID-19 crisis shed light on the multifaceted nature of these global crises and highlight the urgent need for a comprehensive and interdisciplinary approach to tackling these interconnected crises.

Table 2 Clusters' most central keywords and themes

Cluster	Keywords	Theme
C1 (34 items)	climate change, pandemic, adaptation, resilience, science, vulnerability, knowledge, attitudes, communication, education, media	Addressing Global Resilience Challenges
C2 (32 items)	covid-19, sustainability, greenhouse gas emission, energy policy, renewable energy, policy, governance, consumption, performance, green recovery	Transitioning to Sustainable Energy Governance
C3 (22 items)	impact, air pollution, climate, lockdown, air quality, urban, city, aerosols, pm2.5, transport	Urban Air Quality
C4 (21 items)	sars-cov-2, health, temperature, transmission, disease, public health, influenza, biodiversity, virus, epidemic	Viral transmission
C5 (17 items)	management, framework, uncertainty, system, innovation, challenges, technology, china, design, circular economy	Future Economic Management Challenges
C6 (3 items)	agriculture, food security, variability	Agriculture and Food Variability

Lessons Learned from COVID-19 Crisis Response (RQ3)

"Never waste a good crisis" is the main lesson we learned from our study. Humanity is facing two interconnected crises, as revealed by examining scientific literature across various themes. The "Cluster 4" section of the literature investigates climate change as an environmental factor that has an impact on the spread of the (corona) virus. However, although warmer climates tend to slow the spread of viruses, major infection rates have been observed in warm countries like Singapore (Ching and Kajino 2020). Rising temperatures and deforestation also increase disease vectors, potentially leading to the next global pandemic (van der Ven and Sun 2021). The literature's "Cluster 2" and "Cluster 3" sections explore the pandemic's environmental effects, such as air quality improvement, record-low greenhouse gas emissions, and a digitalization boost (Cheval et al. 2020) and how lockdowns have reduced greenhouse gas emissions in the world's top three emitters (Perkins et al. 2021). Though the benefits of these measures in developed countries should be viewed as pandemic-driven initiatives' side effects, not as a "victory for the climate," due to the measures' economic impact and disruption of the renewable energy industry's supply chain (Ibn-Mohammed et al. 2021; Sovacool et al. 2020). According to Yoshino et al. (2021), the post-COVID-19 economic fallout will result in a reduced investment rate toward the 2030 Agenda for Sustainable Development. Hepburn et al. (2020) identified that only 4% of green policies were implemented in G20 rescue-phase packages. "Cluster 5" and "Cluster 6" discuss the challenges and issues following COVID-19 and the climate crisis.

With the foregoing in mind, to answer RQ3 we turned our focus to the "Cluster 1" section of the literature to try to uncover the parallels and contrasts between the pandemic and climate crises in order to draw lessons from the former to address the latter more effectively.

Connecting the Dots Between the Two Crises

The COVID-19 crisis has highlighted our vulnerability to global risks such as a pandemic or climate change, underscoring the need for planning and preparation (Howarth et al. 2020). Botzen et al. (2021) even suggest that the COVID-19 crisis could be viewed as *"a rapid learning experiment about how to cope with climate change,"* given that it demonstrated that changes in lifestyle are feasible and that a slow response to planet-scale threats can have massive health and economic consequences.

On the other hand, the pandemic has also revealed a concerning rise in science skepticism or denial, particularly regarding climate change and vaccination, both of which are subjects of heated debate (Rutjens et al., 2021). Although the attitudes surrounding these issues do not have the same roots, they are both linked to conspiracy thinking. Bessi et al.'s (2015) Facebook study found that conspiracy news is three times more likely to be liked and shared by users than scientific information.

Therefore, in today's globalized and interconnected society, a disease outbreak is remarkably similar to a large-scale emergency such as climate change in the following ways:

Counter-Intuitive Processes

The pandemic crisis has a broad temporal and spatial range. Since people are not always prone to react to longer-term or remote threats, early warnings of coronaviruses' possible outbreaks (Cheng et al. 2007; Ng and Tan 2017) were ignored, and reactions to the early COVID-19 cases were hushed or downplayed, resulting in a planet-wide pandemic (Manzanedo and Manning 2020).

Similarly, despite its increasingly visible impact and scientific calls for action, Moser (2010) identifies a number of challenges inherent in the nature of climate change that is partially responsible for the muted response, including the lack of visibility, as well as the temporally and geographically distant impact between cause and effect. This can be attributed to the "availability" behavioral bias, which refers to the individuals' willingness to respond to a possible threat only after it has happened in their local environment (Botzen et al. 2021).

Point of Irreversible Change

Complex systems, such as the climate or the spread of infectious disease, have tipping points where the systems can abruptly and irreversibly shift (Fan et al. 2021). In the case of COVID-19, controlling it once it has reached a certain level of abundance within a population becomes extremely difficult since the ongoing spread is exponential (Manzanedo and Manning 2020). Regarding climate, the scientific community agrees that crossing certain thresholds, such as global warming of 1,5 °C, can cause abrupt and irreversible effects on biodiversity (Fuentes et al. 2020; IPCC 2014; Trisos et al. 2020). That approach can lead individuals, due to inherent biases, to oversimplify the full risk of a probable disaster when that threshold is not reached. In the case of COVID-19, for example, people may misunderstand the concept of exponential spread, whereas in the climate crisis, they may oversimplify the concept of nonlinear climate dynamics (Botzen et al. 2021).

Heterogeneous Impacts & Inequitable Effects

The climate crisis, like the COVID-19 crisis, is an issue whose underlying causal mechanisms and ramifications cross borders, with consequences felt far beyond the source of the problem. However, despite the fact that such crises have a global impact, they disproportionately affect low-income communities (Oldekop et al. 2020).

Climate change and pandemic crises do not affect all social groups equally, in terms of geography and social factors. Lockdowns, rising unemployment, and access to medical care disproportionately affect low-income people. While global temperatures may rise consistently across the majority of the globe, changes in precipitation, extreme events, and sea level rise will fluctuate geographically and over time, resulting in local, regional, and societal emergencies (Manzanedo and Manning 2020; Heyd 2021; Mattar et al. 2021).

It is Expensive to Wait

When one compares the public health situation between nations that reacted quickly and those that responded after a considerable percentage of the population had been infected with COVID-19, it is clear that waiting has a fatal cost: In the early phases of the epidemic, a single day of lockdown delay raised the number of cumulative cases by 40%, according to estimates. Respectively, a meta-analysis of 16 studies on mitigation policies shows that delaying climate action by 10 years increases the cost of later climate action by 37% (Klenert et al. 2020).

Even if strong climate action, like reducing greenhouse gas emissions, is only done for a few decades, it could give ecosystems the time they need to adapt, which would lessen the severity of the ecological effects (Trisos et al. 2020).

Spotting the Differences Between the Two Crises

Apart from the commonalities, pandemics and climate crises have distinct characteristics. Their fundamental difference is that COVID-19 pandemic is a health crisis, while the climate crisis is an environmental one. Focusing on their differences, Gemenne and Depoux (2020) even suggest that we should stop describing climate change as a crisis because of its irreversibility. Assessing how it affected UN biodiversity and climate conferences, Ortiz et al. (2021) consider COVID-19 to be the proverbial "tip of the iceberg," since the loss of biodiversity and the effects of climate change will overshadow this problem. Recognizing and understanding the differences between these two crises are crucial to the effective mitigation of their results.

Temporal and Spatial Ranges

In terms of space and time, humans have a limited range of perception. That results in an important distinction between climate and pandemic crises; the impacts of the former are less immediately visible than those of the latter. Quantifying climate change is more abstract and indirect (e.g., temperature increase, CO_2 levels) compared to quantifying COVID-19 (number of cases or deaths) (Bouman et al. 2021).

The virus affects human populations more rapidly than climate change, which has slower and less noticeable effects (rapid increases in sickness and mortality). The fast and visible effects of the pandemic create a greater sense of urgency than the slower and less noticeable impacts of climate change, which can make it harder to connect causes and consequences and result in a lower sense of urgency (Heyd 2021). Therefore, because of its very nature, COVID-19 is a hazard for us now, but climate change is not a temporary crisis. This results in different mitigation measures that are perceived and accepted differently by the public. It is easier to conform to temporary measures, like lockdowns or travel restrictions than to implement permanent lifestyle changes in order to live more sustainably (Fuentes et al. 2020; Gemenne and Depoux 2020).

In addition, factors such as the popularity of "tipping point" rhetoric (which has its own merits and advantages concerning climate change discussion) (van der Hel et al. 2018) and the time gap between emissions and their consequences for climate change inaccurately portray climate change as a future issue. COVID-19, on the other hand, compressed space and time "in an instant," making the world a hazardous place (Leal-Filho et al. 2020).

Effectiveness of Individual Action

Mitigation of COVID-19 is very much dependent on individual action (such as social distancing) and individual sense of responsibility (e.g., not transmitting the virus to others). That is not the case for the climate crisis, where changing individuals' habits—no matter how important—has less impact (Howarth et al. 2020; Bouman et al. 2021). While the impacts of climate change are already being felt by the current generation, they will be even more pronounced for future generations. In contrast, COVID-19 is a threat we are currently facing (Gemenne and Depoux 2020). As a result, individuals may feel their contribution is inconsequential, or they may decide to leave the problem for future generations to resolve (Loureiro and Allo 2021).

Who is to Blame?

The role of political leadership in dealing with crises is critical. During the pandemic, several politicians blamed the virus's spread on "the Chinese," "irresponsible young people," and so on. Others tried to shift the blame to "unpredictability." The same rhetoric is frequently employed in the aftermath of climate change-related extreme weather occurrences (Prideaux et al. 2020).

Actually, in the case of COVID-19, there is no specific person to blame. Whereas everyone in a carbon-dependent fossil fuel society is responsible for climate change to differing extents (Lidskog et al. 2020).

Discussion

Lessons for Climate Crisis Through the Lens of the Pandemic

The pandemic has brought to light basic flaws in the global socio-economic system, illuminating the ways in which the predominance of poverty, inadequate health care systems, unequal educational opportunities, and a lack of global collaboration have contributed to the worsening of the issue (Fenner and Cernev 2021).

Here, we present a (by no means comprehensive) list of the lessons that COVID-19's "learning experiment" can teach us about tackling the climate crisis.

Importance of Effective Communication

The pandemic took over public discourse. Some researchers associate the disproportionate media coverage with the theory of a "finite pool of worry" or "worry budget," according to which people's emotional capacity is limited, so they prioritize their worry according to which crisis is imminent (Bostrom et al. 2020; Botzen et al. 2021). On the other hand, as we examined in RQ1, academic and public concern (in some cases of developed countries) about climate change has not decreased during the pandemic. That suggests that some audiences may be in a state of awareness, acceptance of the emergency, and preparedness to respond (Mocatta and Hawley 2020).

One of the most common causes of crisis communication failure is panic. It can be caused by high levels of uncertainty surrounding an unknown threat, leaders' failure to focus people's concerns on specific actions, rumors, and inconsistent public messages (Ruiu 2020).

According to Lipscy (2020), uncertainty concerning the likelihood of a crisis occurring is especially high before such a crisis happens. That limits the likelihood that preventive measures will be implemented. To put it bluntly, officials prefer being credited by voters for dealing with an immediate emergency rather than implementing preventative measures for a crisis that may or may not occur during their lifetimes.

Uncertainty about an unknown threat should be communicated with both political and scientific cohesion (Ruiu 2020) and the information provided should be clear and straightforward, taking into mind the background of the audience (Sovacool et al. 2020). Bouman et al. (2021) partly associate the public's response to crises with the concept of *personal norms* (an individual's sense of responsibility or morality) and propose strengthening the intrinsic motives of the individuals rather than using external motivation (such as threatening rhetoric). For that to be achieved, Bouman et al. (2021) suggest a. communicating to the public in a simplified manner the relationship between actions and effects within a complex system such as the environment, b. making it relatable to the individual's experience, and c. strengthening the feeling of belonging to a community with similar goals. Because individual action is insufficient, government and private sector actors should act in tandem with the above.

Large-Scale Change is Possible

The pandemic served as a catalyst for increased hostility between nations as they competed for limited resources such as masks, vaccines, and ventilators, to mention a few (Manzanedo and Manning 2020; Fenner and Cernev 2021). The reactions that countries have presented to the COVID-19 threat have been very diverse, but they have also shown that the international community is capable of making extremely significant adjustments if it is sufficiently motivated to do so (Manzanedo and Manning 2020; Perkins et al. 2021).

As previously noted, COVID-19 gained wide coverage via a variety of sources, including mass media, social media, posters in various organizations, information sessions, and so on. As a result, people's habits and behaviors changed all over the world. According to Sovacool et al. (2020), this can be attributed to a form of the "Christmas Effect," the "Coronavirus Effect," which involves multiple groups of individuals coming together and speaking consistently and frequently in unison about the outbreak (much like what happens during Christmas time in parts of Western societies). Could the "Climate Crisis Effect" phenomenon be harnessed for long-term sustainability?

A Matter of Priority

Due to the global scale of both crises, it is essential for world nations to respond to both crises in unison: No country can achieve reducing its greenhouse gas emissions on its own, whereas COVID-19 cannot be eradicated unless it is eradicated in all countries (Fuentes et al. 2020). From an economic point of view, both crises would be more effectively addressed if governments prioritized health or ecological well-being over "growth imperative"[4] (Stuart et al. 2022).

Pooling international resources and fostering global cooperation is of the utmost importance, especially since the pandemic paradigm showed that the nations' early-stage response of denialism and isolationism was not enough to tackle the crisis. Instead, the resource-sharing and global cooperation that ensued were far more effective (Klenert et al. 2020).

Prevention is Better Than Cure

A number of studies demonstrate the economic repercussions of international governments choosing not to undertake steps to mitigate climate change and, instead, opting

[4] The "growth imperative" prioritises economic expansion above slow-growth policies.

to adapt to its effects (Manzanedo and Manning 2020). The pandemic management has prompted Sovacool et al. (2020) to propose active planning on climate crisis mitigation by a. providing people with clear instructions on how to minimize their carbon footprints immediately, b. investing in clean energy infrastructure and green policies, and c. making use of emerging technologies.

When putting this into practice, we need to remember to distribute the burden fairly among everyone involved. When putting this into action, we must remember to evenly spread the responsibility among all parties involved. For example, when it came to COVID-19 vaccine protection, a smaller percentage of Africa's population was inoculated, whereas wealthier, industrialized nations "hoarded" vaccines (Fatton 2021). This should work as a cautionary tale, as we should expect of the developed world to carry a bigger portion of the responsibility for the reduction of the effects of climate change. Firstly, because developed countries are the main contributors to global anthropogenic emissions (for the time being) (Ward and Mahowald 2014). Secondly, because it is within their technical ability to establish environmentally friendly practices toward that purpose (Wahaj et al. 2022).

Conclusion

COVID-19 has gone viral in terms of academic and media engagement. However, it did not dominate climate-related research. Indeed, at the junction of the two crises, new issues have emerged, which the current study attempted to identify: crisis response, effective transition strategies, the effects of COVID-19 on climate, the influence of climate on viral transmission, as well as foreseeable challenges. However, the pandemic dominated mainstream and social media discussions over the climate crisis for a time. This could be attributed to the immediacy of the pandemic crisis, the recognizability of its effects compared to the repercussions of the climate crisis, and the coordinated international responses.

We found that the two crises share the following features in a literature review: They are counter-intuitive, they are defined by a point of irreversible change, and rapid action is less expensive than waiting to deal with the issues in the future. They also differ in many ways, including the following: They span a wide range of time and place, they produce a varied sense of urgency due to their timings and impacts, and individual actions have diverse effectiveness.

Finally, we focused on the lessons that may be applied from one crisis to the next.

- Effective communication can be the catalyst for bringing mankind on board while facing a crisis,
- People should feel empowered that we can change since we have already done so, and
- National priorities should shift toward policies that ameliorate the problem rather than waiting to adapt after the tipping point is crossed.

Because there is no vaccine for climate change, the purpose of this study is to emphasize the need for further research into effective climate change communication and contribute to "Flattening the Curve" of climate change.

Appendix A. VOSViewer Thesaurus File

covid-19 pandemic	pandemic
coronavirus	sars-cov-2
sars-cov-2	sars-cov-2
air-pollution	air pollution
climate-change	climate change
air-quality	air quality
covid 19	covid-19
diseases	disease
impacts	impact
influenza-virus	influenza
pandemics	pandemic
viruses	virus
sars	sars-cov-2
aerosol	aerosols
carbon dioxide	co2
cities	city
covid	covid-19
covid-19 crisis	crisis
disasters	disaster
epidemics	epidemic
greenhouse-gas emissions	greenhouse gas emissions
infectious-diseases	infectious diseases
mental-health	mental health
model	models
pollutants	pollution
public-health	public health
systems	system
transitions	transition
zoonoses	zoonosis
climate changes	climate change
co2 emissions	greenhouse gas emissions

(continued)

(continued)

covid-19 pandemic	pandemic
co2 emission	greenhouse gas emissions
emissions	greenhouse gas emissions
risks	risk
technologies	technology
policies	policy
sustainable development	sustainability

References

Bessi A, Coletto M, Davidescu G-A, Scala A, Caldarelli Guido, Quattrociocchi W (2015) Science vs Conspiracy: collective narratives in the age of misinformation. PLoS One 10. Public Library of Science: e0118093. https://doi.org/10.1371/JOURNAL.PONE.0118093

Bostrom A, Böhm G, Hayes AL, O'Connor RE, (2020) Credible threat: perceptions of pandemic coronavirus, climate change and the morality and management of global risks. Front Psychol 11. https://doi.org/10.3389/fpsyg.2020.578562

Botzen W, Duijndam S, van Beukering P (2021) Lessons for climate policy from behavioral biases towards COVID-19 and climate change risks. World Development. https://doi.org/10.1016/j.worlddev.2020.105214

Bouman T, Steg L, Dietz T (2021) Insights from early COVID-19 responses about promoting sustainable action. Nat Sustain 4, 194–200. https://doi.org/10.1038/s41893-020-00626-x

Cheval S, Adamescu C M, Georgiadis T, Herrnegger M, Piticar A, Legates D R (2020) Observed and potential impacts of the COVID-19 pandemic on the environment. Int J Environ Res Public Health 17. Henri Coanda Air Force Acad, Brasov 500183, Romania. https://doi.org/10.3390/ijerph17114140

Ching J, Kajino M (2020) Rethinking air quality and climate change after COVID-19. Int J Environ Res Public Health 17. Meteorol Res Inst, 1–1 Nagamine, Tsukuba, Ibaraki 3050052, Japan. https://doi.org/10.3390/ijerph17145167

de Moor J, De Vydt M, Uba K, Wahlström M (2021) New kids on the block: taking stock of the recent cycle of climate activism. Soc Mov Stud 20. https://doi.org/10.1080/14742837.2020.1836617

Fan J, Meng J, Ludescher J, Chen X, Ashkenazy Y, Kurths J, Havlin S, Schellnhuber HJ (2021). Statistical physics approaches to the complex Earth system. Physics Reports. North-Holland. https://doi.org/10.1016/j.physrep.2020.09.005

Fatton R (2021) The paradoxes of the pandemic and world inequalities. Social Sciences 10. https://doi.org/10.3390/socsci10090332

Fenner R, Cernev T (2021) The implications of the Covid-19 pandemic for delivering the Sustainable Development Goals. Futures 128. https://doi.org/10.1016/j.futures.2021.102726

Ficetola GF (2013) Is interest toward the environment really declining? The complexity of analysing trends using internet search data. Biodivers Conserv 22. https://doi.org/10.1007/s10531-013-0552-y

Fisher DR, Nasrin S (2021). Shifting coalitions within the youth climate movement in the US. Politics and Governance 9. https://doi.org/10.17645/pag.v9i2.3801

Fuentes R, Galeotti M, Lanza A, Manzano B (2020) COVID-19 and climate change: a tale of two global problems. Sustainability 12. Egade Business Sch Tec Monterrey, Zapopan 45201, Jalisco, Mexico. https://doi.org/10.3390/su12208560

Gemenne F, Depoux A (2020) What our response to the COVID-19 pandemic tells us of our capacity to respond to climate change. Environ Res Lett 15:14–16. https://doi.org/10.1088/1748-9326/abb851

Google (2015) Trends Help.

Hepburn C, O'Callaghan B, Stern N, Stiglitz J, Zenghelis D (2020) Will COVID-19 fiscal recovery packages accelerate or retard progress on climate change? Oxf Rev Econ Policy 36. Univ Oxford, Smith Sch Enterprise & Environm, Oxford, England: S359–S381. https://doi.org/10.1093/oxrep/graa015

Heyd T (2021) Covid-19 and climate change in the times of the Anthropocene. Anthropocene Review 8. Sage Publications Inc.: 21–36. https://doi.org/10.1177/2053019620961799

Howarth C, Bryant P, Corner A, Fankhauser S, Gouldson A, Whitmarsh L, Willis R (2020) Building a social mandate for climate action: lessons from COVID-19. Environ Resource Econ 76. https://doi.org/10.1007/s10640-020-00446-9

Ibn-Mohammed T, Mustapha KB, Godsell J, Adamu Z, Babatunde KA, Akintade DD, Acquaye A, et al. (2021) A critical analysis of the impacts of COVID-19 on the global economy and ecosystems and opportunities for circular economy strategies. Resour Conserv Recycl 164. Univ Warwick, Warwick Mfg Grp WMG, Coventry CV4 7AL, W Midlands, England. https://doi.org/10.1016/j.resconrec.2020.105169

IPCC (2014) Climate Change 2014: synthesis report. Contribution of working groups I, II and III to the Fifth Assessment Report of the Intergovernmental Panel on Climate Change. IPCC.

Kamiński M, Łoniewski I, Marlicz W (2019) Global internet data on the interest in antibiotics and probiotics generated by Google Trends. Antibiotics 8. https://doi.org/10.3390/antibiotics8030147

Klenert D, Funke F, Mattauch, L O'Callaghan B (2020) Five Lessons from COVID-19 for advancing climate change mitigation. Environ & Resour Econ 76. European Commiss, Joint Res Ctr, Calle Inca Garcilaso 3, Seville 41092, Spain: 751–778. https://doi.org/10.1007/s10640-020-00453-w

Le Quéré C, Jackson RB, Jones MW, Smith AJP, Abernethy S, Andrew RM, De-Gol AJ et al (2020) Temporary reduction in daily global CO2 emissions during the COVID-19 forced confinement. Nat Clim Chang 10. https://doi.org/10.1038/s41558-020-0797-x

Leal-Filho W, Nagy GJ, Ayal DY (2020) Viewpoint: climate change, health and pandemics—a wake-up call from COVID-19. Int J Clim Chang Strat Manag 12. Hamburg Univ Appl Sci, Fac Life Sci, FTZ ALS, Hamburg, Germany: 533–535. https://doi.org/10.1108/IJCCSM-08-2020-212

Lidskog R, Elander I, and Standring A (2020) COVID-19, the climate, and transformative change: comparing the social anatomies of crises and their regulatory responses. Sustainability 12. Orebro Univ, Sch Humanities Educ & Social Sci, SE-70281 Orebro, Sweden. https://doi.org/10.3390/su12166337

Lipscy PY (2020) COVID-19 and the politics of Crisis. Int Organ 74. Univ Toronto, Dept Polit Sci, Toronto, ON, Canada: E98–E127. https://doi.org/10.1017/S0020818320000375

Loureiro ML, Allo M (2021) How has the COVID-19 pandemic affected the climate change debate on Twitter? Environ Sci & Policy 124. Univ Santiago De Compostela, Ecobas, Santiago De Compostela, Spain 451–460. https://doi.org/10.1016/j.envsci.2021.07.011WE-ScienceCitation IndexExpanded(SCI-EXPANDED)

Loureiro ML, Alló M (2021) How has the COVID-19 pandemic affected the climate change debate on Twitter? Environ Sci Policy 124. Elsevier: 451–460. https://doi.org/10.1016/j.envsci.2021.07.011

Manzanedo R, Manning P (2020) COVID-19: lessons for the climate change emergency. Sci Total Environ 742. https://doi.org/10.1016/j.scitotenv.2020.140563

Masson-Delmotte V, Zhai P, Pörtner H-O, Roberts D, Skea J, Shukla PR, Pirani A, Moufouma-Okia W, Péan C, PidcockR , Connors S, Matthews JBR, Chen Y, Zhou X, Gomis MI, Lonnoy E, Maycock T, Tignor M, Waterfield T (2018) Global warming of 1.5°C. An IPCC Special Report on the impacts of global warming of 1.5°C above pre-industrial levels and related global greenhouse gas emission pathways, in the context of strengthening the global response to the threat of climate change. IPCC - Sr15 2: 17–20

Mattar SD, Jafry T, Schroder P, Ahmad Z (2021) Climate justice: priorities for equitable recovery from the pandemic. Climate Policy 21. Glasgow Caledonian Univ, Ctr Climate Justice, Glasgow, Lanark, Scotland: 1307–1317. https://doi.org/10.1080/14693062.2021.1976095

McAllister JT, Lennertz L, Mojica ZA (2022) Mapping a discipline: a guide to using VOSviewer for bibliometric and visual analysis. Sci Technol Librar 41 (3):319–348. https://doi.org/10.1080/0194262X.2021.1991547

Mocatta G, Hawley E (2020) The coronavirus crisis as tipping point: communicating the environment in a time of pandemic. Media Int Aust 177. Deakin Univ, Burwood, Vic, Australia Pu-Sage Publications Ltd PI - London Pa, England: 119–124. https://doi.org/10.1177/1329878X20950030

Moser SC (2010) Communicating climate change: history, challenges, process and future directions. Wiley Interdisciplinary Reviews: Climate Change 1. Wiley-Blackwell: 31–53. https://doi.org/10.1002/wcc.11

Ng Oi Wing, Yee Joo Tan (2017) Understanding bat SARS-like coronaviruses for the preparation of future coronavirus outbreaks—Implications for coronavirus vaccine development. Hum Vaccines & Immunother 13. Hum Vaccin Immunother: 186–189. https://doi.org/10.1080/21645515.2016.1228500

Oldekop J A, Horner R, Hulme D, Adhikari R, Agarwal B, Alford M, O Bakewell, et al. (2020) COVID-19 and the case for global development. World Dev 134. Univ Manchester, Global Dev Inst, Manchester, Lancs, England. https://doi.org/10.1016/j.worlddev.2020.105044

Ortiz AMD, de Leon AM, Torres JNV, Guiao CTT, La Vina AGM (2021) Implications of COVID-19 on progress in the UN Conventions on biodiversity and climate change. Glob Sustain 4. UCL, Bartlett Sch Environm Energy & Resources, Inst Sustainable Resources, London, England. https://doi.org/10.1017/sus.2021.8

Perkins KM, Munguia N, Ellenbecker M, Moure-Eraso R, Velazquez L (2021) COVID-19 pandemic lessons to facilitate future engagement in the global climate crisis. J Clean Prod 290. State Univ New York, Sch Nat & Social Sci Purchase Coll, Psychol, 735 Anderson Hill Rd, Purchase, NY 10577 USA. https://doi.org/10.1016/j.jclepro.2020.125178

Prideaux B, Thompson M, Pabel A (2020) Lessons from COVID-19 can prepare global tourism for the economic transformation needed to combat climate change. Tour Geogr 22. https://doi.org/10.1080/14616688.2020.1762117

Rauchfleisch A, Siegen D, Vogler D (2021) How COVID-19 displaced climate change: mediated climate change activism and issue attention in the Swiss media and online sphere. Natl Taiwan Univ, Grad Inst Journalism, Taipei, Taiwan, Environmental Communication. https://doi.org/10.1080/17524032.2021.1990978

Ruiu ML (2020) Mismanagement of Covid-19: lessons learned from Italy. J Risk Res 23. Northumbria Univ, Social Sci, Newcastle Upon Tyne, Tyne & Wear, England PU - Routledge Journals, Taylor & FranciS Ltd PI - Abingdon PA - 2–4 Park Square, Milton Park, Abingdon Ox14 4RN, Oxon, England: 1007–1020. https://doi.org/10.1080/13669877.2020.1758755

Rutjens BT, van der Linden S, van der Lee R (2021) Science skepticism in times of COVID-19. Group Process & Intergroup RelatS 24. Univ Amsterdam, Amsterdam, Netherlands: 276–283. https://doi.org/10.1177/1368430220981415

Sabherwal A, Ballew MT, van der Linden S, Gustafson A, Goldberg MH, Maibach EW, Kotcher JE, Swim JK, Rosenthal SA, Leiserowitz A (2021) The Greta Thunberg effect: familiarity with Greta Thunberg predicts intentions to engage in climate activism in the United States. J Appl Soc Psychol 51:321–333. https://doi.org/10.1111/jasp.12737

Schaub S, Huber R, Finger R (2020) Tracking societal concerns on pesticides—A google trends analysis. Environ Res Lett 15. IOP Publishing Ltd. https://doi.org/10.1088/1748-9326/AB9AF5

Sovacool BK, Del Rio DF, Griffiths S (2020) Contextualizing the Covid-19 pandemic for a carbon-constrained world: insights for sustainability transitions, energy justice, and research methodology. Energy Research & Soc Science 68. Univ Sussex, Sch Business Management & Econ,

Sci Policy Res Unit SPRU, Brighton, E Sussex, England. https://doi.org/10.1016/j.erss.2020.101701
Spisak BR, State B, van de Leemput I, Scheffer M, Liu Y (2022) Large-scale decrease in the social salience of climate change during the COVID-19 pandemic. PLoS One 17. https://doi.org/10.1371/journal.pone.0256082
Stuart D, Petersen B, Gunderson R (2022) Shared pretenses for collective inaction: the economic growth imperative, COVID-19, and climate change. Globalizations 19. No Arizona Univ, Sch Earth & Sustainabil, Program Sustainable Commun, Flagstaff, AZ 86011 USA: 408–425. https://doi.org/10.1080/14747731.2021.1943897
Thijs B, Steg L, Dietz T (2021) Insights from early COVID-19 responses about promoting sustainable action. Nature Sustainability. https://doi.org/10.1038/s41893-020-00626-x
Trisos CH, Merow C, Pigot AL (2020) The projected timing of abrupt ecological disruption from climate change. Nature 2020 580:7804 580. Nature Publishing Group: 496–501. https://doi.org/10.1038/s41586-020-2189-9
United Nations (2015) Transforming our world: the 2030 agenda for sustainable development United Nations United Nations transforming our world: the 2030 agenda for sustainable development. United Nations.
van der Hel S, Hellsten I, Steen G (2018) Tipping points and climate change: metaphor between science and the media. Environ Commun 12. https://doi.org/10.1080/17524032.2017.1410198
van der Ven H, Sun YX (2021) Varieties of Crises: comparing the politics of COVID-19 and climate change. Glob Environ Polit 21. McGill Univ, Sch Environm, Montreal, PQ, Canada: 13–22. https://doi.org/10.1162/glep_a_00590
van Eck NJ, Waltman L (2020) VOSviewer manual version 1.6.16. Univeristeit Leiden.
Vincent CCC., Lau SKP,. Woo PC, Kwok YY (2007) Severe acute respiratory syndrome coronavirus as an agent of emerging and reemerging infection. Clinical microbiology reviews 20. Clin Microbiol Rev: 660–694. https://doi.org/10.1128/CMR.00023-07
Wahaj Z, Alam MM, Al-Amin A (2022) Climate change and COVID-19: shared challenges, divergent perspectives, and proposed collaborative solutions. Environ Sci Pollut Res 29. Natl Univ Sci & Technol NUST, NUST Business Sch NBS, Islamabad, Pakistan: 16739–16748. https://doi.org/10.1007/s11356-021-18402-5
Ward DS, Mahowald NM (2014) Contributions of developed and developing countries to global climate forcing and surface temperature change. Environ Res Lett 9. https://doi.org/10.1088/1748-9326/9/7/074008
WHO (2020) Coronavirus (COVID-19) events as they happen. https://Www.Who.Int/Emergencies/Diseases/Novel-Coronavirus-2019/Events-As-They-Happen
Wouter B, Duijndam S, van Beukering P (2021) Lessons for climate policy from behavioral biases towards COVID-19 and climate change risks. World Dev. https://doi.org/10.1016/j.worlddev.2020.105214
Wu X, Nethery RC, Sabath BM, Braun D, Dominici F (2020) Exposure to air pollution and COVID-19 mortality in the United States: a nationwide cross-sectional study. medRxiv: the preprint server for health sciences. https://doi.org/10.1101/2020.04.05.20054502
Yoshino N, Taghizadeh-Hesary F, Otsuka M (2021) Covid-19 and optimal portfolio selection for investment in sustainable development goals. Environ Res Lett 38. Keio Univ, Tokyo, Japan. https://doi.org/10.1016/j.frl.2020.101695

Normalization, Legalization, and Equalization: Toward a Sustainable and Humane EU Border Policy

Henk van Houtum and Rodrigo Bueno Lacy

Abstract In this chapter, we analyze the systemic mismatch between the EU's self-aggrandizing promise of sustainable and humane migration policies and the actual practices characterizing its border regime. The EU's discriminatory border regime (which we dissect into a *pre-border* visa regime, the *in-situ* land and sea borders, and the *post-border* camps) has set in motion a recurrent demarcation of increasingly inhumane, unlawful and deadlier borders. We contend that, in its self-proclaimed attempt to protect its foundational values through a selectively permeable border regime, the EU has triggered an autoimmune disorder that has become the Union's most formidable threat. We conclude by discussing three possible paths out of this downward spiral: normalization, legalization and equalization. This, we believe, is the urgent turn that the EU needs to take in order to escape this suicidal paradox.

Keywords EU border regime · Human migration · Sustainable and humane border policy

Introduction

There is no doubt that the EU subscribes to the UN's Sustainable Development Goals (SDGs) and is formally committed to their realization. Overall, these goals aim at ending poverty and inequality as well as at improving health and education around the world—while tackling climate change by developing sustainable frameworks to preserve our oceans and forests by 2030. The European Commission is unambiguous about its support for such lofty objectives: 'We are committed to implementing the

This chapter is a reworked and updated version of a previously published journal article (Van Houtum and Bueno Lacy 2020b)

H. van Houtum (✉)
Nijmegen Centre for Border Research, Radboud University, Nijmegen, The Netherlands
e-mail: henk.vanhoutum@ru.nl

H. van Houtum · R. B. Lacy
University of Eastern Finland, Kuopio, Finland

A. Klonari et al. (eds.), *Re-visioning Geography*, Key Challenges in Geography,
https://doi.org/10.1007/978-3-031-40747-5_15

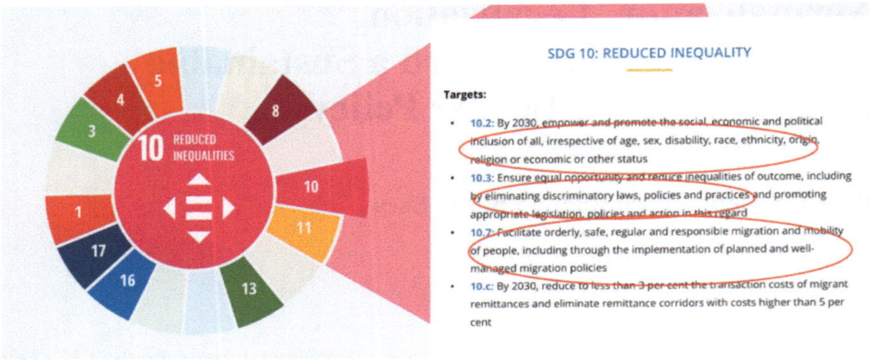

Fig. 1 SGD#10: Reduced inequality

SDGs in all our policies and encourage EU countries to do the same'. In this chapter, we take this self-proclaimed commitment seriously by assessing the progress of goal number 10—'Reduce inequality within and among countries'—against the rising inequality that migrants around the world experience.[1]

Comprised within SGD#10 are the following targets (see Fig. 1)[2]:

> 10.3 Ensure equal opportunity and reduce inequalities of outcome, including by eliminating discriminatory laws, policies and practices and promoting appropriate legislation, policies and action in this regard.
> 10.7 Facilitate orderly, safe, regular and responsible migration and mobility of people, including through the implementation of planned and well-managed migration policies.

We begin by the irrefutable mismatch between official rhetoric and daily practice: the gap between the nobility articulated by such objectives and the actual ruthlessness of the EU migration policy has become truly distressing.[3] Although estimates differ, the most meticulous appraisals have counted over 52,000 deaths among the human beings who have died trying to reach the EU since the early 90's—when Schengen was progressively incorporated into EU law—and many others have been pushed back through beatings, torture, and sexual assault by EU border forces (UNITED 2023; 111 2023).[4]

This sheer horror has turned the EU's external border into the deadliest on the planet—by far (van Houtum 2015; van Houtum and Bueno Lacy 2020a, 2020b; IOM 2022). Despite its appalling magnitude, however, even the number of registered

[1] https://unstats.un.org/sdgs/indicators/Global%20Indicator%20Framework%20after%202022%20refinement_Eng.pdf.

[2] https://unstats.un.org/sdgs/indicators/Global%20Indicator%20Framework%20after%202022%20refinement_Eng.pdf.

[3] https://unitedagainstrefugeedeaths.eu

[4] The Fatal Policies of Fortress Europe.

deaths is an underestimate, for it is impossible to tell how many more anonymous migrants have drowned in the Mediterranean Sea or succumbed along the perilous North African routes—or further afield.

A crucial catalyzer of these fatalities has been the EU-wide trend to criminalize humanitarianism. Not only did the EU discontinue 'Mare Nostrum'—its single most successful rescue operation in the Mediterranean—only one year after its implementation, but NGOs attempting to save lives at sea have been unscrupulously harassed by EU Member States to the absurd excess of being charged with 'human smuggling' and 'human trafficking': a cynically disingenuous policy that seems bound to ensure even higher casualties in the Mediterranean—turning it, as the UNHCR has described it, into a 'sea of blood' (Bueno Lacy and van Houtum 2020).

At the same time, the violence and death that characterize the experience of undocumented migrants trying to make their way to the EU have become such a 'normal' part of current EU border policies, that news about these perilous journeys or the mental and bodily devastation they lay waste to hardly make headlines anymore (Laurent and O'Grady 2018; AiK 2018; AP 2018; Kemp 2017; Malik 2019; van Houtum and Boedeltje 2009; Lucas, Ramsay and Keen 2019; Plenel 2019). This overt callousness poses the conundrum that we analyze in this chapter: Why is the EU purposefully undermining its own committment 'to implementing the SDG's in all [its] policies'?

Although we recognize that the EU is not a homogeneous political entity but a complex supranational organization composed of diverse political institutions, culturally specific member states, antagonistic political parties of all ideological stripes and, overall, a wide range of interests, for the purpose of this chapter we evaluate the overall effect that the EU's border regime exerts on its political community as a whole.

We contend that the ongoing violence, dehumanization, and deterrence with which undocumented migrants are being 'welcomed' are not anomalous practices but integral to the structure and ideology of EU's current border policies (Jones 2017; Kriesi and Pappas 2015; van Houtum and van Naerssen 2002; van Houtum 2021). To this end, we analyze EU's b/ordering and othering regime through Jacques Derrida's notion of *geopolitical autoimmunity*, which he defined as the strange behavior by which an organization, 'in quasi-suicidal fashion, "itself" works to destroy its own protection, to immunize itself against its "own" immunity' (Derrida 2003, p. 94).

We employ the metaphor of 'autoimmunity' in full antagonism with the white-supremacist rhetoric of far-right politicians such as Thierry Baudet, who have distorted it to denounce 'massive immigration' as the cause of the West's 'weakening body'. Instead, we draw on Derrida's conception of autoimmunity to argue the opposite: that the EU's self-destructiveness is not rooted in its openness to the world but in the counterproductive effects of its increasingly closed and xenophobic border regime (Bueno Lacy and Van Houtum 2015). Dating back to its inception, the EU has been inspired by a nativist principle to develop a network of biopolitical filters along its external and internal borders. This architecture has been designed to discriminate against migrants from specific countries—not least by endangering and criminalizing their mobility. We argue that not only has this *border disorder* (Bueno

Lacy and van Houtum 2013) alienated the EU from its self-professed values (i.e., the rule of law and universal human rights) but it also has legitimized and normalized nativist authoritarian populists (Boedeltje and van Houtum 2008; van Houtum and Bueno Lacy 2017; van Houtum 2021; De Jonge and Gaufman 2022). Thus, the autoimmunity that we recognize has its roots inside EUrope and not beyond its borders.

Our analysis breaks down the EU's border regime in three filtering mechanisms: (1) the pre-border of legal entry documents or, as we term it, the *paper border*; (2) the in-situ physical border controls or what we refer to as the *iron border*; (3) and the post-border articulated by the reception and detention camps that keep migrants ostracized even after they have entered the EU. We analyze how these three cogwheels of the EU's b/ordering and othering machinery have developed over time by casting light on how they have become self-reinforcing engines propelling a self-destructive border policy. Finally, we suggest three alternative directions that could take the EU out of this suicidal paradox (van Houtum 2010).

We conclude by stressing the ominous political implication of this EUropean *border disorder* (Bueno Lacy and van Houtum 2013): the harrowing fate of immigrants is inextricably linked to the fate of the EU, for their suffering and deaths are both symptoms and consequences of an autoimmune reaction that might ultimately lead to the EU's demise. Succinctly put, by driving migrants and their children to commit suicide, the EU is opening a black hole that might swallow whole the ethos, values and laws on which much of its post-war peace and prosperity have been built. Ultimately, the aim of our analysis is to issue a warning: we may be witnessing a dangerous authoritarian turn—or even a s*EU*cide—characterized by the gratuitous self-destruction of the post-war project of European integration. A project which was founded on principles indistinguishable from the UN's goals of basic human dignity, safety and equality that it increasingly denies to those who need it the most.

Derrida's Autoimmunity

Autoimmunity as tool for the critical analysis of geopolitics is famously associated to the deconstructive method developed by Jacques Derrida. To him, autoimmunity evoked the backfiring mechanism by which a political hegemon flexes its 'techno-socio-political machine' in order to consolidate its power yet unleashes an unintended reaction that undermines it and eventually threatens the hegemon's survival. Derrida identified a series of symptoms typical to this autoimmune disorder: (1) a *reflex of power* and the *reflection* it produces; (2) a *trauma* that envisions an inauspicious future—barred any action undertaken to prevent its repetition; (3) *invisible* and *anonymous* enemy forces that could hardly be pinned down to a particular state, cartographical location or physical entity; (4) *apocalyptic* descriptions of geopolitical events carrying religious undertones and, perhaps more decisively: (5) a *double incomprehension*: a political organization's inability to grasp the traumatic events to which it responds and to realize that what it deems its reasonable responses to them only aggravate them (Derrida 2003, p. 90, pp. 97–98). Ultimately, such autoimmunity sets in motion a dauntingly counterproductive machinery of self-fulfilling prophecies that are fueled not by a specious 'clash of civilizations' but instead by what Edward Said called 'a clash of ignorance' (2001).

In a famous interview with the philosopher Giovanna Borradori, Derrida resorted to a deconstructive analysis of 9/11 to dissect the autoimmune syndrome that he saw affecting the US' global hegemony (Derrida 2003). He called out the asymmetry between the US commemoration of 9/11 as an unparalleled historical tragedy and the far more atrocious violence orchestrated by the United States around the globe, which was unleashed many times before 9/11 and has been reoccurring many times afterward without arousing a comparable amount of either media epitaphs or political lamentations. Whereas 9/11 enjoys the privilege of arousing pathos in both Europe and the United States, comparable massacres beyond their territories and perpetrated by their own armies do not cause such an intense upheaval in either their media or public opinion (e.g., Vietnam, Chile, Guatemala, Cambodia, Rwanda, Palestine, Iraq, Afghanistan, Yemen, Congo and so on) (Derrida 2003, p. 92).

The Autoimmune Borders of the EU

In what follows, we deploy Derrida's notion of autoimmunity to analyze the b/ordering and othering policies that the EU has developed as a response to undocumented migration (van Houtum and van Naerssen 2002). To this end, we dissect the EU's b/ordering response into three immunizing borders, each characterized by a different materiality and function: the pre-border (i.e., the 'paper border'), the in-situ border (i.e., the 'iron border') and the post-border (i.e., the 'camp border').

The Pre-Border: The EU's Paper Border

Arguably, one of the most significant landmarks in the recent history of EU's b/ordering and othering policy (van Houtum and van Naerssen 2002; van Houtum 2021) has been the creation of a common external border—what we call 'the paper border'. The common paper b/ordering of the EU dates back to the Schengen Agreement of 1985, which envisioned the gradual abolition of internal checks in exchange for the establishment of strict border controls along the EU's external borders—a decision that implied merging Member States' border controls under a joint command. This agreement was further refined by the Dublin Convention of 1990, which harmonized the EU's asylum procedures later enshrined in the Maastricht Treaty of 1992. The Schengen area was effectively established in 1995 (and later incorporated into EU law by the Treaty of Amsterdam that came into effect in 1999). The Schengen Information System (SIS) and Visa Information System (VIS)—implemented in 2006 and 2011, respectively—that derived from these agreements constitute the fundamental architecture of the EU's common external border surveillance system, which is designed to filter out global border crossers lacking the traveling papers required by the Schengen Agreement.

With the demarcation of this paper *bordering* (van Houtum and van Naerssen 2002; van Houtum 2021), the EU, a supranational organization, began to mimic the nation-state's anachronistic political myth: it legalized—and thereby normalized—the apocryphal synonymousness between 'EU citizens' and 'Europeans' (Slootweg et al. 2019). This geo-ideological caging carved up a fracture between the EU and the much broader historical idea of Europe, which is anchored along the whole Mediterranean basin and, as consequence of imperialism, colonial legacies and transculturation, resonates with varying intensity across the world (Bueno Lacy and van Houtum 2015). The understanding of a neatly-demarcated 'Europe' coinciding with the current political boundaries of the EU is unprecedented: Europe has always been a geographically undetermined idea and it has never been either a congruent political organization or a demos (Delanty 1996); nor has the European continent ever been severed from its contiguous Mediterranean geographies by such sharp borders (Braudel 1995, 2002). Yet, since the signing of the Schengen Agreement in 1985, the EU has been gradually fortifying itself by turning the Mediterranean into its moat.

This abduction of Europe by the EU has been progressively reified through a conscious *ordering* (van Houtum and van Naerssen 2002; van Houtum 2021) brought about by the distinct process of *EUropeanization.* This identitary strategy has inculcated a synonymization between the EU—a political project dating back to 1951—and a European heritage that could be traced as far back as to the Kurgan civilization in the 4th millenium BC (Gimbutas 1985)—or whenever a historian might decide these origins lie, for such considerations are unavoidably political decisions too (Foucault 1971; Southgate 2011). Such EUropeanization has been characterized by the manufacture of maps, coins, symbols, narratives and accompanying geopolitical practices that have attempted to shoe-horn European history and culture into the current borders and geopolitical concerns of the EU (Boedeltje and van Houtum 2008; Bueno Lacy and van Houtum 2015). As membership to the EU has become equated with a historical belonging to Europe, EUropeanization has cultivated a tacit imagination of neighboring countries as lacking an intrinsic Europeanness—i.e., a geopolitically constructed *othering* process (van Houtum and van Naerssen 2002; van Houtum 2021).

A prime example of this geopolitical otherization took place in 1987, when the European Economic Community (EEC)—the immediate predecessor of the EU—received a membership application from Morocco which it immediately rejected arguing Morocco's lack of Europeanness. By codifying contingent geopolitical prejudice into law, the EEC prevented a North African country from meeting the basic eligibility criteria to be considered part of a geography of which it has nonetheless been part since, at least, the Graeco-Roman Antiquity which the EU itself considers the cradle of its civilization. It is worth noticing that this reasoning amounted to more than an innocent incursion into basic physical geography: with this decision, the EEC implicitly asserted itself as the institutionalized embodiment of European civilization and, as such, it claimed its prerogative to legally define and confer, in a discretionary manner, its arbitrary acknowledgment of Europeanness. Such an exercise in what could perhaps be termed 'grand geopolitical anachronism' repeated itself in 2004, when the EU demarcated the Bosporus as yet another clear-cut boundary of Europe by indefinitely delaying Turkey's accession (pending since 1987, which makes Turkey the only candidate to which EU membership has been promised yet never granted).

Through these legal, territorial and discursive reifications of its narrow idea of Europe, the EU has suggested that it regards the borders of European culture as roughly coinciding with those of a Christian—or, at least, of an essentially non-Muslim—European civilization. These geopolitics have allowed the present to invent the past by b/ordering EUropeanness in a way that consciously leaves out large swathes of land whose people, cultures and heritage have played a critical role in the making of what the EU itself glorifies as quintessential European culture: North Africa, Asia Minor, Russia and also the scattered worldwide geographies with which Europeans share so much transculturation and vice versa.

We contend that a troubling consequence of carving up this hard external border—on which the EU's invention of Europe is predicated—has been the resurrection of *traumatic* prejudices about Europe's others: the non-Europeans who have been traditionally imagined as backward, violence-prone and overall *inferior* intruders (Vitkus 1997; Davidson 1980: 24–25). Although such civilizational threat is mostly confined to sensationalized accounts or downright fabrications magnified by murkily manipulated media (Juhász and Szicherle 2017; Horaczek 2019), the *invisibility* and *anonymity* inherent to such non-existent boogeymen have made their signifiers—i.e., the flesh-and-bone human beings immigrating to the EU—look like legitimate targets of ever more vicious social demonization and thus of state surveillance and brutal repression (Fekete 2022).

The striking culmination of this paper b/ordering regime was the common Schengen list of visa-required countries introduced in 2001. This significant—yet still remarkably under-researched—'black and white list' (later re-branded as the 'negative and positive list') makes a sharp distinction between countries whose citizens require a visa to enter the EU—largely Muslim, African and overall less affluent countries—and those exempted from it—largely OECD members as well as a few countries in South America and Asia (Mau et al. 2015; Neumayer 2006; Salter 2003, 2006; van Houtum 2010; van Houtum and Lucassen 2016, van Houtum and van Uden, 2021; van Houtum, 2021; see Fig. 2).

This list is based on nativist discrimination—a principle that is forbidden by law in all Member States of the EU and which runs against the EU's own Copenhagen criteria and Lisbon Treaty. This global apartheid has, in effect, almost entirely closed off all legal migration channels to the EU for the large majority of the world (van Houtum 2010; van Houtum and van Uden 2021). The legalization of such discrimination and prejudice has nurtured a selective dehumanization of refugees, which is a vivid illustration of what has been termed *borderism*: the discriminatory politics of spatial segregation that essentialize—and politicize—the value of human beings on

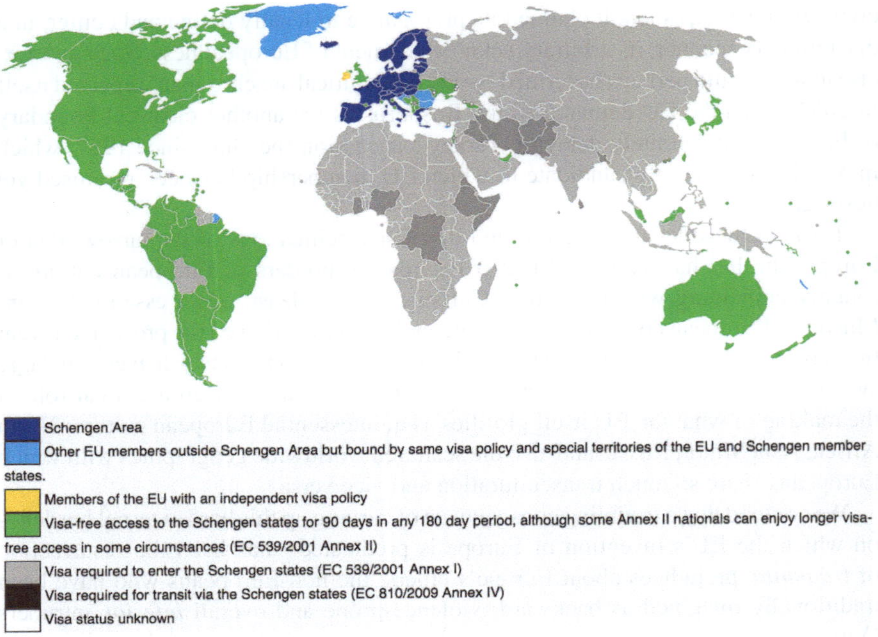

Fig. 2 The paper fortress of the EU (*Source* https://commons.wikimedia.org/wiki/File:Schengen_visa_requirements.png)

the basis of the bordered (id)entity they are born into, reside in, travel from or are associated with (van Houtum 2021).

This pre-border divides EUropeans from non-EUropeans on the basis of arbitrary geographical discrimination even before the actual fences, border guards, and detention camps are even able to exert their own b/ordering constraints (Bueno Lacy and van Houtum 2022). Moreover, such pre-border policies have outsourced the EU's border control to governmental offices far away from the EU's actual border (Bialasiewicz 2012a). Thus, this paper border should not be conceived as a line on a map dividing one country from another but as a techno-political mechanism of global reach meant to b/order the EU at remote control (Zaiotti 2016). Rather than guards with guns, this first border of the EU is staffed by bureaucrats entrenched in faraway embassies and armed to the teeth with unobtainable stamps and impenetrable forms. This political technology, which could be termed *tele*-bordering, is employed to blacklist entire nations—barred few exceptions such as the wealthy, who can afford golden passports (Carrera 2014; van Houtum and van Uden 2021); Russian and US oligarchs (Collins 2022); as well as the families of tyrants and other kleptocrats the world round who have made EUrope their home (Ragget 2020). This means that, in practice, most of the citizens of these blacklisted countries cannot acquire visas to legally travel to the EU—a reality subverted by an unscientific far-right discourse that insists on portraying undocumented migrants as 'tsunamis' of willful law-breakers. The implication is that the paper border of the EU keeps people remotely caged in the inequitable lottery of birth (see Rawls 1971, pp. 118–123; van Houtum and van Uden 2021).

The result of this tele-bordering has been as counterproductive as it has been grim (Miller 2019), for a first suicidal paradox inherent to the EU's paper fortress is captured by the following Escher-like tragedy: even if someone is fleeing life-threatening circumstances or an oppressive regime, they can rarely get a visa for the EU precisely because they stem from countries whose disrepute, ironically, the EU extrapolates to the individual character of those who denounce it with their exodus. By refusing potential refugees the possibility to obtain legal entry to its territory, the EU's paper fortress punishes the victims of unenviable fate for being born in the wrong place and for the all-too-human inclination to leave behind political persecution, generalized violence, economic despair, or natural disaster. This constitutes both a violation of international refugee law as well as a factual rejection of both the humanist ethos and legal custom on which the internationally-recognized right to ask for another country's protection has been built. Such custom includes an express exhortation to governments for the 'issue and recognition of travel documents', which 'is necessary to facilitate the movement of refugees, and in particular their resettlement' (UN 1951).

The repercussion of the EU's willful non-compliance with such international obligations is that—and this is the second paradox of this paper-border regime—access to the EU's *regular* asylum system can only be gained *irregularly*: through smugglers and other illicit peddlers. The safe alternative of air travel is also unavailable to undocumented migrants because, since 2001, air carriers can be fined for boarding passengers lacking the required visa (Directive 2001/51/EC)—a policy that amounts to the erection of an effective b/ordering dome over the EU's airspace. Such paradoxical policy welcomes refugees yet illegalizes the channels that would allow them to legally and safely travel to the EU, what FitzGerald has described as 'the Catch-22' of the rich world's asylum policies (2019).

By forcing asylum seekers to undertake a reckless odyssey—which criminalizes a large portion of the world as well as those who assist undocumented migrants in their journeys—the EU has boosted a large-scale smuggling industry that profits from the legal void that the EU itself has made sure to enforce. Instead of the humanitarian philanthropy which the EU so dishonestly characterizes as the guiding value behind its border regime, it is its anti-refugee obsession disguised as an anti-smuggling preoccupation that has become the legal framework on which a billion-dollar industry of refugee smugglers and border enforcers (e.g., Frontex) has been built (Lyman and Smale 2015; Spijkerboer 2018).

This *border industrial complex* embodies the third paradox of the paper-border regime: the EU has decided—against its own principles and international obligations—to voluntarily create a border system that guarantees the production of 'illegality', corruption, and human insecurity. Ergo, this paper border should be credited with turning the routes to seek asylum in the EU—a supposedly safe destination—into a sordid and perilous survival of the fittest. Since this is precisely the kind of distress that international refugee law is meant to preclude, the so-called migration crisis of 2015 in the EU would be better described as a ‘refugee-protection crisis’. Scandalously, by erecting such an insurmountable paper border, the EU has advocated a politics of death—a *necropolitics* (Mbembé 2003).

Seen through Derrida’s conceptualization of autoimmunity, one could argue that the *reflex of power* manifested as the EU’s territorial strategy to protect itself from unwanted foreigners through Schengen’s paper b/ordering has been predicated on an inexistent *apocalyptic* threat of *invisible* and *anonymous* non-EUropeans—a fear ceaselessly stoked by much of the EUropean press and opportunistic politicians, who wantonly associate asylum seekers with all sorts of crime and moral decay (Albahari 2018; Burrell and Hörschelmann 2019; Trilling 2019). Furthermore, the EU has reinforced its paper-border regime by waging a facetious ‘war against human trafficking’ (Frontex 2022a): a real peril faced by refugees which nonetheless is heightened—not diminished—by both the EU’s border regime and Frontex, its callous border enforcer. Such a tortured ‘humanitarian’ narrative relies on a cunning misrepresentation of the causality between asylum seekers and human smuggling and trafficking (Cuttitta 2018; Dadusc and Mudu 2020): as though smugglers and traffickers were the cause and not the consequence of the EU’s border regime and thus the key symptom of its autoimmunity (Bueno Lacy and van Houtum 2020). Such conscious mischaracterization inverts the causality between migrants who are forced to travel without documents and a EU border regime of militarized coast guards and criminal gangs who exploit the legal limbo to which such system dooms them, thus criminalizing the most vulnerable of migrants while legitimizing far-right prejudices against them (De Genova 2013). By manufacturing such a spectacle, the EU border regime has helped to normalize—and popularize—a stream of EUrosceptic and xenophobic political movements that are trying to pose as the preservers of Europe’s imagined native culture, which they define in terms antithetical to the EU’s ethos yet reminiscent of the heyday of European imperialism.

Overall, such misrepresentation of global mobility has led to a politics of death and to the criminalization of solidarity: a fraying rule of law that is being presented as the regrettable but unavoidable collateral damage that EUrope has to accept in order to preserve the ‘enlightened’ European civilization which the Union itself has essentialized (Plenel 2019). What Derrida identifies as autoimmunity’s *double incomprehension* lies in the EU’s inability to grasp the suicidal paradox in which it has trapped itself: in a short period—since the Schengen Agreement was signed in 1985—the EU’s has triggered a border politics that, in an inexorably self-defeating manner, has been attempting to shield EUrope’s humanist heritage from the very universal scope that makes such heritage worth of preservation. The result is autoimmune border regime that is isolating the EU not only from refugees but also from the humanist values that safeguard the fundamental freedoms of its own citizens.

The In-Situ Border: The EU's Iron Border

The second b/ordering—or immunization—strategy of the EU that we wish to address is the construction of all kinds of material deterrences that have been erected over time along the external borders of the EU and which we metonymically classify as the 'in-situ border' or 'iron border'. This border complements the 'gate at a distance' erected by the paper border with all sorts of terrain-related obstacles including towers, walls and barbed wire.

This violent infrastructure is typically guarded by men and women in uniform who are equipped with guns, handcuffs, surveillance vehicles and sophisticated military gear. The spatial shape of this iron b/ordering regime is *arterial* (Vogt 2017; Campos-Delgado and Coté-Boucher 2022): it is not confined to either the EU's external borders or its Member States' internal borders but it also encompasses a thinly ramified network of border controls extending throughout Schengen territory. Such a system comprises on-the-spot passport controls at airports, trains and highways; as well as surveillance patrols along the EU's maritime borders, which are tasked with stopping refugees from either reaching the EU or remaining in it (Minca and De Rijke 2017; van Houtum 2010).

In contrast to the remotely controlled legal procedures characterizing the largely invisible paper border, a defining aspect of the iron border around—and within—the EU's territory is its visibility to the public eye (Campos-Delgado 2022). It suffices to google 'fences' and 'EU' (or anything of the sort) to come across thousands of pictures featuring the heterogenous materiality of the iron border, such as the iconic and violently guarded fences separating the African continent from the Spanish enclaves of Ceuta and Melilla or the recently built fences along the Hungarian-Serbian border. The gruesome human-rights abuses characterizing the iron border have mushroomed as its infrastructure has expanded along the Belarusian, Ukrainian, Croatian and Greek borders, thus turning the EU's Balkan and eastern routes into hallways of horror. There is a growing body of evidence showing that military and paramilitary forces have been robbing, beating, torturing, raping and either murdering undocumented migrants or leaving them to die all along the external borders of the EU (Tondo 2018b; BVMN 2020; Deeb et al. 2021; Mbayé 2022; Kassam 2022; Guterres 2022; HRW 2022). This purposeful terrorization of vulnerable populations represents neither a mistake nor a rarity but instead a systematic EU strategy to deter undocumented migrants from requesting the refugee protection to which they are righteously entitled.

On the basis of this evidence, the iron borders of the EU should be understood as a consciously performative power play: a geopolitical spectacle in which the official humanist rhetoric serves mainly as a public-relations' façade intended to conceal the callousness of EU's migration-filtering infrastructure. This border show is intended to project safety and security for domestic electoral consumption by displaying an intimidating system of migration-deterring barriers which, in turn, is meant to lionize the government as being 'tough' on unwanted outsiders. The straightforward message that this vulgar political theater is meant to convey is that the government is protecting its people by keeping a close eye on threatening foreigners as well as a clenched fist to thwart their efforts to enter the country by 'illegal' means. More than a line of control, the iron border constitutes a camera-happy spectacle of a drama pre-scripted by the paper border. This performative *mise en scène* relies on an unwritten but foreseeable plot in which barbed wire—which epitomizes the division between EUrope and a threatening world of incompatible and undesirable strangers—casts unsuspecting migrants into the threatening stereotypes on which xenophobic EUrosceptics feed.

The iron border's attention-grabbing visibility intensified with the outbreak of the refugee-protection crisis in the summer of 2015. This spectacle included the sensationalized arrivals of undocumented immigrants disembarking from fragile dinghies, trying to climb fences, cutting their way through barbed wire or running away from the border police. By comparison: although governments and migration scholars estimate that the number of the largely invisible visa-overstayers—who entered the EU *legally*—is at least as large as the number of undocumented migrants, the latter's trespassing of the EU's outer borders receives far more media attention and is the target of much more political aversion. Undoubtedly, what has heightened this sense of crisis is the surge of far-right politicians who have relied on such images to frame undocumented migration as an invasion and a menace to sovereignty. Ultimately, this narrative constitutes the core rationale of the EUropean far-right's ethno-exclusionary demand: an existential threat as justification for merciless borders, i.e., selective dehumanization aracist violence.

The political sway of these border aesthetics should not be underestimated: this 'spectacular' theater of trespassing undocumented migrants has justified the EU's expansion of its own iron border (DeGenova 2017). It is estimated that the EU has constructed almost 1,000 km of iron borders in the last two decades—more than six times the total length of the Berlin Wall (Ruiz Benedicto and Brunet 2018)—which are soon to be equipped with digital surveillance systems at sea and on land. Not surprisingly perhaps, although still ironically, the costs of physical border controls have gone up at about the same speed—and in a similar proportion—as the turnover in the smuggling industry (The Migrants' Files 2014).

Since its foundation in 2004, Frontex's budget has exponentially increased—from €6.2 million in 2004 to €754 million in 2022 (Frontex 2022b), thus making it one of the best funded agencies in the EU (Grün 2018). Between 2000 and 2014 (one year before the refugee-protection crisis), the EU spent almost €13 billion on border control: a number that bulged to €19.7 billion for the 2014–2020 budget cycle and which has swollen to €43.9 billion for the current period (2021–2027) (Jones et al. 2022). This zeal for border militarization has conferred the EU the dishonorable distinction of having one of the costliest border regimes on the planet. One thing should be clear: as long as the EU's visa-based paper border keeps working as the main manufacturer of irregular migrants, we should expect the iron border and its costs to keep rising accordingly.

In order to understand the significance of this bulging border budget and the geopolitics of the EU's iron fence, one should look at the reach of its arteries beyond the European continent. The EU has struck a growing number of bilateral deals with neighboring and faraway dictatorships to which it is outsourcing its border controls so that it can stop undocumented migrants with the help of faraway henchmen who are not shy to employ atrocious violencewhile exculpating the EU from any responsibility. This means that the EU is incorporating contiguous and distant autocracies into its border regime by hiring them as its *de facto* immigration enforcers in exchange for large sums of money—thereby raising the costs of external border controls and boosting the border security industry even further. This neo-colonial outsourcing of migration control to poorer countries, warlords and dictators is factually stretching the EU's own iron border far beyond its EU's physical boundaries (Lahav 1998; Lavenex 2006; Rijpma and Cremona 2007; Levy 2010; Ferrer-Gallardo and van Houtum 2014; Zaiotti 2016; Carrera et al. 2018). Such geopolitical machinations betray the same kind of 'dictator-empowering'-policy that the EU decried as shameful back in 2010, when Berlusconi and Gaddafi struck a deal that committed Libya to stopping migrants in return for money (Bialasiewicz 2012a).

Today, the EU pact with Libya has given rise to a full-fledged slave market run by cold-blooded human traffickers who, incentivized by the EU's crackdown on irregular migration and the resulting business downturn of would-be profitable passengers, are now auctioning refugees as slaves (Asongu and Kodila-Tedika 2018). How times have changed: only one self-manufactured 'refugee crisis' later, the EU is striking collaborations with downright tyrannies—such as Rwanda, Egypt, Morocco, Sudan and Tunisia (Jakob and Schlindwein 2019)—by means of outright bribes (Verhofstadt 2015).

The infamous deal between the EU and Erdogan's despotic administration is a prime case in point: since 2016, Turkey has been cutting short the journeys of asylum seekers traveling to the EU in return for €6 billion and the conditional promise of visa-free access for its citizens (DW 2021). Although the official EU narrative is that such contribution is intended to support the humanitarian reception of asylum seekers in Turkey, it is evident that the same aid could also be provided by EU Member States within their jurisdictions, a fact that reveals the policy's true purpose: preventing refugees from setting foot into EU territory. Incontrovertibly, the EU's magnanimous embrace of over 5 million Ukrainian refugees—the self-constructed 'good' or 'deserving' migrants who 'look like us'—has unmasked the insincerity of such narrative (Bueno Lacy and van Houtum 2022).

The autoimmunization of this 'iron' b/ordering strategy is reflected in the incongruence between the EU's desire to gain more control over its borders in order to safeguard its democracy, human rights, rule of law, and diplomatic power of attraction, on the one hand; and its pursuit of such honorable objectives through the recruitment of autocratic and unsafe neighboring countries, on the other. By outsourcing its border policies to smugglers and repressive regimes with the aim of tightening its grip on migration, the EU is, *incomprehensibly*: (1) losing control over its expanding, shadowy and ever more distant physical border, thus undermining its own sovereignty and making itself liable to blackmail while becoming morally complicit in the exploitation of refugees by pitiless regimes elsewhere; (2) widening the global mobility divide by fostering human misery, criminal economic activities and political instability in its periphery, which in turn is nurturing the desire to migrate among its neighboring populations; (3) hollowing out the EU's core values and

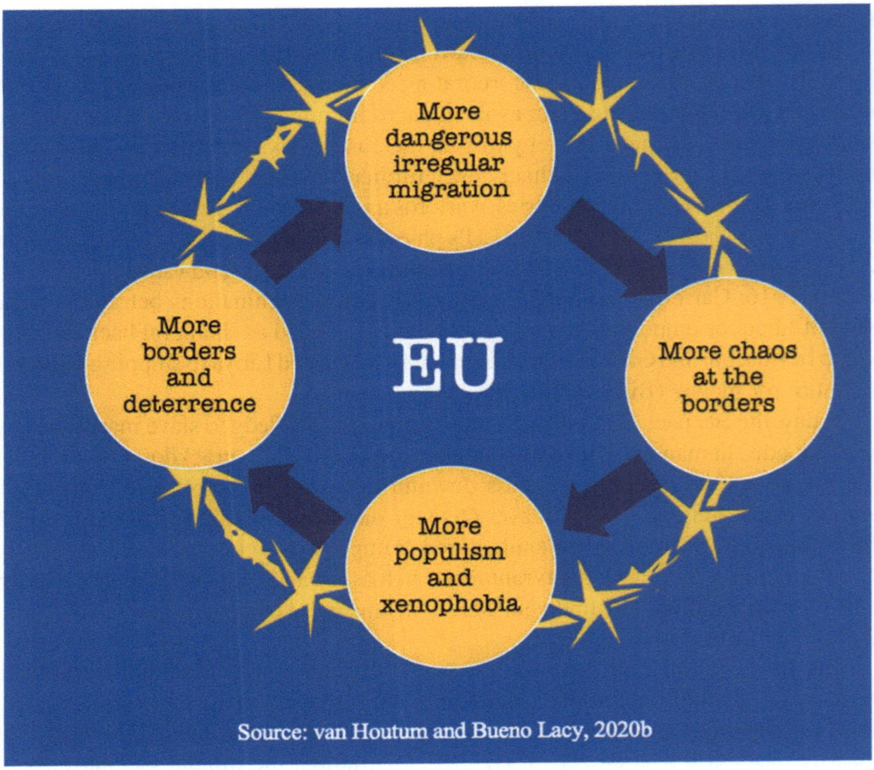

Fig. 3 The vicious cycle of EU's autoimmune b/ordering regime (*Source* authors)

contributing to legitimize the discourse on which authoritarian EUrosceptic populists draw their strength (van Houtum and Bueno Lacy 2017). The result is a narrow tunnel vision that keeps the EU obsessed with stopping undocumented migrants at all costs, even though the sensationalized chaos and manufactured 'insecuritization' at its borders are undermining solidarity with refugees while strengthening the hand of EUrosceptics—who exploit the threat inherent to the aesthetics of the iron border to push their demands for even higher walls and an ever more vicious border regime (see Fig. 3).

The Post-Border: The Camp

The third immunizing b/ordering pillar of the EU's border architecture that we identify is the post-border, aka 'the refugee camp'. Undocumented migrants who have the fortune to make it across the paper and iron borders must endure yet another procedure of exclusion, which takes the shape of concentrated segregation in reception and detention camps strewn along the EU's external and internal borders (Agamben 1998). Migrants have to wait in such secluded reception until their case is 'processed', a hostile policy that coincides with what Derrida called 'hos*ti*pitality' (2000): a portmanteau of hostility and hospitality formulated as a critique on the enslaving inclinations inherent to Christian charity—and an appeal to solidarity instead. Like the iron border, the hostile hospitality of EUrope's refugee camps abnormalizes and exceptionalizes migrants in space, society, and representation in the hope to discourage them from seeking the EU's protection.

These camps correspond to what Vaughan-Williams described as 'zoo-like spaces' (2015): refugees are caged yet ceaselessly exposed to the inquisitive eye of either cameras or the intimacy-depriving layouts that characterize refugee- reception and detention camps. Such unnecessary exposure amounts to a spatial confirmation of their social undesirability as well as to a forced animal-like performance that, we argue, contributes to stoke the already prevalent disdain—or plain fear—for racialized refugees in the EU. Furthermore, the segregation and maltreatment of people who share a bodily resemblance or cultural affinity with already-discriminated ethnic minorities in EUropean societies send an ominous message to the EU's diasporic citizens: it tells them that the fundamental rights which the EU allegedly guarantees do not fully apply to people who look like them. By legitimizing such discrimination, the immunization of the border camp fails to ensure the very rule of law it is designed to safeguard. Instead, it ends up emboldening authoritarian leaders and political movements inside the EU who have signaled their preference to employ similarly discriminating practices against easily identifiable minorities with scant political representation (van Houtum and Bueno Lacy 2017).

Although refugee camps have elicited a barrage of critique due to their recurrent human-rights violations, Camp Moria in Lesbos being the most emblematic due to its dreadful conditions (McElvaney 2018; Minca 2005), the preventable suffering of their inmates has nevertheless become a normalized policy across the EU.

We cannot stress this vicious border cycle enough: by creating a hostile environment for undocumented migrants, the EU is nodding to rapacious ethno-nationalists who see in the Union's lawless border violence a blueprint to employ the state's apparatus and legitimacy to trample upon the fundamental rights of their opponents (i.e., racialized national minorities, traditionally oppressed sexual minorities, liberated women, communitarians, environmental activists, unions, etc.). No wonder the far-right has become the EU's dominating political force (Mudde 2019b): the brutal EU's b/ordering regime has been manufacturing a theater in which undocumented migrants are forced to play the role of dangerous hordes while vicious border police forces are cast into the role of civilization's bulwark against barbarism.

Certainly, there are Europeans willing to rescue refugees from the claws of the sea or help them find their way into EUropean societies. However, their humanitarian deeds are largely offset by border tactics that ensure the failure of undocumented migrants to integrate before they even get a chance to find a foothold in the EU. Such tactics involve warehousing migrants of blacklisted countries under inhumane conditions while making them dependent on aid for both their livelihood and freedom. This amounts, in Derrida's terms, to an *incomprehensible* deterrence politics that is detrimental for undocumented migrants as well as for cohesion across the EU (Fernández 2014; Kingsley 2018; Leape 2018; Smith 2018).

As though the social animosity that such policies are carving along racial lines was not volatile enough, states like Hungary, Poland and Slovakia have compounded such ethnocentrism with religious intolerance by expressly stating their willingness to host *only* Christian refugees (Bastide 1968; Cienski 2017; Reuters 2015). As though the EU was not hemorhaging solidarity in all directions, the EU's expectations for Greece to manage a disproportionate number of asylum arrivals after the austerity-fueled misery that the European Commission together with the European Central Bank and the IMF imposed on it begs belief: It looks as though the EU were running a stress test designed to find its own point of breakage. Ultimately, these self-defeating processes betray a blind incomprehension that characterizes what the EU's autoimmune disorder looks like to us: as though the EU were pursuing a border strategy bound to nurture resentment against itself and undocumented migrants alike.

Given the cold shoulder that EU Member States and institutions showed Greece when it needed their solidarity the most, the country does not seem very receptive to the EU's calls to improve the inhumane conditions in which it keeps the detained migrants languishing on its Mediterranean islands. Unsurprisingly, Greece has been accused of misusing EU funds meant for the critically overcrowded and underfunded refugee camps in its Aegean islands (Howden and Fotiadis 2017), while its far-right government has indulged in the senseless destruction of extraordinary refugee-support networks like Exarcheia and introduced laws to deport thousands of asylum seekers without concern for their rights under international refugee law (King and Manoussaki-Adamopoulou 2019; Smith 2018, 2019). Greece's animosity toward the EU—and undocumented migrants—is critical to understand how this excruciating autoimmune cycle is being fueled by the mutual reinforcement between the Union's internal austerity policies and its external border regime: by defaulting on its asylum obligations, the EU is simultaneously degrading its larger promise to the EUropean population, namely that of a political community guided by the rule of law and solidarity.

It is worthwhile to reflect on a particular question that Derrida posed in his deconstruction of geopolitical autoimmunity: 'Can't "letting die", "not wanting to know that one is letting others die"—hundreds of millions of human beings, from hunger, AIDS, lack of medical treatment and so on—also be part of a "more or less" conscious and deliberate terrorist strategy'? (Derrida 2003, p. 108). Derrida's reflection poses a harrowing question for the EU's border regime: is it less cruel because it repels potential refugees at a distance by preventing them from even legally applying to migrate to the EU in the first place? Is it less violent because it premeditatedly builds hurdles that preclude asylum seekers from safely entering the EU and purposefully creates ever more inhumane hosting conditions once they have reached what they imagined would be a safe territory?

Perhaps, by pushing undocumented migrants—many of whom have been displaced against their will—seeking the EU's protection into a hopelessness so intolerable that an alarming number of children in refugee camps have chosen to take their own lives (Tondo 2018a), the suicidal autoimmunity of the EU's b/ordering strategy is coming to an abhorrent full circle.

Toward a Sustainable and Humane Border Policy

'For the first time in 30 years, I really believe that the European project can fail' (Lefranc 2016). This alarming message came from no less than the vice president of the European Commission, Frans Timmermans, who believes that the 'refugee crisis' of 2015 strained solidarity across the EU to the brink of rupture. The continuation of this crisis, Timmermans frets, poses an existential threat to the project of EUropean integration.

Employing Derrida's notion of autoimmunity, we have argued that the EU has been cornering itself into a dead end. Not only has it been unable to muster support for a comprehensive migration and asylum system across its supranational community—although a compromise on an asylum policy that degrades refugee protections seems to have been reached as far-right leaders become aligned across the EU (Bautista et al. 2013)—but also the Union is increasingly taking the self-destructive road toward dirty deals with autocracies, and either imprisoning legitimate asylum seekers, or brutally pushing them back to countries where they might die or suffer severe harm.

The politicization of undocumented migration, which itself is the result of the EU's inability to assume its role in the longstanding geopolitical inequality that causes it, has been exacerbated by the adoption of extreme policies by establishment parties. Under the pretense of 'normality', political forces of a self-confessed liberal inspiration have copied the far-right's predatory worldview as a counterproductive strategy to stop its rise—or, perhaps, as an understated confession of their ideological proximity to it (van Houtum and Bueno Lacy 2017; Mudde 2019a; Kundnani 2023). In the bigger picture, since the structural causes that keep pushing people away from their countries (e.g., inequality and poverty, financial plundering, armed conflict and widespread violence, droughts and agricultural collapse, overfishing and the depletion of ancestral fisheries, and overall livelihood-destroying global ecocide, etc.) are unlikely to be addressed anytime soon and the worsening effects of climate change are surely going to keep magnifying them (Franzen 2019; Nordås and Gleditsch 2007; McGuire 2022; Provost and Kennard 2023), the question the EU should be asking itself is not whether the next existential crisis will come, but rather *when*.

To break this self-defeating political path, the EU urgently needs a drastic revision of both its violent b/ordering regime and the essentialist EUropean discourse that supports it (Bueno Lacy and Van Houtum 2015; Jones 2017). To this end—and as a conclusion—we offer three different paths that the EU could take in order to correct course and keep its own promise of 'committing to implementing the SDGs': normalization, legalization and equalization (van Houtum 2015; van Houtum and Lucassen 2016).

Normalization would require accepting migration as a non-negotiable reality of today's globalizing world, which would represent a first step toward the design of a border policy informed by scientific assessments instead of prejudice. The dominant pattern of world migration shows that global mobility is still very much the exception rather than the rule: 97% of the world's population is not a migrant. Refugees represent less than 1% of the world's population and more than 85% of all refugees on the planet are hosted outside the EU—mostly in less affluent countries (de Haas 2008; UNHCR 2022). Moreover, the EU's neighboring countries are hosting a higher number of refugees than the EU—in absolute and relative terms. Although this does not mean that hosting an increasing number of refugees does not represent a difficult task for EUropean societies, it shows that such a challenge does not warrant transferring responsibility to dictatorships.

The numbers also make clear that the panic-stricken depiction of an 'invasion' of migrants coming to the EU is not only scientifically unfounded, but also dehumanizing and thus contemptible. It is a worrying sign of our times to realize that all kinds of phobic metaphors used to refer to undocumented migrants have become normalized in the EU over the last decade. Think of the threatening descriptions and (cartographic) imaginaries of undocumented migrants conjured up by hydrophobic metaphors evoking flows, streams, floods, waves and tsunamis; zoophobic metaphors suggesting swarms, flocks, rats, cockroaches and insects; as well as bellicose and criminalizing metaphors that bring to mind invasions, armies, illegal and criminal activities, hordes, and violent conflict (see van Houtum and Bueno Lacy 2020a, 2020b). When dehumanization is normalized and unchallenged, untamed extremism goes rampant and physical violence becomes ever more likely (Smith 2020).

Meaningful reform would mean *legalization*: the creation of legal channels for migrants to safely travel to the EU would require its laws to allow for migration's natural circularity (Clemens et al. 2019). This specific policy path would also require the EU to get rid of the boogeyman of 'the economic migrant' (Althaus 2016; Carling 2023). Moreover, the fear of economic migrants reveals perhaps one of the biggest flaws of Schengen: the criminalization of people whose biggest threat to the prosperity of the EU polity seems to be their ambition to work in order to earn the kind of living standards that their countries of origin cannot offer them. It is a testament to the extreme nature of our times that such unremarkably liberal ideas as respect for those who seek fairness of opportunities through the right to work are today seen as extreme proposals within the EU, which prides itself on its universal rights, rule of law and market economy (Holmes 1993, pp. 3–4).

Opening more legal migration channels is not merely morally right but it would be in the interest of everyone: migrants themselves, their countries of origin—where they send much of the money they earn—and, finally, the EU's economy, particularly regarding the preservation of its welfare states amidst ageing populations.

Furthermore, the legalization of migratory movements would not only protect refugees and drastically disrupt the illicit chaos and high death rates at the gates of the EU but it would also protect and strengthen the EU's rule of law by disrupting the supply-and-demand chains on which smugglers, slave traders and even violent extremists depend. Such legalization would also reduce the informal economy by giving undocumented migrants a chance to stand again on their own feet by contributing to the state's coffers; and by setting clear rules for them to acquire citizenship and social security rights, which could depend on their years of participation. With legalization we also mean that the EU should abide by its own rule of law: although all EU Member States have signed the Refugee Convention and its protocols—which means that their pledge to aid people escaping their countries is a commitment of their own volition—the increasingly vicious border regime they have put in place is vigorously hollowing out the protections that such international agreements afford asylum seekers. Trampling upon such international obligations stands in direct contradiction with the EU's own rule of law and is weakening its promise to uphold fundamental rights while legitimizing xenophobia as well as the arbitrary abuse of power across EUropean politics.

Finally, a comprehensive reform of the EU's border regime should encompass an *equalization*—i.e., an equal distribution of refugees across the EU and among the neighboring regions on the basis of shared responsibility and eventual resettlement guided by refugees' autonomy. Such a course would involve dismantling the Dublin regulation and implementing provisions akin to those stipulated by the Temporary Protection Directive afforded to Ukrainian refugees (Bueno Lacy and van Houtum 2022). Instead of ad hoc funds such as the current Asylum Migration and Integration Fund (2013–2027)—which replaced the European Refugee Fund (2000–2013)—a more reliable funding tool could be sourced from a common fiscal policy first restricted to this very purpose. Crucially, such a policy could eventually be extended to fund public services such as those typically demanded by both refugees and communitarian EU migrants alike (i.e., social housing, education, healthcare, public transport and everyday utilities). A common fiscal policy for migrants would not merely promote the seamless integration of refugees into EUropean societies but it would also bolster the EU's freedom of movement by making it easier for its citizens to relocate in other Member States. Moreover, by decoupling national public services from national taxes, a common fiscal policy could delegitimize xenophobic rabble-rousers whose electoral success depends on misrepresenting migrants as freeloaders of public services.

Critically, equalization would also require the immediate deployment of a EU Rescue Guard that would reflect the EU's *raison d'être*: by preventing more migrant deaths. This would imply an immediate end to the concentrated-segregation politics of the EU's refugee camps and replace them for dignified housing monitored by the democratic scrutiny of media and society. That such a scenario is not an impossibility but instead a matter of political will has been demonstrated by the manifest willingness —even enthusiasm—to accommodate over 4 millions of Ukrainian refugees, who represent about four times the number of Syrian refugees whose arrival to the EU, by contrast, was framed as a refugee crisis that could threaten the very survival of the EU (Bueno Lacy and van Houtum 2022).

In the longer run, an equalization agenda would imply a wider series of tasks that would take the EU outside of its perceptual isolation by assuming itself as the significant global actor it is. That would demand, first of all, the overall adoption of a EUropean political identity based on a humanist supranational historiography that rejects the racist tropes, imperialist symbolism, and colonial narratives evoked by the EU's current identitary paraphernalia (Bueno Lacy 2020; Bialasiewicz 2012b).

Secondly, it would imply the EU's pursuit of fair trade, a global green deal, and a concrete multilateral mechanism for the peaceful resolution of conflicts—at least around its immediate neighborhood or, at the very least, in the regions and countries that are the sources of its largest refugee populations. Hence, such course should involve the replacement of savagery-prone border police forces and co-opted foreign dictatorships with a veritable Euro-Mediterranean cohesion policy designed to root out the wealth disparities between the EU and its neighborhood. Such disparities underlie the poverty, violence, and despair pushing refugee populations to EUrope, factors whose roots often hark back to a longstanding European colonialism that has not disappeared but merely transformed (Gregory 2004).

Eventually, such equalization agenda would also need to take into account what is perhaps the most important measure: a drastic revision of the discriminatory EU visa regime to root out the nativist principle built into the design of the Union's political community. Of the three borders discussed in this chapter, this paper border is arguably the most untreatable root of the EU's refugee-protection crisis Overall, the autoimmune policies devised to address the most vulnerable form of human mobility have not only magnified the challenge posed by an increased number of asylum seekers in the EU but they have also exacerbated other geopolitical problems to the point that this self-made migration crisis has become the most threatening existential threat the EU has ever faced. As we have argued, the visa regime has created a global caste system of elite travelers whose mobility is welcomed, and a cohort of wretches whose mobility is banned, criminalized, and deterred as consequence of having the wrong place of birth—to the extent that they could die not only trying to flee but also even after they have arrived to an allegedly safe port (see also Bueno Lacy and van Houtum 2022). The deliberate intention to keep the less affluent, bodily contrasting, and religiously different trapped at a distance—or spatially and representationally invisibilized—simply because they were born unlucky is an act of discrimination that is at odds with the equal moral worth of human beings laid down in the Universal Declaration of Human Rights as well as in the Treaty of Lisbon that the EU is supposed to uphold. Surely, the world's discriminatory visa system—already a hundred years old—may today seem unbreakable, but so did once the trans-Atlantic slave trade, South Africa's *apartheid* and the divine right of kings (Le Guin 2014).

The discussion—on how to achieve a responsible, sustainable and just border policy surely should not stop at our proposals. Worryingly, however, the EU seems poised to keep medicating itself with increasingly self-poisoning remedies. Such an appalling course, we argue, should not be regarded as a momentary lapsus but as a train wreck happening in slow motion instead. Since the EU closed its external borders with the introduction of Schengen, its political community has followed an ever-deadlier path of discriminatory self-enclosure that excludes a large portion of the world. Today, the EU is experiencing the limits of this border model: the current politicization of migration characterized by ghastly measures to curtail the movement of refugees is shaking the EU to its foundations and endangering Schengen as well as the non-discrimination principle underlying the protection of human rights, solidarity and the rule of law. Alarmingly, these are the liberal-democratic principles of the Copenhagen criteria and, ultimately, they constitute the very ethos of the EU.

Barring a drastic change in the EU's trajectory, the death and suicide of undocumented migrants and their children will not stop. What is more foreboding, perhaps, is that the EU might share their fate—at least as we know it. Perhaps Frans Timmermans is right: for the first time, the project of European integration that has brought historically unseen prosperity and peace to a continent characterized by its historical bloodshed seems as though it might fail. Ironically, it might fail because the European Union has become its own most formidable threat.

References

Agamben G (1998) The Camp as the'Nomos' of the Modern. In the Omnibus Homo Sacer. Stanford University Press, pp 166–180

Agencies in Kecskemét (2018, June 14) Four men jailed over deaths of 71 migrants locked in lorry. The Guardian. https://www.theguardian.com/world/2018/jun/14/people-smuggling-gang-jailed-over-hungary-migrant-lorry-deaths

Akkerman M (2016) It is time to end illegal pushbacks and abolish Frontex. ROAR Magazine. https://roarmag.org/essays/end-pushbacks-abolish-frontex/

Albahari M (2018) From right to permission: asylum, mediterranean migrations, and Europe's war on smuggling. J Migr Hum Secur 6(2):121–130. https://doi.org/10.1177/2331502418767088

Althaus A (2016) The false dichotomy between 'Economic Migrants' and refugees | IOM Blog. IOM. https://weblog.iom.int/false-dichotomy-between-economic-migrants-and-refugees (August 4, 2023).

AP in Athens (2018, October 10) Three female migrants found murdered near Greece-Turkey border. The Guardian. https://www.theguardian.com/world/2018/oct/10/three-women-found-dead-near-greece-river-border-with-turkey

Asongu SA, Kodila-Tedika O (2018) "This one Is 400 Libyan Dinars, This one is 500": Insights from cognitive human capital and slave trade. Int Econ J 32(2):291–306. https://doi.org/10.1080/10168737.2018.1480643

Bastide R (1968) Color, racism, and Christianity. Color and Race 13:34–49

Bautista J et al. (2023, June 27) Europe's new asylum policy: 'The First Thing They Do Is Imprison You.' Der Spiegel. https://www.spiegel.de/international/europe/europe-s-new-asylum-policy-the-first-thing-they-do-is-imprison-you-ad7396d76-f53f-4d81-8f71-82c759c6e27d

Benedicto AR, Brunet P (2018) Building walls: fear and securitization in the European Union (War & Pacification). Transnational Institute

Bialasiewicz L (2012a) Off-shoring and out-sourcing the borders of EUrope: Libya and EU border work in the Mediterranean. Geopolitics 17(4):843–866. https://doi.org/10.1080/14650045.2012.660579

Bialasiewicz L (2012b) Spectres of Europe: Europe's past, present and future. In D. Stne (Ed.), The oxford handbook of postwar European history. Oxford University Press, pp 98–119. https://doi.org/10.1093/oxfordhb/9780199560981.013.0005

Boedeltje F, van Houtum H (2008) The abduction of Europe: a plea for less 'Unionism' and more Europe. Tijdschr Econ Soc Geogr 99(3):361–365. https://doi.org/10.1111/j.1467-9663.2008.00467.x

Braudel F (1996) The Mediterranean and the Mediterranean world in the age of Philip II: vol I. University of California Press

Braudel F (2002) The Mediterranean in the ancient world. Penguin UK

Bueno Lacy R, van Houtum H (2013) Europe's border disorder. E-journal for international relations. https://www.e-ir.info/pdf/44738

Bueno Lacy R, van Houtum H (2015) Lies, damned lies & maps: the EU's cartopolitical invention of Europe. J Contemp Eur Stud 23(4):477–499. https://doi.org/10.1080/14782804.2015.1056727

Bueno Lacy R, van Houtum H (2020) Death as policy: the EU's criminalisation of solidarity with undocumented migrants. In Expanding Boundaries. Routledge, pp 72–84

Bueno Lacy R, van Houtum H (2022) The proximity trap: how geography is misused in the differential treatment of Ukrainian refugees to hide for the underlying global apartheid in the EUropean border regime. ASILE (https://www.asileproject.eu/the-proximity-trap-how-geography-is-misused-in-the-differential-treatment-of-ukrainian-refugees-to-hide-for-the-underlying-global-apartheid-in-the-european-border-regime/), 27 June 2022

Bueno Lacy Rodrigo (2020) In the image of Kronos or how Europe is devouring itself: The iconological construction of EUropean identity, its geopolitical implications for the project of European integration and why it needs to be re-imagined [Radboud University]. http://hdl.handle.net/2066/225617

Burrell K, Hörschelmann K (2019) Perilous journeys: visualising the racialised "refugee crisis." Antipode 51(1):45–65. https://doi.org/10.1111/anti.12429

BVMN (2020) The black book of pushbacks, vol 1 & 2. Border Violence Monitoring Network

Campos-Delgado A, Côté-Boucher K (2022) Tactics of empathy: the intimate geopolitics of Mexican migrant detention. Geopolitics 1–24. https://doi.org/10.1080/14650045.2022.2039633

Campos-Delgado A (2022) Bordering theatricality: Mexico's mass expulsion of Indian citizens. AntiAtlas J, 5

Carling J (2023) The phrase 'Refugees and Migrants' undermines analysis, policy and protection. Int Migr 61(3):399–403

Carrera S (2014) The price of EU citizenship: the Maltese citizenship-for-Sale affair and the principle of sincere cooperation in Nationality matters. Maastricht J Eur Comp Law 21(3):406–27

Carrera S, El Qadim N, Fullerton M, Garcés-Mascareñas B, York Kneebone S, López Sala AM, Chun Luk N, & Vosyliuté L (2018) Offshoring asylum and migration in Australia, Spain, Tunisia and the US: Lessons learned and feasibility for the EU [Working Document]. CEPS

Cienski J (2017, May 21) Why Poland doesn't want refugees [News]. POLITICO. https://www.politico.eu/article/politics-nationalism-and-religion-explain-why-poland-doesnt-want-refugees/

Clemens M, Dempster H, Gough K (2019) Promoting new kinds of legal labour migration pathways between Europe and Africa. Centre for Global Development. https://www.cgdev.org/publication/promoting-new-kinds-legal-labour-migration-pathways-between-europe-and-africa

Collins C (2022, March 17) Cracking down on Russian oligarchs means cracking down on U.S. Tax Havens. CounterPunch.Org. https://www.counterpunch.org/2022/03/17/cracking-down-on-russian-oligarchs-means-cracking-down-on-u-s-tax-havens/

Control. ACME: Int J CritAl Geogr, 13(2): 295–304

Cuttitta P (2018) Delocalization, humanitarianism, and human rights: the Mediterranean border between exclusion and inclusion. Antipode 50(3):783–803. https://doi.org/10.1111/anti.12337

Dadusc D, Mudu P (2020) Care without control: the humanitarian industrial complex and the criminalisation of solidarity. Geopolitics 1–26. https://doi.org/10.1080/14650045.2020.1749839

Davidson B (1980) The African slave trade. James Currey Publishers

De Genova N (2013) Spectacles of migrant 'Illegality': the scene of exclusion, the obscene of inclusion. Ethn Racial Stud 36(7):1180–98

De Genova N (2017) The borders of "Europe": Autonomy of migration, tactics of bordering. Duke University Press

De Genova N (2018) The "migrant crisis" as racial crisis: Do black lives matter in Europe? Ethn Racial Stud 41(10):1765–1782. https://doi.org/10.1080/01419870.2017.1361543

de Haas H (2008) The Myth of invasion: The inconvenient realities of African migration to Europe. Third World Q 29(7):1305–1322 de Jonge L, Gaufman E (2022) The normalisation of the far right in the Dutch media in the run-up to the 2021 general elections. Discourse Soc 09579265221095418. https://doi.org/10.1177/09579265221095418

Deeb B, Sapoch J, Vögele N, Lüdk S, Christides G, Simeonidis P. Beer A, Govedrica S, Popoviciu A. Statius T, Keizer P, van Driel E, Bakotin J, Derifaj D, Laghai S, Sabic L, & van Dijken K (2021) Unmasking Europe's Shadow Armies. Lighthouse Reports. https://www.lighthousereports.nl/investigation/unmasking-europes-shadow-armies/

Delanty G (1996) The frontier and identities of exclusion in European history. Hist Eur Ideas 22(2):93–103. https://doi.org/10.1016/0191-6599(95)00077-1

Derrida J (2003) Autoimmunity: real and symbolic suicides / A dialogue with Jacques Derrida. In Borradori G, Philosophy in a time of terror: Dialogues with Jurgen Habermas and Jacques Derrida. University of Chicago Press

DW (2021, June 23) EU eyes billions in extra funds for Turkey migration deal | DW | 23.06.2021. DW.COM. https://www.dw.com/en/eu-eyes-billions-in-extra-funds-for-turkey-migration-deal/a-58019261

European Commission (2022) EU budget: Commission proposes major funding increase for stronger borders and migration. European Commission - European Commission. https://ec.europa.eu/commission/presscorner/detail/en/IP_18_4106

European Union (2009) Treaty of Lisbon. https://eur-lex.europa.eu/legal-content/EN/TXT/?uri=celex%3A12007L%2FTXT

Farmer A, Human Rights Watch (Organization). (2012) Boat ride to detention: Adult and child migrants in Malta. Human Rights Watch.

Fekete L (2022) Racism, radicalisation and Europe's 'Thin Blue Line.' Race & Class 64(1):3–45

Fernandez B (2014, May 18) Detention in Malta: Europe's migrant prison [News]. Al Jazeera. https://www.aljazeera.com/opinions/2014/5/18/detention-in-malta-europes-migrant-prison

Ferrer-Gallardo X, Houtum H. van. (2014) The Deadly EU Border Frontex. (2022a) Frontex Analysis: Countering trafficking in human beings at EU's borders. The European Border and Coast Guard Agency. https://frontex.europa.eu/media-centre/news/focus/frontex-analysis-countering-trafficking-in-human-beings-at-eu-s-borders-NA3Dvd

FitzGerald DS (2019) Refuge beyond Reach: how rich democracies repel asylum seekers. Oxford University Press. https://doi.org/10.1093/oso/9780190874155.001.0001

Foucault M (1971) Nietzsche, la généalogie, l'histoire. In Bachelard S, Dagognet F, Canguilhem G, Foucault M, Guéroult M, Henry M, Laplanche J, Pariente J.-C., Serres M (Eds.), Hommage à Jean Hyppolite. Presses universitaires de France

Franzen J (2019, September 8) What if we stopped pretending the climate apocalypse can be stopped? The New Yorker. https://www.newyorker.com/culture/cultural-comment/what-if-we-stopped-pretending

Frontex (2022b) Key facts. The European border and coast guard agency. https://frontex.europa.eu/about-frontex/faq/key-facts/

Gleditsch NP, Nordås R (2009) Climate change and conflict: acritical overview. Die Friedens-Warte 84(2):11–28

Gimbutas M (1985) Primary and secondary homeland of the Indo-Europeans. Comments on Gamkrelidze-Ivanov articles in recent Russian papers on the Indo-European problem and on the ethnogenesis and original homeland of the slavs. J Indo-Eur Stud 13(1 & 2):185–201

Gregory D (2004) The Colonial Present: Afghanistan. Wiley, Palestine. Iraq

Grün G-C (2018, February 15) Follow the money: What are the EU's migration policy priorities? | DW | 15.02.2018. DW.COM. https://www.dw.com/en/follow-the-money-what-are-the-eus-migration-policy-priorities/a-42588136

Guterres A [@antonioguterres]. (2022, June 29) I am shocked by the violence on the Nador-Melilla border on Friday which resulted in the deaths of dozens of migrants & asylum seekers. The use of excessive force is unacceptable, and the human rights and dignity of people on the move must be prioritized by countries. [Tweet]. Twitter. https://twitter.com/antonioguterres/status/1541992352046370817

Holmes S (1993) The anatomy of antiliberalism. Harvard University Press

Horaczek N (2019) Propaganda war in Europe: the far-right Media. European Press Prize. https://www.europeanpressprize.com/article/propaganda-war-europe-far-right-media/ (July 18, 2023).

Houtum H. van (2021) Beyond Borderism: Overcoming discriminative B/ordering and Othering, TESG https://onlinelibrary.wiley.com/doi/10.1111/tesg.12473

Houtum H van, Uden A.M. van (2021), The birth of the paper prison. The global inequality trap of visa borders. Environment and Planning C, pp 20–23 https://henkvanhoutum.nl/wp-content/uploads/2021/02/2021-Rethinking-biopolitcs-pagina-20-23_pages.pdf

Houtum H van, Bueno Lacy R (2020a) The migration map trap. On the invasion arrows in the cartography of migration. Mobilities, 15(2):196–219. https://doi.org/10.1080/17450101.2019.1676031

Houtum H van, Lacy RB (2020b) The autoimmunity of the EU's deadly b/ordering regime. Overcoming its paradoxical paper, iron and camp borders, Geopolitics 25(3):706–733. https://doi.org/10.1080/14650045.2020.1728743

Houtum H van (2010) Human blacklisting: the global apartheid of the EU's external border regime. Environment and planning d: Society and Space 28(6):957–976. https://doi.org/10.1068/d1909

Houtum H van (2015, January 7) Dood door beleid: EU-grens dodelijkste grens op aarde. de Volkskrant. https://www.volkskrant.nl/gs-be784791

Houtum H van, Boedeltje F (2009) Europe's shame: death at the borders of the EU. Antipode 41(2):226–230. https://doi.org/10.1111/j.1467-8330.2009.00670.x

Houtum H van, Bueno Lacy R (2017) The political extreme as the new normal: the cases of Brexit, the French state of emergency and Dutch Islamophobia. Fennia - International Journal of Geography, 195(1): 85–101. https://doi.org/10.11143/fennia.64568

Houtum H van, Lucassen L (2016) Voorbij Fort Europa: Een nieuwe visie op migratie. Atlas Contact, Uitgeverij

Houtum H van, van Naerssen T (2002) Bordering, ordering and othering. Tijdschr Econ Soc Geogr 93(2):125–136. https://doi.org/10.1111/1467-9663.00189

Howden D, Fotiadis A (2017, March 9) Where did the money go? How Greece fumbled the refugee crisis. The Guardian. https://www.theguardian.com/world/2017/mar/09/how-greece-fumbled-refugee-crisis

Human Rights Watch (2022, June 29) Morocco/Spain: horrific migrant deaths at Melilla border. Human Rights Watch. https://www.hrw.org/news/2022/06/29/morocco/spain-horrific-migrant-deaths-melilla-border

International Organization for Migration, Missing Migrants Project, https://missingmigrants.iom.int/, accessed 25 January 2022

Jakob C, Schlindwein S (2019) Dictators as Gatekeepers: Outsourcing EU Border ¿controls to Africa. Daraja Press.

Jones R (2017) Violent Borders: refugees and the right to move. Verso Books, p 224

Jones C, Kilpatrick J, Maccanico Y (2022) At what cost? Funding the EU's security, defence, and border policies, 2021–2027. Transnational Institute

Juhász A, Szicherle P (2017) The political effects of migration-related fake news, disinformation and conspiracy theories in Europe. Friedrich Ebert Stiftung, Political Capital, Budapest.

Kassam, A (2022, June 30) 'A bloodbath': Refugees reel from deadly Melilla mass crossing. The Guardian. https://www.theguardian.com/world/2022/jun/30/a-bloodbath-refugees-reel-from-deadly-melilla-mass-crossing

Kemp R (2017) Ross Kemp: Libya's Migrant Hell. Sky. https://www.sky.com/watch/title/programme/ca063fef-330c-4245-b948-1ea2c8a56570/ross-kemp-libyas-migrant-hell

King A, Manoussaki-Adamopoulou I (2019) Inside Exarcheia: the self-governing community Athens police want rid of." The Guardian. https://www.theguardian.com/cities/2019/aug/26/athens-police-poisedto-evict-refugees-from-squatted-housing-projects (November 16, 2021)

Kingsley P (2018, October 3) 'Better to Drown': A Greek refugee camp's epidemic of misery. The New York Times. https://www.nytimes.com/2018/10/02/world/europe/greece-lesbos-moria-refugees.html

Kriesi H, Pappas TS (2015) European populism in the shadow of the great recession. ECPR Press. https://cadmus.eui.eu//handle/1814/36489
Kundnani A (2023) What is Antiracism? And why it means Anticapitalism, 1st edition. Verso Books
Lahav G (1998) Immigration and the state: The devolution and privatisation of immigration control in the EU. J Ethn Migr Stud 24(4):675–694. https://doi.org/10.1080/1369183X.1998.9976660
Laurent O, O'Grady S (2018, June 28) Thousands of migrants have been abandoned in the Sahara. This is what their journey looks like. Washington Post. https://www.washingtonpost.com/news/worldviews/wp/2018/06/28/thousands-of-migrants-have-been-abandoned-in-the-sahara-this-is-what-their-journey-looks-like/
Lavenex S (2006) Shifting up and out: The foreign policy of European immigration control. West Eur Polit 29(2):329–350. https://doi.org/10.1080/01402380500512684
Lefranc F-X (2016, November 7) Frans Timmermans: "The European project can fail." https://www.euractiv.com/section/euro-finance/interview/frans-timmermans-the-european-project-can-fail/
Le Guin U (2014) Ursula K Le Guin's speech at national book awards: 'Books Aren't Just Commodities.' The Guardian. https://www.theguardian.com/books/2014/nov/20/ursula-k-le-guin-national-book-awards-speech (August 4, 2023)
Leape S (2018) Greece has the means to help refugees on Lesbos–but does it have the will? The Guardian. https://www.theguardian.com/global-development/2018/sep/13/greece-refugees-lesbos-moria-campfunding-will (November 16, 2021)
Levy C (2010) Refugees, Europe, camps/state of exception: "Into the zone", the European Union and Extraterritorial Processing of Migrants, Refugees, and Asylum-seekers (Theories and Practice). Refug Surv Q 29(1):92–119. https://doi.org/10.1093/rsq/hdq013
Lucas A, Ramsay P, Keen L (2019) No end in sight: the mistreatment of asylum seekers in Greece. StateWatch
Lyman R, Smale A (2015, September 3) Migrant smuggling in Europe is now worth 'Billions.' The New York Times. https://www.nytimes.com/2015/09/04/world/europe/migrants-smuggling-in-europe-is-now-worth-billions.html
Malik K (2019, November 30) When refugees in Libya are being starved, Europe's plan is working. The Guardian. https://www.theguardian.com/commentisfree/2019/nov/30/when-refugees-in-libya-are-being-starved-europes-plan-is-working
Mau S, Gülzau F, Laube L, Zaun N (2015) The Global Mobility Divide: How Visa Policies Have Evolved over Time. J Ethn Migr Stud 41(8):1192–1213
Mbayé S (2022, July 5) "No son muertes, son asesinatos." Público. https://blogs.publico.es/otrasmiradas/61224/no-son-muertes-son-asesinatos/
Mbembe A (2003) Necropolitics. Publ Cult 15(1):11–40
McElvaney K (2018, January 19) Rare look at life inside Lesbos' largest refugee camp [News]. Al Jazeera. https://www.aljazeera.com/gallery/2018/1/19/rare-look-at-life-inside-lesbos-moria-refugee-camp
McGuire B (2022) Hothouse earth: an inhabitant's guide. Icon Books
Miller T (2019) Empire of borders: the expansion of the US border around the world. Verso Books
Minca C (2005) The return of the camp. Prog Hum Geogr 29(4):405–412
Minca C, Rijke A (2017) Walls! walls! society and space, Walls, p 4
Mudde C (2019a, May 14) Why copying the populist right isn't going to save the left. The Guardian. https://www.theguardian.com/news/2019a/may/14/why-copying-the-populist-right-isnt-going-to-save-the-left
Mudde, C (2019b, May 28) The far right may not have cleaned up, but its influence now dominates Europe. The Guardian. https://www.theguardian.com/commentisfree/2019b/may/28/far-right-european-elections-eu-politics
Neumayer E (2006) Unequal access to foreign spaces: How states use visa restrictions to regulate mobility in a globalized world. Trans Inst Br Geogr 31(1):72–84. https://doi.org/10.1111/j.1475-5661.2006.00194.x
Plenel E (2019, September 12) Cette Europe qui nous fait honte. Mediapart. https://www.mediapart.fr/journal/international/120919/cette-europe-qui-nous-fait-honte

Provost C, Matt K (2023) Silent coup: how corporations overthrew democracy. Bloomsbury AcademIc
Raggett C (2020, March 3) Networks of impunity: corruption and European foreign policy—European Council on foreign relations. ECFR. https://ecfr.eu/publication/networks_of_impunity_corruption_and_european_foreign_policy/
Rawls J (1971) A theory of justice (Revised). Harvard University Press
Reuters Staff (2015, August 20) Slovakia says it prefers Christian refugees under resettlement scheme. Reuters. https://www.reuters.com/article/us-europe-migrants-slovakia-idUSKCN0QP1HN20150820
Rijpma JJ, Cremona M (2007) *The extra-territorialisation of EU migration policies and the rule of law* [Working Paper]. European University Institute. https://cadmus.eui.eu//handle/1814/6690
Said EW (2014) The clash of ignorance. Routledge, In Geopolitics
Salter, MB (2003) Rights of passage: the passport in international relations. Boulder
Salter MB (2006) The global visa regime and the political technologies of the international self: borders, bodies. biopolitics. Alternatives 31(2):167–189. https://doi.org/10.1177/030437540603100203
Slootweg A, van Reekum R, Schinkel W (2019) The raced constitution of Europe: The Eurobarometer and the statistical imagination of European racism. Eur J Cult Stud 22(2):144–163. https://doi.org/10.1177/1367549418823064
Smith H (2018, September 26) Lesbos refugee camp at centre of Greek misuse of EU funds row. The Guardian. https://www.theguardian.com/world/2018/sep/26/lesbos-refugee-camp-at-centre-of-greek-misuse-of-eu-funds-row
Smith H (2019, November 1) Greece passes asylum law aimed at curbing migrant arrivals. The Guardian. https://www.theguardian.com/world/2019/nov/01/greece-passes-asylum-law-aimed-at-curbing-migrant-arrivals
Smith D L (2020) On inhumanity: dehumanization and how to resist It. Oxford University Press
Southgate B (2011) History meets fiction. Taylor & Francis
Spijkerboer T (2018, June 20) High risk, high return: How Europe's policies play into the hands of people-smugglers. The Guardian. https://www.theguardian.com/world/commentisfree/2018/jun/20/how-europe-policies-accelerate-people-smuggling
The Migrants' Files (2015) http://www.themigrantsfiles.com/
Tondo L (2018a, October 3) "We have found hell": Trauma runs deep for children at dire Lesbos camp. The Guardian. https://www.theguardian.com/global-development/2018/oct/03/trauma-runs-deep-for-children-at-dire-lesbos-camp-moria
Tondo L (2018b, November 14) "They didn't give a damn": First footage of Croatian police "brutality." The Guardian. https://www.theguardian.com/global-development/2018b/nov/14/didnt-give-a-damn-refugees-film-croatian-police-brutality-bosnia
Trilling D (2019, August 1) How the media contributed to the migrant crisis. The Guardian. https://www.theguardian.com/news/2019/aug/01/media-framed-migrant-crisis-disaster-reporting
UN Convention and Protocol Relating to the Status of Refugees (1951) https://www.unhcr.org/protection/basic/3b66c2aa10/convention-protocol-relating-status-refugees.html
UNHCR (2022, May 26) Figures at a glance. UNHCR. https://www.unhcr.org/figures-at-a-glance.html
UNITED (2023) Death by policy—Time for Change! UNITED. http://www.unitedagainstrefugeedeaths.eu/wp-content/uploads/2014/06/ListofDeathsActual.pdf
Verhofstadt G (2015, December 1) Europe bribes Turkey. POLITICO. https://www.politico.eu/article/europe-bribes-turkey/
Vitkus DJ (1997) Turning Turk in Othello: the conversion and damnation of the moor. Shakespear Q 48(2):145–176. https://doi.org/10.2307/2871278
Vogt W (2017) The arterial border: Negotiating economies of risk and violence in Mexico's security regime. Int J Migr Bord Stud 3(2–3):192–207. https://doi.org/10.1504/IJMBS.2017.083244
Zaiotti R (Ed.). (2016) Externalizing migration management: Europe, North America and the spread of "remote control" practices. Routledge. https://doi.org/10.4324/9781315650852

Houtum van, Henk Prof Dr, Head of the Nijmegen Centre for Border Research & Professor Political Geography and Geopolitics at Radboud University Nijmegen research interests: Bordering, Migration, Global Justice, Visas, Maps.

Bueno Lacy, Rodrigo Dr, YUFE post-doctoral researcher in European Identity at the Karelian Institute, University of Eastern Finland; research interests: border studies, geopolitics, radical cartography, state-building, and global history.

Gepspatial Technologies and Application in Agriculture

Pest Management with Precision Farming Tools: The Case of the Olive Fly (*Bactrocera Oleae*)

Giorgos Katsikogiannis, Vasilis Kopsachilis, Viron Ignatios Michalakis, Dimitris Kavroudakis, Michalis Vaitis, Thomas Tscheulin, Giorgos Stavrianakis, Stratis Sentas, and Thanasis Kizos

Abstract The management of pests is an area where different objectives meet: the need to ensure that agricultural production covers the needs of a growing population, the need to ensure safety and low levels of health risks for people and the need to conserve wildlife and biodiversity are not always compatible. In this chapter, we present precision farming tools that are used to manage and control the most important pest of the olive tree, the olive fly (*Bactrocera Oleae*) at the landscape level. A geodatabase, linked to two android applications (one for recording insect populations in traps and one for recording spraying routes) and a WebGIS application, is used to monitor populations and design management options. System architecture is presented, along with findings from its application on Samos and Lesvos Islands. These tools can be used to monitor and predict population movement and changes.

G. Katsikogiannis · V. Kopsachilis · V. I. Michalakis · D. Kavroudakis · M. Vaitis · T. Tscheulin · G. Stavrianakis · S. Sentas · T. Kizos (✉)
University of the Aegean, Mytilene, Greece
e-mail: akizos@aegean.gr

G. Katsikogiannis
e-mail: geod16001@aegean.gr

V. Kopsachilis
e-mail: vkopsachilis@geo.aegean.gr

V. I. Michalakis
e-mail: v.michalakis@aegean.gr

D. Kavroudakis
e-mail: dimitrisk@aegean.gr

M. Vaitis
e-mail: vaitis@aegean.gr

T. Tscheulin
e-mail: t.tscheulin@aegean.gr

G. Stavrianakis
e-mail: gmstavrianakis@geo.aegean.gr

S. Sentas
e-mail: s.sentas@aegean.gr

A. Klonari et al. (eds.), *Re-visioning Geography*, Key Challenges in Geography,
https://doi.org/10.1007/978-3-031-40747-5_16

Keywords GIS · Insecticides · Olive fly · Olive tree · Precision farming · Spraying efficiency

Introduction

Olive cultivation is an important economic and cultural activity for Greece, the third largest olive producing country in the world. The most important enemy of the tree is the olive fly (*Bactrocera Oleae*, Diptera, Tephritidae, referred as "OLF" here), as its infestation results in both quantitative and qualitative damage to olive oil production (Daane and Johnson 2010; Malheiro et al. 2015). To manage OLF in Greece, a national program based on bait spraying is implemented, coordinated by the Ministry of Agriculture, and carried out by the Offices for Rural Development (DAOK in Greek) of olive-growing prefectures. The program is based on monitoring OLF populations with traps (of the McPhail type), which are placed in the canopy of olive trees (1 every 2000 trees) and checked by trained personnel (called "trap setters"), every 5 days. The attractant solution commonly used is ammonium sulfate at a ratio of 2% (Varikou et al. 2013, 2014). Depending on OLF population levels and other technical data such as the ratio of males to females, fertility, weather conditions, etc. bait spraying is applied, to attract and kill adults before they lay their eggs in the fruit. Killing adults at this stage is intended to reduce OLF abundance in the next generation and generally to reduce infestations. These bait sprays are today—and after the ban of the use of organophosphate insecticides—administered with a protein attractant solution and pyrethroid insecticides, or with the ready-to-use bait Success 0.24cb (spinosin) (Varikou et al. 2016). The efficacy of bait sprays is monitored by monthly sampling of olive fruit from random trees in the sprayed area and by the estimation of viable oviposition on the olive fruit. Pest population dynamics and infestation on the olive fruit depend on many biotic and abiotic parameters, such as weather conditions, the availability and condition of the olive fruit, the increasing resistance to insecticides, etc. (Kampouraki et al. 2018).

The effectiveness of the spraying program, in addition to the above-mentioned factors, directly depends on the early detection of the appearance of the first adults (referred as the "base generation") as well as the recording of population outbreaks for timely spraying. Since population measurements in the traps are carried out by the "trap setters" on specific dates each month (four to five times), count OLF numbers and renew the attractant solution in McPhail traps, this is a key process for timely and spatially explicit recording of overall OLF presence and identification of hot spots. Moreover, population movements can also be deducted from the trap data with relatively good temporal resolution.

Another key process of the effectiveness of the program is spraying itself, performed in almost all cases by agricultural tractors with mounted sprayers. These sprayers spray on both sides of the tractor reaching a radius that exceeds up to 50 m from the movement of the tractor. Therefore, the position and course of the tractor are vital in recording the extent of the spraying and understanding its effectiveness and avoid overlapping or omitting areas. Overlaying with population monitoring data, the relief and land cover in the spraying area, these data can provide the key to better planning and management of spraying and reduce the amount of substances used.

The emerging food crisis has highlighted the need to increase agricultural production. At the same time, an emerging trend focuses alternatively on the use of less or no chemicals and the management of crop pests with natural means whenever this is feasible. Precision agriculture is a tool that can be used to address this new reality. As Pierce and Nowak (1999) assert, precision agriculture is defined as the application of technologies and principles to manage spatial and temporal variability associated with all aspects of agricultural production. Elaborating on the role of GIS in agriculture is about allocative efficiency, profitability and record keeping.

In this paper, we present the conceptualization, planning, development and implementation of an integrated management system of the OLF Program on the Samos Island in the Aegean Sea, Greece. The system consists of two Android applications, available in Greek language, that cover two critical processes, trap monitoring and spraying and an interface that informs the coordinators in real time about OLF population hotspots, assisting them to decide on ordering spraying and monitoring its effectiveness. All data are automatically stored in a geodatabase and displayed spatially in cartographic backgrounds. All components of the system were developed by the University of the Aegean. In the following section, the system is described in detail and then some results of its implementation are presented.

Methods and Data

System Contents and Architecture

The system consists of the following subsystems and applications (Fig. 1):

- A Geospatial database, where all data is stored and managed. The database (DB) is the core of the system, as all the other parts communicate with it for entering and extracting data. Its implementation was carried out in the PostgreSQL/PostGIS database.
- An application for mobile devices that allows recording and transmission of OLF population for each McPhail trap (Dakos mobile app). The application allows the cartographic display of the traps, the recording of the data of the trap population and cleaning. Internet connection (WiFi or 3G/4G) is necessary, however, the trap data from the DB to the device can be stored before the departure of the spraying crews and the OLF population data can be stored in the device and transferred to the DB when a connection is available. This makes the application functional even in case of incomplete or problematic internet connection. The application checks the user's location: they are not allowed to enter data unless they have approached the respective trap within a certain distance (by default 20 m, or after setting for

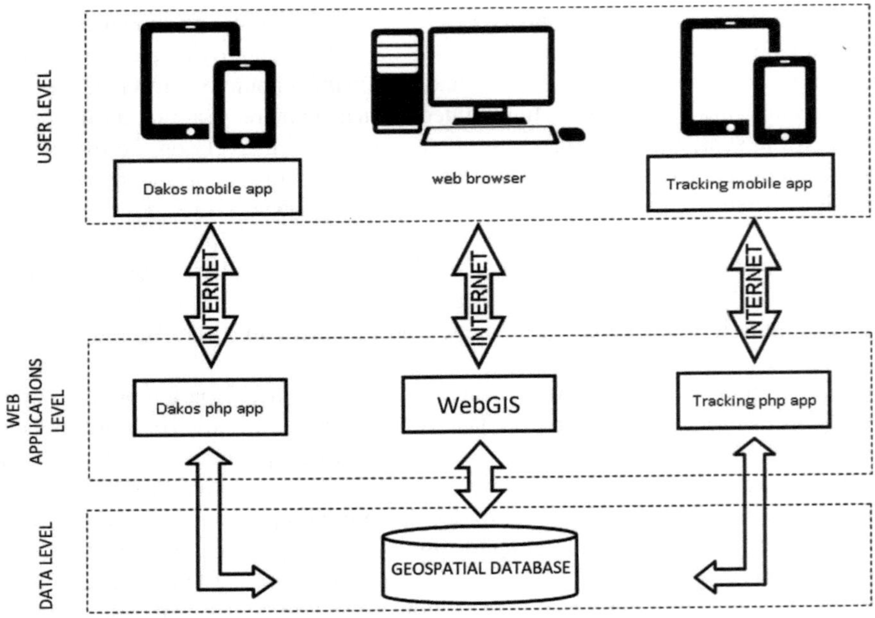

Fig. 1 System architecture

each trap separately, considering any deviations of the GPS of the device and the position of each trap). This ensures that users actually visit the traps and thus the measurements are reliable. Figure 2 shows some screenshots of the application, developed with the Java programming language and running on Android devices.

- An application for mobile devices that allows recording and transmission of the routes of spraying tractors, during spraying in real time (Tracking mobile app). This application allows the vehicle coordinates to be recorded and transmitted to the DB during spraying. In particular, the application records the position of the vehicle every 5 s using the GPS of the device, which together with the date, time and identity of the vehicle, are transmitted to the database. The vehicle's identity is declared once by the supervisor on each device using a code. The application works with the device in sleep mode, to save battery. It was developed with Java programming language and works on devices with Android operating system.
- A Web-based geographic information system (WebGIS) where data is displayed and managed. The goal of the WebGIS is to monitor and support the monitoring program by the supervisor, achieved through two independent applications: (i) a monitoring application, which allows the creation of reports, charts and cartographic representations of the data and (ii) a management application, which provides a user-friendly environment for the supervision and processing of the raw data maintained in the database. For the implementation of the subsystem, the programming languages Java and Javascript, HTML and some libraries were used

a b c

Fig. 2 Screenshots of "Dakos mobile" app **a** Traps **b** Trap and user position **c** Trap data input (the app is available only in Greek language. As in "Έλεγχος θέσης" = Check your position in map, "Σύνολο δάκων" = Total of adults in trap, "Θηλυκά" = Female pests in trap, "Φάρμακο Θειική αμμωνία" = Solution - ammonium sulfate, "Καταχώρηση» = Register numbers, "Πίσω στο χάρτη" = Go back to map)

(indicative: OpenLayers3, DataTables, JFreeChart, Timeline), while the distribution of the maps is carried out through the Geoserver map server and WMS services (Web Map Services).

The system architecture consists of three layers: the user layer, the web application layer and the data layer (Fig. 1).

- The user level includes three types of users:
- Supervisors, who interact with WebGIS via web browsers and can visualize the data in tables, figures and maps, to monitor the progress of the program and to order sprayings.
- Trapp setters that use a tablet's or smartphone's in-built satellite navigation receiver in combination with the installed "Dakos mobile" app. Access to the application is controlled by a password. The log data is stored in the database.
- Spraying operators, that use the satellite navigation of a phone, where the "Tracking mobile" app is installed. The log data is stored in the database.
- The web applications layer mainly includes the web-based geographic information system (WebGIS) that allows cartographic display and management of the data stored in the database. It also gives access to free background data.
- The data layer includes the database management system (DBMS) and an appropriately designed geospatial database to support system operations.

The system has been developed using free software/open-source software.

Visualization Tools

Tables and Charts

The system allows reporting of insect counts per trap and per area. In the first case, the report contains the code of the trap, the date of measurement, the total number of insects and the numbers of males and females. In the second case, the report contains the name of the area, the date of measurement and the averages of the total number of insects, the number of males and the number of female insects of the traps of the area. The supervisor can customize the reports for the traps or areas and time period desired. Reports can be exported as CSV files for further processing.

The system also allows plotting of insect counts per trap and per area, where the horizontal axis refers to dates of counts and the vertical axis refers to insect count values. The supervisor can customize the chart by selecting the traps or areas, time interval and type of count (total, males, females) desired. Optionally, a time series of meteorological data can be integrated into the chart. Figure 3 shows, for example, the measurements of the total number of insects for the area of "Ag. Konstantinou", together with the time series of maximum temperature values for the meteorological station of "Karlovasi" (Samos Island).

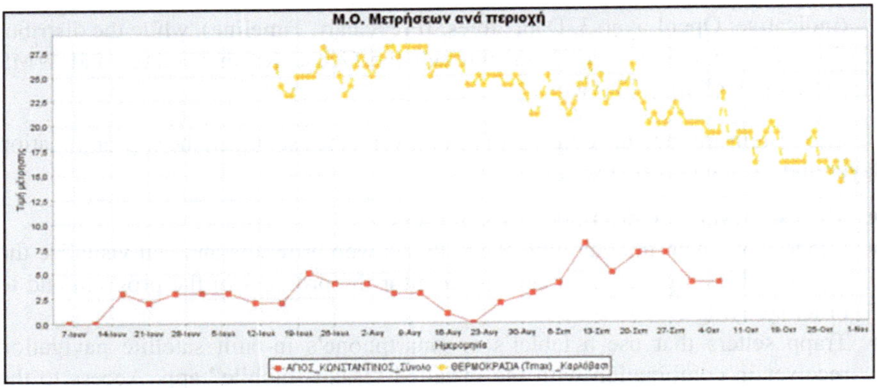

Fig. 3 Screenshot of the web application: Plot of measurements of the total number of insects together with a time series of maximum temperature values (the app is available only in Greek language. As in "Μ.Ο. Μετρήσεων ανά περιοχή" = average of pest in traps per area, Vertical axis = Number of adults in traps, Horizontal axis = Date of measurement, "ΑΓΙΟΣ_ΚΩΝΣΤΑΝΤΙΝΟΣ" = is the name of the selected area, "ΘΕΡΜΟΚΡΑΣΙΑ" = Temperature - Tmax)

Mapping Traps and Routes of Spraying Vehicles

The system allows the creation of cartographic representations of OLF counts in the traps. A dynamic interactive cartographic environment displays trap locations and an interactive timeline with measurement/cleanup dates available in the DB (Fig. 4). Specifically, cartographic representations of: (a) the total number of insects, (b) the male/female ratio (M/F) and (c) the total number of insects and the M/F ratio simultaneously are available. Each trap is colored according to the number of insects counted (green, orange, red), with its symbol representing the M/F ratio. More detailed information is given by selecting a specific trap. There are options for different map backgrounds and the ability to save/print the current map display.

The system allows the generation of cartographic representations of the sequential positions of vehicles performing bait sprays (which simulate a route). In a dynamic interactive cartographic environment, the sequential positions of the vehicles are displayed, with a different color for each vehicle. The supervisor can customize the display by selecting the desired spraying date and vehicles (Fig. 5). There are also options for different map backgrounds and the ability to save/print the current map display.

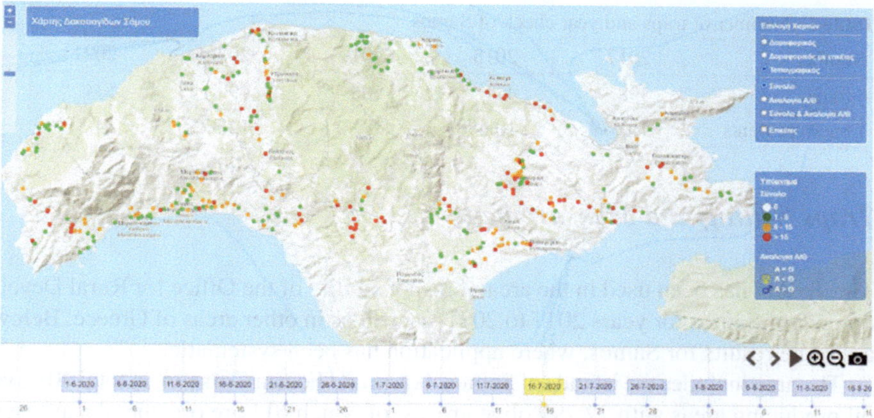

Fig. 4 Screenshot of the web application: Mapping of OLF counts from McPhail traps (the app is available only in Greek language. As in "Χάρτης Δακοσυλλήψεων Σάμου" = Map of OLF in traps for Samos island, "Επιλογή χαρτών" – "Δορυφορικός", "Δορυφορικός με ετικέτες" "Τοπογραφικός" = Background map, Satellite map, Satellite map with labels, Topographical map, "Υπόμνημα" – "Σύνολο" = Index – Total, "Αναλογία Α/Θ" = Sex ratio, "Α" = mail, "Θ" = femaile)

Fig. 5 Screenshot of the web application: Mapping of spraying vehicle routes (the app is available only in Greek language. As in "Χάρτης Online Παρακολούθησης" = real time tracking map, "Επιλογή υποβάθρου" = Background selection, "Δορυφορικός", "Δορυφορικός με ετικέτες", "Τοπογραφικός" = Background map, Satellite map, Satellite map with labels, Topographical map, "Επιλογή ημερομηνίας" = Date selection, "Υπόμνημα" = Index)

Table 1 Number of traps and year check of Samos

Number	2017	2018	2019	2020	2021
# traps	398	421	500	499	534
# measurements	10.679	10.688	13.911	14.373	14.319

Traps Setting and Measurements

The system has been used in the area of responsibility of the Office for Rural Development of Samos for years 2017 to 2021, as well as in other areas of Greece. Below are some results for Samos, where application has been systematic.

The area includes the islands of Samos, Ikaria and Fourni Corseon, in total 40 olive oil producing areas with 22,799 olive groves, of which 611 are organic. An average of 470 traps were installed per year, while an average of 12,800 measurements were performed using the system, with an increasing trend (detailed data are listed in Table 1).

Results and Discussion

Tables 2 and 3 present the M/F ratio as well as the totals of insects measured in the traps for the five years of implementation (2017–2021) per month, giving an overall sense of the change in the insect population.

Table 2 Sex ratio Male/Female of pest adults measured in traps for 5 years period

	Male/Female olive flies ratio				
	2017	**2018**	**2019**	**2020**	**2021**
June	1.93	1.42	1.83	1.15	1.56
July	1.35	1.30	1.77	1.21	1.38
August	1.21	1.30	1.32	1.31	1.40
September	1.19	1.24	1.17	1.23	1.15
October	0.94	1.11	1.18	1.27	1.19

Table 3 Totals of pest adults, measured in traps for 5 years period

	Total olive flies				
	2017	**2018**	**2019**	**2020**	**2021**
June	7,647	18,458	3,976	6,071	9,173
July	19,245	21,456	6,176	21,452	21,403
August	15,242	19,817	23,135	22,186	11,887
September	23,062	25,022	37,973	35,796	20,398
October	25,196	20,103	34,883	30,640	36,952
Totals	**90,392**	**104,856**	**106,143**	**116,145**	**99,813**

The recording, transmission and storage in the DB of the measurements of the OLF population, as well as the storage of the coordinates of the OLF traps and meteorological variables, enables calculations to be carried out for a better understanding of the population dynamics and spread of OLF and monitoring program. The depiction of monthly counts with the use of a Kriging spatial interpolation method was applied using the insect measurements from the set of traps. Surface areas were calculated only in olive grove areas, as recorded by the declarations of their owners (Fig. 6). This depiction can reveal seasonal patterns of higher population presence and assist in planning for next season.

The results show some common patterns of OLF population change during the monitoring season (northern and mountainous areas at the beginning, southern and lowland toward the end of the season), but also appear to show population movements. An exception seems to be 2019, where the weather conditions were very different compared to the other two years (initially intense heat and later low temperatures and very high humidity). The highest population values recorded are in the north in September, but apart from that one-year maximum number, typically the highest values are recorded in the warmer south in October, when higher temperatures and relative humidity values are combined, favoring PLF population explosions.

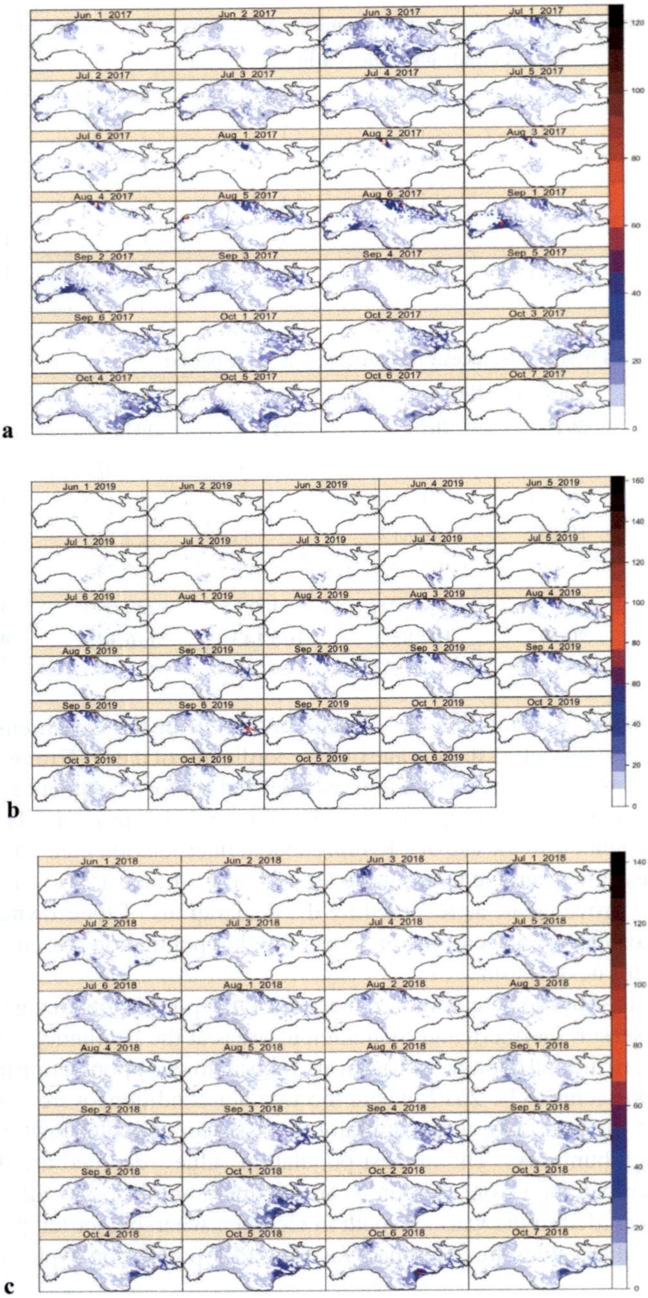

Fig. 6 10-day measurements of OLF in Samos for 3-year period (2017–2019): **a** 2017 **b** 2018 **c** 2019

Conclusion

The management of the OLF is a very important issue in many olive oil producing areas. Common management Programs such as the ones practiced in Greece can offer landscape level solutions to its management rather than single farm level management. The Integrated System that is described here can provide input and assistance to the supervisors of the Programs at two levels: (a) the monitoring of the population level, in which the supervisors can see in real-time changes and schedule spraying; (b) the monitoring of spraying application, in which the supervisors can monitor in real-time sprayed areas, avoid overlapping and with the combination of OLF population monitoring estimate spraying efficiency.

Although a detailed study has not yet been carried out to evaluate the economic and environmental benefits of the application of the Integrated System, supervisors assert that when practiced in full, it has facilitated their everyday work and has eased the realization of the Program overall and on a day to day basis. New functions in the WebGIS, such as spatial analysis calculations, would provide an extra assistance in the future. Also, the aim is to enrich this application and make it user-friendly for agricultural cooperations and individual olive farmers who will use it in their fields, obtaining information about their region.

The combination of the application of this system along with the implementation of climate smart and environmentally friendly farming practices could lead to an integrated management of OLF in olive groves. This type of cultivation requires that the farmers will implement low input agriculture by managing crop pests with natural means, for example by allowing understory plants to grow and act as nesting sites for predator and parasitoid arthropods. With this type of olive grove cultivation practices, farmers will enable natural enemies of OLF in order to manage its population. In the case, the OLF population reaches an alarming level, the system will inform the supervisors of the program who will proceed to precise spraying. With these tools, we will create an olive cultivation system which manages the OLF with integrated methods and less or no chemicals, according to the trends that are being promoted by both EU and the new CAP.

References

Daane KM, Johnson MW (2010) Olive fruit fly: managing an ancient pest in modern times. Annu Rev Entomol 55(1):151–169. https://doi.org/10.1146/annurev.ento.54.110807.090553

Kalaitzaki A, Alexandrakis V, and Varikou K (2005) Integrated protection of olive crops = Protection intégrée des olivaies. Dijon: IOBC/WPRS

Kampouraki A, Stavrakaki M, Karataraki A, Katsikogiannis G, Pitika E, Varikou K, Vlachaki K (2018) Recent evolution and operational impact of insecticide resistance in olive fruit fly Bactrocera Oleae populations from Greece. J Pest Sci 91(4):1429–1439. https://doi.org/10.1007/s10340-018-1007-8

Malheiro R, Casal S, Baptista P, Pereira JA (2015) A review of Bactrocera Oleae (Rossi) impact in olive products: from the tree to the table. Trends Food Sci Technol 44(2):226–242. https://doi.org/10.1016/j.tifs.2015.04.009

Pierce FJ, Nowak P (1999) Aspects of precision agriculture. Adv Agron 67:1–85. https://doi.org/10.1016/S0065-2113(08)60513-1

Varikou K, Alexandrakis V, Gika V, Birouraki A, Marnelakis C, Sergentani C (2013) Estimation of fly population density of Bactrocera Oleae in olive groves of crete. Phytoparasitica 41(1):105–111. https://doi.org/10.1007/s12600-012-0270-0

Varikou K, Garantonakis N, and Birouraki A (2014) Comparative field studies of Bactrocera Oleae Baits in olive orchards in crete. Crop protection 65(November): 238–43. https://doi.org/10.1016/j.cropro.2014.08.005

Varikou K, Garantonakis N, Birouraki A, Ioannou A, Kapogia E (2016) Improvement of bait sprays for the control of Bactrocera Oleae (Diptera: Tephritidae). Crop Prot 81(March):1–8. https://doi.org/10.1016/j.cropro.2015.11.007

Sustainable Networking Solutions in Remote IoT Environments: Use Cases, Challenges, and Solutions for Smart Agriculture

Stefanos Plastras, Sofia Polymeni, Dimitrios N. Skoutas, Georgios Kormentzas, and Charalabos Skianis

Abstract Smart agriculture has laid a solid foundation for the development of society and the growth of national economies. However, due to the vast variety of cultivable lands, some of which may be in inaccessible regions, numerous requirements and challenges have emerged over time, necessitating the development of innovative technological solutions. Despite the fact that technological improvements have substantially facilitated the expansion of agricultural services, more effort is required to fully integrate novel solutions based on heterogeneous network architectures. In addition, the incorporation of new technology to achieve high performance, while preserving high quality and a minimal environmental footprint, remains a challenge for sustainable agricultural development. Therefore, researchers continue to work on designing and developing new technologies and techniques to meet these objectives. Consequently, the management of Internet of Things (IoT) data, the expansion of terrestrial connectivity, and the development of a sustainable infrastructure for the future generation of agriculture are critical research challenges. In this context, new technological approaches and solutions, such as Non-Terrestrial telecommunications Networks, Machine Learning (ML)-based algorithms, and Green IoT protocols, are being proposed and investigated. The aim of this chapter is to familiarize readers with the current agricultural landscape, its requirements, and the most promising technologies being developed to meet those needs.

S. Plastras (✉) · S. Polymeni · D. N. Skoutas · G. Kormentzas · C. Skianis
Department of Information and Communication Systems Engineering, University of the Aegean, Karlovasi, Samos, Greece
e-mail: s.plastras@aegean.gr

S. Polymeni
e-mail: spolymeni@aegean.gr

D. N. Skoutas
e-mail: d.skoutas@aegean.gr

G. Kormentzas
e-mail: gkorm@aegean.gr

C. Skianis
e-mail: cskianis@aegean.gr

A. Klonari et al. (eds.), *Re-visioning Geography*, Key Challenges in Geography,
https://doi.org/10.1007/978-3-031-40747-5_17

Keywords Remote IoT · Sustainable Smart Agriculture (SA) · Non-Terrestrial Networks (NTNs) · Machine Learning (ML) · Green IoT

Introduction

The global production and distribution of agricultural products, as well as the growth of national economies, are strongly reliant on the agricultural sector, elevating the importance of intelligent agriculture in the contemporary world. By adopting an extroverted strategy, modern smart agriculture has found a position in a world-wide consumer-producer ecosystem marked by the spread of novel and sustainable products. As a result, farmers and the agro-industrial environment as a whole have adopted novel farming techniques and, with the aid of technology, have delivered substantial benefits to the world over the years (Khan et al. 2021). However, as the world's population is expected to reach 10 billion by 2050, and in response to the needs of the twenty-first century for sustainable agricultural solutions, the United Nations (UN) has established a set of regulations aimed at creating a more sustainable world in the areas of green energy, natural resource conservation, and climate change mitigation (Abbasi et al. 2022).

Moreover, developments in agricultural technology demonstrate that research is advancing. The Internet of Things (IoT) technology based on 5G networks is anticipated to bring about a massive change by enabling the internet connection of numerous physical objects and devices. In the near future, it is anticipated that these technologies will dramatically increase industrial and agricultural productivity (Sobin 2020). Ericsson's study on IoT-enabled cellular networks predicts that the number of IoT devices linked via the Narrow-band IoT (NB-IoT) and Long-Term Evolution Category-M (LTE-M) technologies would overtake 2G/3G connected IoT devices in 2023 and account for 51% of all cellular IoT connections by 2027 (Ericsson 2022). In light of this, the development of diverse IoT ecosystems within the context of smart agriculture has created both new challenges and opportunities.

Notably, a number of smart agriculture applications, such as distributed water management services, agricultural land irrigation, and even crop production, are largely implemented in remote areas (Ullo and Sinha 2021). For example, island clusters are a use case scenario in which the implementation and development of various IoT-enabled agricultural services present inherent connectivity issues, such as the interconnection between IoT objects and the communication with data storage and analysis services, which are primarily implemented on a cloud data center (Tao et al. 2021). In such locations, the absence of an effective terrestrial communication infrastructure makes the deployment and operation of advanced agricultural technologies extremely difficult. Incorporating new technologies able to provide sufficient and secure communication with the core network and remote servers is necessary for data collection in a remote agricultural region. Consequently, it is anticipated that remote geographic regions will drive the development of Remote IoT solutions to overcome the aforementioned challenges.

Non-Terrestrial Networks (NTNs), which consist of satellite systems and Unmanned Aerial Vehicles (UAVs), are gaining attention as a possible technological basis for the integration of terrestrial networks. Therefore, NTNs could potentially be used to provide connectivity to advanced agricultural services in remote areas (Vaezi et al. 2022). Moreover, Machine Learning (ML) can offer communication infrastructures with the ability to self-manage and self-repair, which could be incredibly useful for both routine maintenance and emergency scenarios demanding rapid, large-scale response, such as a massive network failure (Bhat and Huang 2021). In this context, Green IoT is also introduced, which provides more sustainable and energy-efficient communications while respecting the environment and promoting the vision of a sustainable society (Popli et al. 2022).

The main driving force guiding this research is the emphasis on the importance of sustainability in smart agricultural systems, especially in remote areas, by analyzing both key factors that aid the deployment of smart agricultural scenarios and the main aspects that influence the design of guidelines for their sustainable development. For a smart agricultural system to be considered sustainable and be implemented in remote areas, the key factors that need to be addressed for its development include low energy consumption to prolong the system's battery life, reliability, and mobility. As already mentioned, energy consumption and battery life are two interdependent concepts. In order to develop a system for remote monitoring while also maintaining its energy consumption to a minimum, long-range, low-power communications are among the key solutions that will allow for large coverage and a larger number of sensors to be deployed. However, in addition to energy consumption, the proposed system must also be reliable, meaning it must perform correctly during a specific time period, and mobile, so that it can be implemented in different locations. In this study, in order to emphasize the significance of sustainable smart agricultural systems, we discuss the key challenges of current Remote IoT systems and then present novel solutions for sustainability in smart agricultural applications of the future.

The remainder of this chapter is organized as follows. Section 2 offers a detailed description of the key fields investigated in this study, while Sect. 3 presents the challenges that have surfaced in various Remote IoT agricultural applications, such as remote connectivity, data collection, and sustainable network infrastructure. In Sect. 4, sustainable networking technologies are presented, namely NTNs, ML, and Green IoT protocols, which can provide a solution to the majority of the issues discussed in Sect. 3. Finally, Sect. 5 summarizes the chapter's conclusions and findings.

Internet of Things in Smart Agriculture: Toward a Sustainable Development

This section provides a detailed description of the key fields investigated in this study, namely the IoT, smart agriculture, and advanced Information and Communication Technology (ICT) enablers, as well as how they can be integrated.

Internet of Things Environments

By expanding the boundaries of the known Internet, the realm of IoT constitutes a tremendous transformation in the daily lives of many people worldwide by proposing numerous intelligent and real-time applications, including smart homes, smart cities, industrial control systems, smart health systems, and autonomous transportation. IoT is considered a unified service layer that combines multiple research fields, such as business logic, sensor technology, middleware, communication technologies, and Application Programming Interfaces (APIs), and delivers results while also taking into consideration the quality of the offered services (QoS) (Sinha and Dhanalakshmi 2022). Therefore, to ensure that an IoT service delivers the required QoS, metrics like data integrity, access control, interoperability, and communication should be taken into account throughout the design process.

In the field of smart agriculture, such metrics include the support of massive device connectivity (e.g., number of sensors supported by the overall infrastructure), precision sensing, as well as appropriate communication technologies allowing for fast and reliable data transmissions. Message Queuing Telemetry Transport (MQTT) and Constrained Application Protocol (CoAP) are two key protocols that aid the development of data transfer services for livestock farming and crop monitoring. In addition, fog and cloud computing have enabled applications such as real-time animal health and plant disease monitoring by bringing the cloud closer to the field and data collection area, thereby reducing processing latency and network congestion.

To conclude, integrating IoT technologies in the field of agriculture can enhance production, available resources and final revenue, and help farmers transform their business into a digital entity by providing sufficient storage, processing, and transmitting capabilities, as well as timely forecasts.

Modern Communication Standards

As mentioned previously, IoT offers a variety of communication technologies and standards depending on the infrastructure and the corresponding needs, such as the implementation cost, the size of the deployed network, as well as its power consumption (Tao et al. 2021), (Polymeni et al. 2022). The distance of the communication link, the expenses involved with network maintenance and installation, the frequency bands accessible for licensed or unlicensed use, and the specific requirements of each application are fundamental considerations when selecting a communication standard (Wang et al. 2022). Currently, 5G networks have a significant impact on many people's lives, elevating the digital realm to the forefront. In addition, the number of IoT devices is expected to increase in the near future, developing new networking challenges (Li et al. 2018; Chettri and Bera 2020).

Additionally, different IoT contexts have distinctive requirements. For instance, a smart home is comprised of small places, so the telecommunication standards that can be utilized in it must have a comparable range, such as Wi-Fi (IEEE 802.11), Bluetooth (IEEE 802.15.1), and Bluetooth Low Energy (BLE). In the context of a smart city or in the case of agriculture fields, however, it is required to employ long-range telecommunication technologies capable of covering a large geographical region with adequate data transmission capabilities. Here, cellular standards provide significant support for these contexts, as they enable a wide range of data rates, while the usage of licensed frequency spectrum limits interference and maximizes coverage area. In addition, the advent of new 5G technologies within the IoT enables the support of mass connectivity scenarios, with up to one million devices per square kilometer and extremely low latency (Li et al. 2018).

The third-generation partnership project (3GPP) has standardized the NB-IoT and massive Long-Term Evolution Category-M (LTE-M) specifications, which are the two versions of the wide-area 5G-IoT systems supported by cellular networks (Wang et al. 2022). For the research community, the question of which technology is regarded as ideal for serving the various IoT environments, particularly for smart agriculture applications, is indeed debatable due to the fact that its answer is a multi-parametric decision that, as stated previously, depends on the specifics of each application's requirements for mass connectivity, mobility, data transmission rates, and the ability of the deployed devices to use a particular range of frequency spectrum.

Smart Agriculture

Agriculture in its traditional sense refers to the production of livestock and agricultural products through entirely physical labor by farmers, without the use of digital data or the existence of computing devices, making it challenging to develop new services in this sector as production demands begin to rise. Smart farming is the transformation of traditional agriculture into a sustainable, intelligent, and digitized agriculture that integrates IoT technology in its entirety and other innovative technologies, having undergone a significant technological upgrade and therefore providing advanced methods and tools for the production and exploitation of agricultural production data (Tao et al. 2021; Sinha and Dhanalakshmi 2022; Abbasi et al. 2022). Smart agriculture is anticipated to make substantial use of 5G-IoT, from initial sensor deployment to cloud-based data storage and processing. The technology stack in the field of smart farming typically consists of the following: the level of smart devices with sensing or data collection capabilities; the level of the network allowing these devices to exchange useful information both ways to the Internet or a local node; the level of middleware offering a unified interface for devices; and lastly, the level of the business model providing value to producers and farmers.

At the device level, different manufacturers have their own integrated systems or circuit boards with specialized data gathering, processing capacity, speed, and networking capabilities. In agriculture, manufacturers generally include a variety of mechanisms of resistance against elements such as humidity, temperature, and precipitation. Energy consumption, development costs, reliability, battery life, and mobility are crucial aspects that influence the design of smart agriculture scenarios and the development of guidelines for their sustainable development. At the level of communication, there are short and long-distance communication standards that cover a wide variety of scenarios. There are instances where it would be acceptable to deploy multiple Wi-Fi-enabled sensors in a limited area. To cover larger areas with a larger number of sensors, however, it is necessary to employ communications technologies with the required capabilities. As mentioned previously, such technologies include NB-IoT and LTE-M, which utilize the infrastructure of cellular networks to provide reliable and secure coverage over wide areas.

In the same category of low–power, wide-area networks are networks like the Long Range Wireless Access Network (LoRaWAN), that use unlicensed frequency bands for transmission. These systems do not require the end-user to pay a subscription fee to a telecommunications service provider because they use a free (unlicensed) frequency spectrum. On the other hand, they necessitate the initial development of the relevant network infrastructure up to the area of interest, which increases the cost of the initial installation. Moreover, compared to IoT systems supported by cellular networks, these systems are susceptible to interference because the frequency bands they employ are open to use by other systems in the same area (Li et al. 2022).

In smart agriculture applications, the level of data storage and analysis is crucial and is typically implemented in the cloud. Moreover, due to the nature of precision agriculture applications, the collected data volume is typically very large, creating processing and storage challenges. The resulting data sets are commonly complex and large and fall under the category of "Big data". Big data analysis necessitates the application of advanced analytic techniques, and it is a field of research that has experienced significant growth in recent years, particularly through the employment of ML-based approaches (Qazi et al. 2022; Misra et al. 2022). ML can also be used to generate forecasts for many aspects of the agricultural production and supply chain that require optimal and valid decisions in order to increase productivity and safety and decrease or even eliminate emergencies. The final level is the business level, which builds on the previous levels to present the farmer or producer with a final business scenario that can be applied and generate further profit and value.

The Fourth Industrial Revolution, which is still taking place as a result of technological progress, is propelling the agricultural industry forward as well as agricultural research and development (Khan et al. 2021; Nasirahmadi and Hensel 2022; Abbasi et al. 2022). Furthermore, technological trends are already influencing the next generation of communication networks, which is anticipated to revolutionize agriculture (Imoize et al. 2021; Zhang et al. 2022).

Information and Communication Technology Enablers for Smart Agriculture

It is an undeniable fact that supply chain fluctuations and global demand for agricultural products, along with climate change and rising food prices, have led to a decline in farm income and agricultural labor force. In addition, as consumer expectations tend to rise, it is essential to have a unified technological ecosystem system that can support agriculture production and business value, guarantee food safety, and be a market-competitive solution. Recent advancements in ICTs have resulted in a rise in the number of farmers and agricultural businesses that have adopted modern cloud computing environments and edge computing technologies, which provide even faster connectivity (Bhat and Huang 2021). Cloud-based approaches help farmers analyze their production by aggregating data from a variety of sensors, terrestrial weather stations, and satellite images, thereby providing real-time analyses of crop, soil, and meteorological data. As shown in Fig. 1, key ICT enablers include IoT, 5G and B5G networks, cloud and edge computing, machine learning, and blockchain.

With the rapid adoption of the IoT, networking devices have become fully integrated in a variety of industries, including agriculture (Chettri and Bera 2020). Due to their ability to collect data and act proactively or in response to a decision, they are considered as the center of the agricultural supply chain and impose a dynamic influence on the agricultural environment. Using IoT devices and technology, farmers gain greater control over livestock husbandry and crop cultivation, enabling them to close the gap between demand and supply. However, the true potential of agriculture

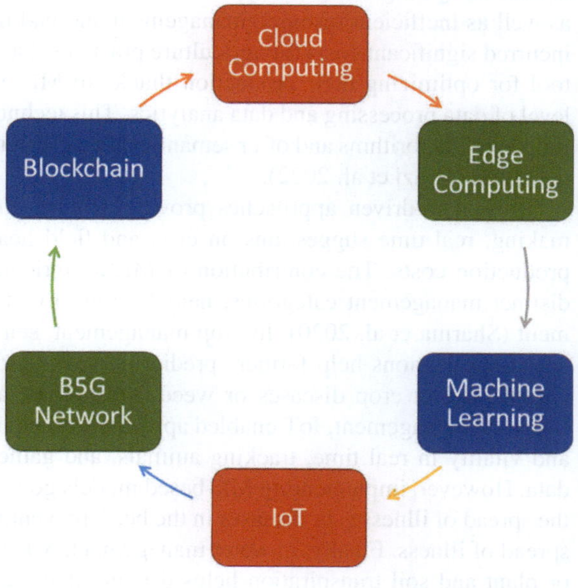

Fig. 1 Modern ICT enablers for smart agriculture

is contingent upon a reliable, high-speed network with low latency. These requirements are met by 5G mobile communication networks, which provide a capable infrastructure for establishing a communication environment (Sinha and Dhanalakshmi 2022). In this context, technologies such as IoT, edge cloud, and robotics can thrive. The 5G network handles all connections between an application and the cloud in both directions. On the other hand, the 6G next-generation network is anticipated to establish new standards and requirements in agriculture, accelerating its evolution (Vaezi et al. 2022; Zhang et al. 2022; Polymeni et al. 2023).

Cloud and edge computing are another set of enabling ICTs. Cloud computing describes a computing paradigm that provides on-demand access to a large pool of computing systems connected to each other in a private, public, or hybrid setting (Shi et al. 2016). Edge computing is a network architecture in which the network's edge acquires enhanced storage and computing capabilities, allowing it to provide the corresponding services to the end-users with reduced latency (Shi et al. 2016). Some of the main features of cloud computing in smart agriculture include remote data acquisition and storage, weather forecasting, and online expert consultation. On the other hand, edge computing is mainly used in combination with cloud or fog computing, helping to enable the evolution of 5G communications and bring cloud capabilities near to end-users by offering intelligent crop management, unmanned agricultural machinery, and safety traceability (Kalyani and Collier 2021).

The area of Artificial Intelligence (AI) known as Machine learning (ML) is an essential technology with several applications in people's everyday lives. The procedures of farm management, hazardous scenario prediction, profit maximization prediction, and resource optimization can become a reality if data can be processed by a smart infrastructure and ML is applied (Sharma et al. 2020). It is true that a lack of knowledge regarding the types of soil, crop, fertilizer, and weather conditions, as well as inefficient resource management and bad market trend predictions, have incurred significant costs for agriculture practices. Farmers are in control of a potent tool for optimizing their production thanks to ML, which takes advantage of the level of data processing and data analytics. This technology's data should supply the proper ML algorithms and offer semantics for a greater understanding of agricultural operations (Qazi et al. 2022).

Thus, data-driven approaches provide a more effective strategy for decision-making, real-time suggestions on crop and field health, and the minimization of production costs. The contribution of ML to agriculture can be divided into four distinct management categories, namely crop, livestock, water, and soil management (Sharma et al. 2020). In crop management, sensors in conjunction with ML-based applications help farmers predict harvest yields, provide crop quality evaluations, detect crop diseases or weed infestations, and identify plant species. In livestock management, IoT-enabled applications help in monitoring livestock health and vitality in real time, tracking animals, and gathering and analyzing historical data. However, implementing ML-based models goes one step further by preventing the spread of illnesses or diseases in the herd, preventing animal loss, or tracking the spread of illness. Finally, in water management, ML-based analysis of precipitation or plant and soil transpiration helps design effective smart irrigation systems that are crucial for optimizing crop production and permitting the development of water-saving methods, while in soil management, ML algorithms can be used for forecasting and identifying agricultural soil parameters, such as composition, dryness, temperature, and wetness, offering less expensive and quicker solutions.

Finally, blockchain is a promising technology for secure transactions that has piqued the attention of the agricultural community (Lin et al. 2017). It can reduce the number of potential failure points, increase data transparency and immutability, and, as a consequence, modernize farm infrastructure, as well as ensure concepts such as party trust and data security. It is also capable of tracing the origin of food, allowing for the reliability of food supply chains and the establishment of trust between producers and consumers. However, as a secure method of data distribution, it is able to leverage data-driven solutions to further enhance agriculture. In conjunction with smart contracts, blockchain permits financial transactions between parties in response to changes in the data presented on the chain. The diverse smart agriculture applications along with the type and the ICTs can be seen in Table 1.

Remarkably, the sustainability of contemporary systems and infrastructures has a substantial impact on their energy efficiency and lifetime. Therefore, it serves as the primary driver of development so that future intelligent agricultural ecosystems have minimal environmental impact and carbon dioxide (CO_2) emissions. In addition to their eco-friendly orientation, these systems can be powered by renewable energy sources such as the sun and wind using advanced techniques such as solar panel energy harvesting and wind energy collection for partial or, in certain circumstances, complete energy sufficiency, thereby reducing electricity consumption by a substantial amount. It is anticipated that these new directions through ICTs and global sustainability guidelines will be infused into the contemporary agricultural world, implementing environmentally and energy-independent innovative practices.

Table 1 ICT-enabled smart agriculture applications

Application	ICTs	Type
Agriculture Marketing	Blockchain, ML	Business
Advanced Crop, Soil Management	ML, IoT, Edge-Cloud Computing	Management
Super High Precision Autonomous Agricultural Robots	ML, B5G, IoT, Edge-Cloud Computing	Monitoring, Sensing
Advanced Irrigation, Water Quality Management	ML, IoT	Management
Remote Sensing	Edge-Cloud Computing, IoT	Sensing
Agricultural Data Analytics, Forecasts	ML, Cloud Computing, Blockchain	Business
Super High Plants-Soil-Crop Disease	ML, Edge Computing, IoT	Monitoring, Sensing
Smart Greenhouse	ML, IoT	Monitoring

Remote IoT Agriculture Challenges

Although the IoT has existed for some time, the concept of Remote IoT has only recently emerged as a result of technology improvements that have expanded IoT's application areas (Goel et al. 2021). As the IoT has expanded to reach the oceans, outer space, and other inaccessible regions of the world, multiple new challenges have surfaced (Triantafyllou et al. 2019). Common applications of Remote IoT systems in agriculture include remote monitoring, control, and activation of a wide range of agricultural equipment in regions with limited or no network access.

The Northern Aegean Islands Paradigm

Samos, Mytilene, Ikaria, and Chios are Aegean Sea islands that make up the North Aegean Region (NAR) of Greece. As a perfect example of how Remote IoT logic may be implemented in practice, this region is the epicenter of ongoing efforts to develop distributed IoT applications for smart agriculture. On these islands, the use of smart agricultural services by farmers, producers, and organizations supports the local production of mastic, olive oil, livestock, and wine. Monitoring services, which include the collection, processing, and transmission of production-related data, utilize spatially distributed data flows. Several monitoring and control devices (traps) for the Dacus oleae insect population contribute to an interesting stream of data from the NAR islands. These devices measure the population of these harmful insects in order to determine the actions required each time to improve the oil's quality. In parallel with the monitoring of the Dacus oleae population, IoT sensors that monitor temperature and humidity provide a second geographically distributed data stream from the islands of the NAR, providing a more complete picture of the olive ecosystem.

A challenge that is immediately apparent is the problem of remote data collection from areas that are typically difficult to access, especially in the case of small islands and islets that lack the appropriate interconnection with the central network infrastructure of the main islands. Thus, in several cases, remote data collection cannot be performed automatically as it is not economically viable to incorporate modern telecommunication technologies in this process due to the location of the sensors, and a manual data gathering process is used (Ullo and Sinha 2021).

In the following sections, a more detailed analysis of these challenges is offered in relation to a variety of remote environments where data monitoring and collection, as well as subsequent transmission in a timely, energy-efficient, and secure manner, are essential.

Remote Connectivity and Data Collection

To this day, some regions of the world lack access to mobile network infrastructure. Due to their distance from population centers, setting up an IoT network in remote areas such as deserts, island complexes, and oceans can be challenging. Currently, the cellular network provides approximately 90% of the required global coverage for broadband and M2M IoT communications. However, as smart agriculture and IoT services advance, it is anticipated that the number of IoT devices will increase to the point that 5G networks will need to be enhanced with new and innovative technologies to meet the demand for these services, particularly in remote areas. In agriculture, applications for precise and real-time monitoring of soil, livestock, climate, and animal health can be deployed in remote areas and open new perspectives for precision agriculture services, enabling the development of new business opportunities and high-value investments. However, in these areas, it is challenging to establish and maintain a communication infrastructure that can interface a huge number of smart devices and sensors and enable them to communicate with the Internet (Callebaut et al. 2021).

Appropriate remote data collection is crucial because once data is collected, it must be transmitted to the cloud for processing. In this process, a key factor to consider is the total transmission delay, which must be reduced within strictly defined time constraints. Utilizing edge computing to process some time-sensitive data locally may be a viable option in certain situations. However, due to the inherent limitations of deployments in remote areas, this approach is not always feasible. Furthermore, when mass connectivity of a large number of IoT devices is required, these limitations are magnified. In any case, the availability and operation of a local network capable of interconnecting all relevant IoT devices is required in such a remote farming system, as only this allows the farmer to manage farm operations efficiently and timely.

In conclusion, it is acknowledged that the issue of connectivity has a significant impact on the agricultural production process, thereby affecting the final product's quality as well as the farmers' overall profit. Real-time IoT monitoring systems used by smart agricultural applications generate a great deal of useful data for farmers, but also present challenges in terms of their timely and efficient management. Furthermore, Internet connectivity also promotes the upgrading of services offered in difficult-to-reach locations and serves as the foundation for the development and delivery of new services.

Sustainable Networking Infrastructure

IoT devices installed in remote or difficult-to-reach locations are powered by batteries. In order to regularly maintain and recharge the batteries of Remote IoT devices, it is required to continuously monitor their energy status. However, since reaching to an isolated location can be expensive and difficult, the designers of such systems are looking for ways to save energy while also utilizing renewable energy sources.

In general, energy is consumed in each component at each architectural level of smart agriculture systems. However, in order to develop a long-term and sustainable development perspective for these systems, an effort must be made to reduce the energy consumed in each single part of the system. Thus, efficient energy management emerges as a new evolving field of research that aims to construct IoT systems in remote environments that fulfill the standards for sustainable development through protocol upgrades or the use of novel management algorithms (Tuysuz and Trestian 2020). Thus, by utilizing design and development frameworks that promote an environmentally friendly and climate change-aware approach, it is possible to reduce the total energy consumption required for data transmission, storage, and analysis, as well as network infrastructure maintenance (Popli et al. 2022).

In addition, the viability of a Remote IoT system is directly related to the need to minimize human intervention as much as possible by automating the necessary operations and being able to remotely activate all of the employed equipment. Indicative examples include the self-organization and self-healing design approaches, which aim to create systems that can sense and react to external changes. In this case, it is essential to exploit the gathered data in order to generate new knowledge that would aid the adaptation processes (Rangel-Martinez et al. 2021). The challenges of smart agriculture in Remote IoT environments are depicted in Fig. 2.

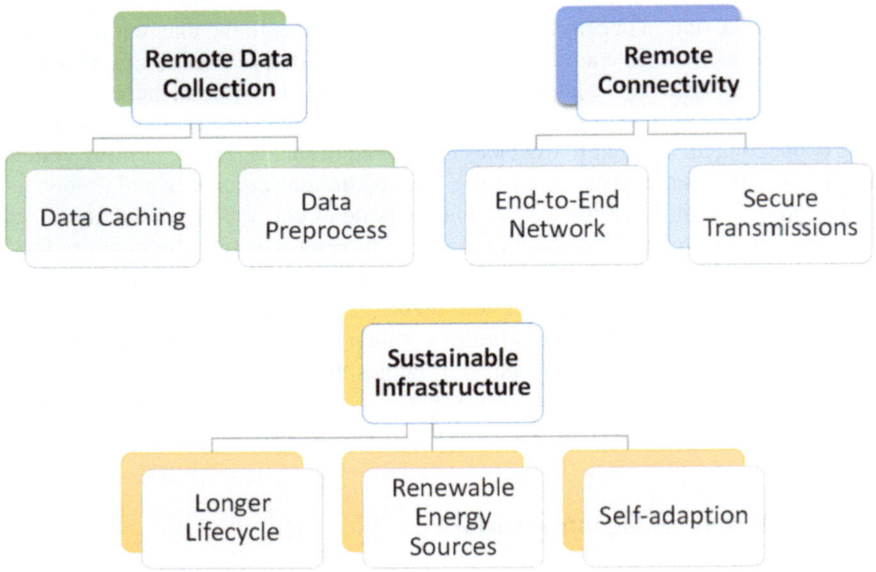

Fig. 2 Remote IoT agriculture challenges

Sustainability and Emerging Networking Solutions

In this section, emerging technologies are presented and classified into two subcategories, namely non-terrestrial networking and sustainability solutions.

Emerging Networking Solutions

The emerging networking technology of Non-Terrestrial Networks NTNs that is expected to enhance the global interconnection of mobile communication networks (Lin et al. 2021; Vaezi et al. 2022) and mitigate the connectivity issues encountered in current Remote IoT applications (Lin et al. 2021), (Rinaldi et al. 2020). In addition, economic considerations for the development of communication infrastructure in remote regions are taken seriously and are an additional reason for the development of NTNs. It is expected that using NTNs will solve the problem of connectivity in these areas and make it possible to create new business models for smart farming. In general, the operating altitude of NTNs is a determining factor in regard of the communication services and coverage they provide. (Azari et al. 2022).

Unmanned Aerial Vehicles (UAVs)

The 3GPP standards organization defines NTN as a network based on the deployment of Unmanned Aerial Vehicles (UAVs) or satellites that operate in the air or space (Michailidis et al. 2022). UAVs can be deployed more flexibly than satellites and at altitudes ranging from a few hundred meters to several kilometers, while their movement can be continuous or fly over a fixed reference point depending on the requirements of the system. Therefore, UAVs can provide low-cost dynamic coverage of remote areas on demand (Delavarpour et al. 2021).

Unmanned Aerial Vehicles (UAVs) based on the altitude at which they operate can support Low Amplitude Platforms (LAPs) as well as High Amplitude Platforms (HAPs). UAVs may function as mobile base stations or as repeaters (Ranjha et al. 2022). In addition, a UAV swarm operation is possible, in which a number of these vehicles fly over a remote area of interest while coordinating with one another to ensure maximum coverage (Li et al. 2022).

Satellites

The satellite can serve a variety of roles in the network's architecture, including that of a base station or repeater. At orbital level, satellite systems typically operate in three altitude regions (Lin et al. 2021). Geostationary Earth Orbit (GEO) is a circular orbit in which satellites reach a height of 35 000 km and cover an approximate surface area of 3500 km2. The lower orbits are classified as Non-Geostationary Earth Orbit (NGEO) including Medium Earth Orbit (MEO) and Low Earth Orbit (LEO). MEO satellites travel between 7000 and 25,000 km in altitude, whereas LEO satellites travel between 300 and 1500 km, covering a surface of 1000 km and continually shifting position around the earth (Vaezi et al. 2022).

The use of satellites is anticipated to result in significant enhancements to existing and emerging smart agriculture applications, particularly in remote regions. In addition, agricultural applications requiring the secure transmission of multimedia data in real time or the monitoring of critical parameters will become feasible (Centenaro et al. 2021).

Finally, satellites and UAVs can collaborate thereby introducing Hybrid-NTN (HNTN). In this sense, NTNs combine all these features into a single, coherent system that Remote IoT systems can use. Future updates to these technologies will be delivered via the current 5G network standard, with the research community expected to improve them even further via the next generation of mobile networks (6G) (Reddy Maddikunta et al. 2021).

Sustainability Solutions

Machine Learning (ML)

ML has evolved into a powerful method for knowledge production systems, with applications in telecom, economics, and healthcare, among others. Algorithms are capable of training learning models and generating a generalized model for problem-solving. ML can enable self-organizing and self-healing capabilities within smart agricultural infrastructures. Typically, the three progressive processes that encompass these possibilities are the detection of an event (e.g., a fault), the identification of the category it falls under, and the implementation of a solution (Qazi et al. 2022).

The categories of machine learning correspond to four types of learning based on the nature of the data and the support of a guiding agent. Unsupervised learning strives to associate and recognize paths and patterns from data without labels, whereas supervised learning attempts to predict an output function based on an input set (labeled data). As for the existence of an agent, reinforcement learning employs a guidance agent that aims each time to explore the best possible outcome and avoid repeating the incorrect decision. Consequently, each round generates feedback that functions as the optimal learning experience.

In the context of smart agriculture applications, intelligent production-increasing or -decreasing decisions and the optimal management of supply chain resources are two areas where machine learning methods can flourish and significantly enhance the performance of the services offered. In addition, the detection of problematic products is now considerably easier, as unsupervised learning models make it possible to identify faulty product characteristics with extreme precision. Furthermore, intelligent robotic agricultural systems can monitor diverse field and plant metrics autonomously. In an ecosystem where IoT devices can malfunction, the soil can become infected and its quality can deteriorate, and the air quality can suffer chemical changes, several difficulties can arise. By adjusting the system parameters that meet the required QoS, a learning model with historical data can address the aforementioned issues, guarantee QoS, and reduce capex/opex costs. In this direction, the human factor that would be essential in similar circumstances can be removed, thereby making the infrastructure smart and sustainable. In the energy sector, it is also possible to discover patterns of energy consumption under specific conditions and to create energy imprints. ML-based models can then use these imprints to predict how much energy will be needed for each usage scenario. This makes the system infrastructure more energy efficient and sustainable.

In the context of sustainable agriculture, however, machine learning can provide solutions for a more intelligent management of the technological infrastructure itself. At this point, learning models can take advantage of robust historical correlations between electrical consumption and then predict daily or weekly system consumption. In addition, machine learning makes it possible for the infrastructure to learn through fault conditions such as routine failure, software failure, and hardware failure. Therefore, self-management of infrastructure, predictive maintenance, and the ability to make local decisions automatically regarding whether or not to activate software and hardware modules based on a set of rules derived from past experience are possible. In this direction, it is possible for new technological trends to emerge that will render the agricultural infrastructure and its subsystems energy sustainable, with minimal environmental impact and self-management, healing, and adaption capabilities. In conclusion, while the aforementioned features are obviously useful in any IoT system, in the context of Remote IoT systems, these features are necessary and can help to reduce or even eliminate the problems that these systems have, particularly in terms of effective and targeted maintenance, fault prediction, energy consumption reduction, and in general dynamic adaptation in changing conditions without human intervention.

Green IoT

Green IoT is a set of rules that an Internet of Things system must adhere to in order to conserve as much energy as possible by fostering a sustainable, consumption-friendly environment (Popli et al. 2022). These rules are enforced at all system levels in an effort to integrate renewable energy sources seamlessly.

In addition, rules and policies for managing device energy consumption can lead to more energy-efficient communication when combined with ML optimization models. (Sakshi Popli et al. 2021). For example, the hardware may contain sleep schedulers that put the device to sleep when it is not transmitting or processing data, thereby substantially decreasing its energy consumption. This concept is also known as the "awakening mechanism". Furthermore, using supervised learning, the system may recognize when an actual data transfer is required and enable it, decreasing unnecessary data transfers to the network and reducing the associated energy consumption.

Furthermore, the integration of infrastructure with renewable energy sources, such as solar panels and wind turbines, can significantly aid in the achievement of this goal by producing a significant amount of the required energy on a daily basis.

Toward this direction, Green IoT enables IoT devices to form small D2D networks, thereby offloading the central gateway or infrastructure responsible for managing their communication and resulting in a reduction in IoT network latency (Hemanand et al. 2021).

Finally, renewable energy sources such as the sun and wind are two prominent examples of how the IoT can collect energy and power a portion of a network, thereby reducing energy consumption significantly. Low-level energy collection is performed by solar panels and wind turbines, which convert solar and kinetic energy into electric energy. In addition, novel energy harvesters, in which researchers are becoming more and more interested, capture heat from the IoT hardware and the environment (thermal harvesters) as well as energy from radio signals (RF harvesters), therefore it is anticipated that energy self-sufficiency can be achieved. The sustainability and the NTN networking solutions can be seen in Fig. 3.

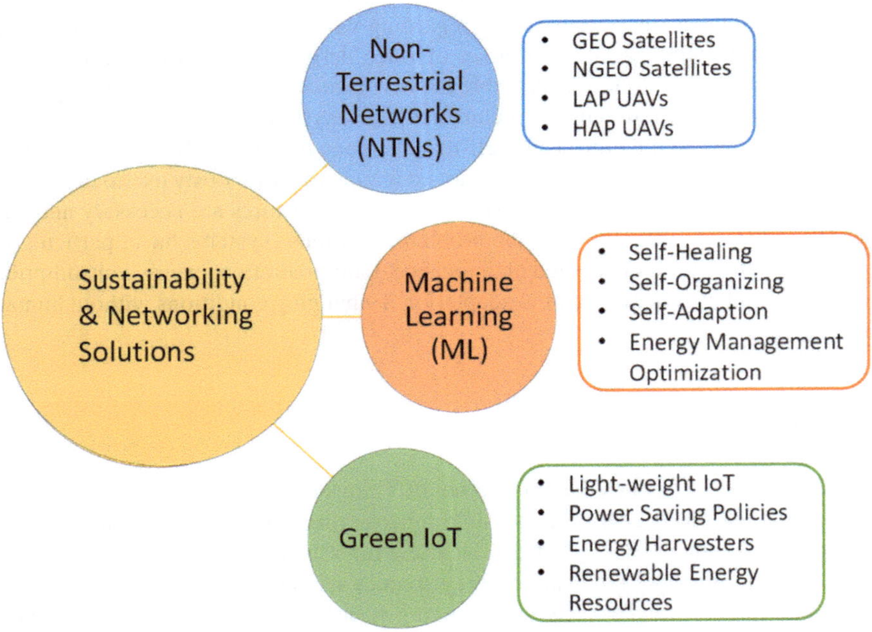

Fig. 3 Sustainability and Non-terrestrial networking solutions

Conclusions

In this chapter, an effort was made to emphasize the importance of sustainability in Remote IoT systems. The chapter begins by presenting the key research fields investigated in this study, including modern communication standards, sustainable smart agricultural scenarios, and ICT enablers. After setting the research context, our work then proceeds to present the main challenges detected in current Remote IoT environments, which include remote connectivity, data collection, and sustainable infrastructure, taking into consideration the system's battery life, reliability, and mobility. Finally, after describing the aforementioned key challenges, the chapter continues to present some new key sustainable solutions that, if properly implemented, could complete the deficiencies in this research field. In this context, NTNs, ML, and Green IoT are three new topics proposed as solutions to the aforementioned challenges. More specifically, NTNs allow for continuous provision of connectivity in locations where an Internet connection is required to support agricultural services, while ML supports intelligent infrastructure management, reducing human labor and faults to a minimum. In addition, the incorporation of Green IoT protocols into IoT-based agricultural systems will provide a level of intelligent and automatic power management through energy policies as well as by integrating renewable energy sources, leading to a sustainable and intelligent infrastructure.

Acknowledgements Support for this study was provided by the research program AGRICA II: AGrifood Research and Innovation Network of ExCellence of the Aegean, ESPA 2014–2022, in the context of the call of the Operational Program "Competitiveness, Entrepreneurship and Innovation", Action "Support of Regional Excellence" (MIS 5046750).

References

Abbasi R, Martinez P, Ahmad R (2022) The digitization of agricultural industry—a systematic literature review on agriculture 4.0. Smart Agric Technol 2:100042. https://doi.org/10.1016/j.atech.2022.100042

Azari MM, Solanki S, Chatzinotas S et al (2022) Evolution of non-terrestrial networks from 5G to 6G: a survey. IEEE Commun Surv Tutor 24:2633–2672. https://doi.org/10.1109/COMST.2022.3199901

Bhat SA, Huang NF (2021) Big data and AI revolution in precision agriculture: survey and challenges. IEEE Access 9:110209–110222. https://doi.org/10.1109/ACCESS.2021.3102227

Callebaut G, Leenders G, Van Mulders J et al (2021) The art of designing remote iot devices—technologies and strategies for a long battery life. Sensors (switzerland) 21(3):913. https://doi.org/10.3390/s21030913

Centenaro M, Costa CE, Granelli F et al (2021) A survey on technologies, standards and open challenges in satellite IoT. IEEE Commun Surv Tutor 23(3):1693–1720. https://doi.org/1109/COMST.2021.3078433

Chettri L, Bera R (2020) A comprehensive survey on internet of things (IoT) toward 5G wireless systems. IEEE Internet Things J 7(1):16–32. https://doi.org/10.1109/JIOT.2019.2948888

Delavarpour N, Koparan C, Nowatzki J et al (2021) A technical study on UAV characteristics for precision agriculture applications and associated practical challenges. Remote Sens (basel) 13(6):1204. https://doi.org/10.3390/rs13061204

Ericsson (2022) Ericsson Mobility Report June 2022. Available from: https://www.ericssonlg.com/en/events/emr-june-202206

Goel RK, Yadav CS, Vishnoi S, Rastogi R (2021) Smart agriculture—Urgent need of the day in developing countries. Sustain Comput: Inform Syst 30:100512. https://doi.org/10.1016/j.suscom.2021.100512

Hemanand D, Jayalakshmi DS, Ghosh U et al (2021) Enabling sustainable energy for smart environment using 5G wireless communication and internet of things. IEEE Wirel Commun 28:56–61. https://doi.org/10.1109/MWC.013.2100158

Imoize AL, Adedeji O, Tandiya N, Shetty S (2021) 6g enabled smart infrastructure for sustainable society: Opportunities, challenges, and research roadmap. Sensors 21(5):1709. https://doi.org/10.3390/s21051709

Kalyani Y, Collier R (2021) A systematic survey on the role of cloud, fog, and edge computing combination in smart agriculture. Sensors 21(17):5922. https://doi.org/10.3390/s21175922

Khan N, Ray RL, Sargani GR et al (2021) Current progress and future prospects of agriculture technology: Gateway to sustainable agriculture. Sustainability (switzerland) 13(9):4883. https://doi.org/10.3390/su13094883

Li J, Kacimi R, Liu T et al (2022) Non-Terrestrial networks-enabled internet of things: UAV-Centric architectures, applications, and open issues. Drones 6(4):95. https://doi.org/10.3390/drones6040095

Li S, Da XuL, Zhao S (2018) 5G internet of things: a survey. J Ind Inf Integr 10:1–9. https://doi.org/10.1016/j.jii.2018.01.005

Lin X, Rommer S, Euler S et al (2021) 5G from space: an overview of 3GPP non-terrestrial networks. IEEE Commun Stand Mag 5:147–153. https://doi.org/10.1109/MCOMSTD.011.2100038

Lin YP, Petway JR, Anthony J et al (2017) Blockchain: The evolutionary next step for ICT e-agriculture. Environments - MDPI 4:1–13. https://doi.org/10.3390/environments4030050

Michailidis ET, Maliatsos K, Skoutas DN et al (2022) Secure UAV-aided mobile edge computing for IoT: a review. IEEE Access 10:86353–86383. https://doi.org/10.1109/ACCESS.2022.3199408

Misra NN, Dixit Y, Al-Mallahi A et al (2022) IoT, big data, and artificial intelligence in agriculture and food industry. IEEE Internet Things J 9:6305–6324. https://doi.org/10.1109/JIOT.2020.2998584

Nasirahmadi A, Hensel O (2022) Toward the next generation of digitalization in agriculture based on digital twin paradigm. Sensors 2292:498. https://doi.org/10.3390/s22020498

Polymeni S, Skoutas DN, Kormentzas G, Skianis C (2022) FINDEAS: a fintech-based approach on designing and assessing IoT systems. IEEE Internet Things J 9:25196–25206. https://doi.org/10.1109/JIOT.2022.3195770

Polymeni S, Plastras S, Skoutas DN, Kormentzas G, Skianis C (2023) The Impact of 6G-IoT Technologies on the Development of Agriculture 5.0: A Review. Electronics 12(12): 2651. https://doi.org/10.3390/electronics12122651

Popli S, Jha RK, Jain S (2022) Green IoT: a short survey on technical evolution & techniques. Wirel Pers Commun 123:525–553. https://doi.org/10.1007/s11277-021-09142-3

Qazi S, Khawaja BA, Farooq QU (2022) IoT-equipped and ai-enabled next generation smart agriculture: a critical review, current challenges and future trends. IEEE Access 10:21219–21235. https://doi.org/10.1109/ACCESS.2022.3152544

Rangel-Martinez D, Nigam KDP, Ricardez-Sandoval LA (2021) Machine learning on sustainable energy: A review and outlook on renewable energy systems, catalysis, smart grid and energy storage. Chem Eng Res Des 174:414–441. https://doi.org/10.1016/j.cherd.2021.08.013

Ranjha A, Kaddoum G, Dev K (2022) Facilitating URLLC in UAV-assisted relay systems with multiple-mobile robots for 6G networks: a prospective of agriculture 4.0. IEEE Trans Industr Inform 18:4954–4965. https://doi.org/10.1109/TII.2021.3131608

Reddy Maddikunta PK, Hakak S, Alazab M et al (2021) Unmanned aerial vehicles in smart agriculture: applications, requirements, and challenges. IEEE Sens J 21:17608–17619. https://doi.org/10.1109/JSEN.2021.3049471

Rinaldi F, Määttänen HL, Torsner J et al (2020) Non-terrestrial networks in 5G & beyond: A survey. IEEE Access 8:165178–165200. https://doi.org/10.1109/ACCESS.2020.3022981

Popli S, Jha RK, Jain S (2021) A comprehensive survey on Green ICT with 5G-NB-IoT: Towards sustainable planet. Comput Netw 199:108433. https://doi.org/10.1016/j.comnet.2021.108433

Sharma R, Kamble SS, Gunasekaran A et al (2020) A systematic literature review on machine learning applications for sustainable agriculture supply chain performance. Comput Oper Res 119. https://doi.org/10.1016/j.cor.2020.104926

Shi W, Cao J, Zhang Q et al (2016) Edge computing: vision and challenges. IEEE Internet Things J 3:637–646. https://doi.org/10.1109/JIOT.2016.2579198

Sinha BB, Dhanalakshmi R (2022) Recent advancements and challenges of Internet of Things in smart agriculture: A survey. Futur Gener Comput Syst 126:169–184. https://doi.org/10.1016/j.future.2021.08.006

Sobin CC (2020) A survey on architecture, protocols and challenges in IoT. Wirel Pers Commun 112:1383–1429. https://doi.org/10.1007/s11277-020-07108-5

Tao W, Zhao L, Wang G, Liang R (2021) Review of the internet of things communication technologies in smart agriculture and challenges. Comput Electron Agric 189:106352. https://doi.org/10.1016/j.compag.2021.106352

Triantafyllou A, Sarigiannidis P, Bibi S (2019) Precision agriculture: A remote sensing monitoring system architecture. Information (switzerland) 10. https://doi.org/10.3390/info10110348

Tuysuz MF, Trestian R (2020) From serendipity to sustainable green IoT: Technical, industrial and political perspective. Comput Netw 182. https://doi.org/10.1016/j.comnet.2020.107469

Ullo SL, Sinha GR (2021) Advances in IoT and smart sensors for remote sensing and agriculture applications. Remote Sens (basel) 13(13):2585. https://doi.org/10.3390/rs13132585

Vaezi M, Azari A, Khosravirad SR et al (2022) Cellular, wide-area, and non-terrestrial IoT: a survey on 5G advances and the road toward 6G. IEEE Commun Surv Tutor 24:1117–1174. https://doi.org/10.1109/COMST.2022.3151028

Wang SY, Chang JE, Fan H, Sun YH (2022) Comparing the performance of NB-IoT, LTE Cat-M1, sigfox, and LoRa for IoT end devices moving at high speeds in the air. J Signal Process Syst 94:81–99. https://doi.org/10.1007/s11265-021-01660-4

Zhang F, Zhang Y, Lu W et al (2022) 6G-Enabled smart agriculture: a review and prospect. Electronics (switzerland) 11(18):2845. https://doi.org/10.3390/electronics11182845

MIX
Papier aus verantwortungsvollen Quellen
Paper from responsible sources
FSC® C105338

If you have any concerns about our products,
you can contact us on
ProductSafety@springernature.com

In case Publisher is established outside the EU,
the EU authorized representative is:
Springer Nature Customer Service Center GmbH
Europaplatz 3, 69115 Heidelberg, Germany

Printed by Libri Plureos GmbH
in Hamburg, Germany